Collins

Edexcel A-Level Maths Year 2

Revision Guide

Phil Duxbury, Rebecca Evans and Leisa Bovey

About this Revision & Practice book

Revise

These pages provide a recap of everything you need to know for each topic.

You should read through all the information before taking the Quick Test at the end. This will test whether you can recall the key facts.

Quick Test

1. $x = 3t$, $y = \frac{5}{t}$, $\{t \in \mathbb{R} : 0 \leqslant t \leqslant 5\}$
   a) Express the parametric equations as a Cartesian equation $y = f(x)$.
   b) Sketch the graph of $y = f(x)$.
2. Sketch the curve given by the parametric equations $x = 2\cos(t) + 1$, $y = 3\sin(t) - 1$, $\{t \in \mathbb{R}: 0 \leqslant t \leqslant 2\pi\}$.

Practise

These topic-based questions appear shortly after the revision pages for each topic and will test whether you have understood the topic. If you get any of the questions wrong, make sure you read the correct answer carefully.

Review

These topic-based questions appear later in the book, allowing you to revisit the topic and test how well you have remembered the information. If you get any of the questions wrong, make sure you read the correct answer carefully.

Mix it Up

These pages feature a mix of questions for all the different topics, just like you would get in an exam. They will make sure you can recall the relevant information to answer a question without being told which topic it relates to.

Test Yourself on the Go

Visit our website at **collins.co.uk/collinsalevelrevision** and print off a set of flashcards. These pocket-sized cards feature questions and answers so that you can test yourself on all the key facts anytime and anywhere. You will also find lots more information about the advantages of spaced practice and how to plan for it.

Workbook

This section features even more topic-based questions as well as practice exam papers, providing two further practice opportunities for each topic to guarantee the best results.

ebook 

To access the ebook revision guide visit

collins.co.uk/ebooks

and follow the step-by-step instructions.

Contents

Statistics and Mechanics

Composite and Inverse Functions

You must be able to:

- Identify and understand the domain and range of a function
- Understand and use composite functions and associated notation
- Recognise and use inverse functions and graphs of inverse functions.

Functions

- A **function** is a **mapping** of one or more objects in one set to a unique object in another set. Functions can be mappings of one-to-one or many-to-one.
- Functions can be expressed using the notation f(x) or f: $x \mapsto$.
- The **domain** of a function f(x) is the set of possible values of the input.
- The **range** of a function f(x) is the set of possible values of the output.
- In this topic, all functions are defined only on the real numbers ($\mathbb{R}$).

Key Point

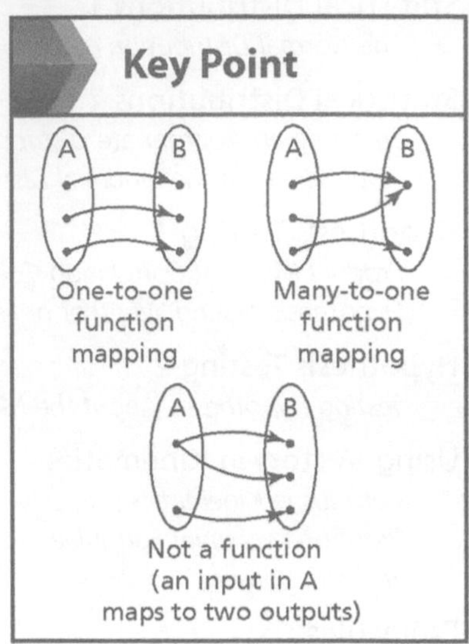

1) Represent the function $y = x^2$ as a mapping diagram and identify the type of mapping.

$y = x^2$ is a many-to-one mapping of the values of the input, x, onto the output, x^2.

2) Identify the domain and the range of the function $y = x^2$ and express the function, domain and range using set notation.

The domain of the function is all real numbers and the range is all real numbers greater than or equal to zero.

The function can hence be written as $f(x) = x^2, x \in \mathbb{R}$ (or as f: $x \mapsto x^2$, $x \in \mathbb{R}$). The range can be written as $\{y \in \mathbb{R}: y \geqslant 0\}$.

Remember, for real numbers $\mathbb{R}$ the values of x^2 are never negative.

Composite Functions

- A **composite function** is a combination of functions in which the output of one function becomes the input of another function.
- The notation fg means that the output of the function g(x) becomes the input to the function f(x).

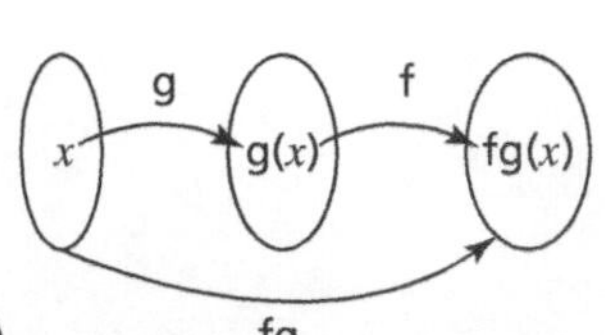

Key Point

fg(x) means do the function g(x) first, then the function f(x).

Given $f(x) = x^2 + 2$ and $g(x) = 2x - 1$

a) Find the function fg(x).

$fg(x) = f[g(x)] = f(2x - 1)$ — First apply the function g.

$= (2x - 1)^2 + 2$ — Now apply the function f.

$= 4x^2 - 4x + 3$ — Expand and simplify.

b) Find the function gf(x).

$gf(x) = g(x^2 + 2)$

$= 2(x^2 + 2) - 1$

$= 2x^2 + 3$

c) Find the values of x such that gf(x) = 53.

$53 = 2x^2 + 3$ — Set the function gf(x) equal to 53 and solve.

$50 = 2x^2$

$x = \pm 5$

d) Find the values for which fg(x) = gf(x).

$4x^2 - 4x + 3 = 2x^2 + 3$ — Set the function fg(x) equal to the function gf(x) and solve.

$2x^2 - 4x = 0$

$x = 0, x = 2$

Inverse Functions

- Given f(x), the **inverse function** $f^{-1}(x)$ is the reverse of f(x) using the inverse operations. Only one-to-one functions have inverse functions.
- $f^{-1}f(x) = ff^{-1}(x) = x$
- The graph of $y = f^{-1}(x)$ is a reflection of $y = f(x)$ in the line $y = x$.
- The domain of f(x) is the range of $f^{-1}(x)$ and the range of f(x) is the domain of $f^{-1}(x)$.

Key Point

$f^{-1}(x)$ performs the inverse operations of f(x) in the reverse order.

1) Find $f^{-1}(x)$ given $f(x) = 2x^2 + 4$, $\{x \in \mathbb{R}: x \geqslant 0\}$.

Finding $f^{-1}(x)$ is equivalent to changing the subject of a formula.

Let $y = f(x)$

$$y = 2x^2 + 4$$

$$y - 4 = 2x^2$$

$$\frac{y}{2} - 2 = x^2$$

$$\sqrt{\frac{y}{2} - 2} = x$$

$$f^{-1}(x) = \sqrt{\frac{x}{2} - 2}$$

f(x): x → square → x^2 → ×2 → $2x^2$ → +4 → $2x^2 + 4$

$f^{-1}(x)$: x → −4 → $x - 4$ → ÷2 → $\frac{x}{2} - 2$ → $\sqrt{\ }$ → $\sqrt{\frac{x}{2} - 2}$

2) Identify the domain and the range of both f(x) and $f^{-1}(x)$.

The domain of $f(x) = 2x^2 + 4$ is $\{x \in \mathbb{R}: x \geqslant 0\}$ (given in the question) and the range is $\{y: y \geqslant 4\}$.

Since the domain is $x \geqslant 0$, the minimum value of y is 4.

The domain of $f^{-1}(x) = \sqrt{\frac{x}{2} - 2}$ is $\{x: x \geqslant 4\}$.

The domain of $y = f^{-1}(x)$ is the range of $y = f(x)$.

The range of $f^{-1}(x)$ is $\{y \in \mathbb{R}: y \geqslant 0\}$.

The range of $y = f^{-1}(x)$ is the domain of $y = f(x)$.

3) Sketch a graph of $y = f(x)$ and $y = f^{-1}(x)$.

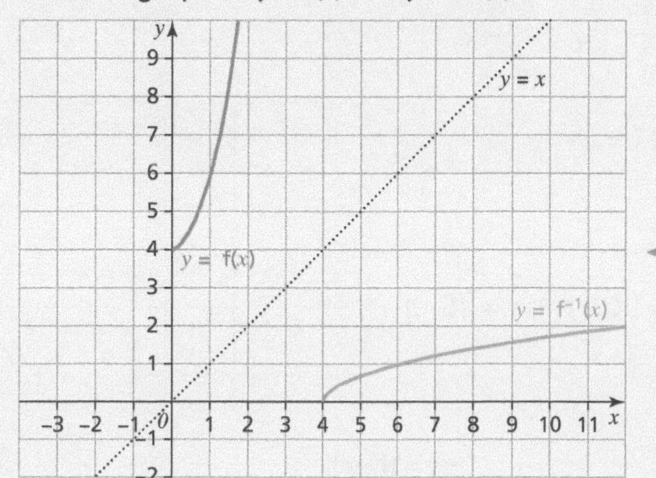

$y = f^{-1}(x)$ is a reflection of $y = f(x)$ across the line $y = x$. Don't forget to restrict the graph to the domain and the range of each function.

Quick Test

f: $x \mapsto \frac{5}{x-2}$, $\{x \in \mathbb{R}: x \neq 2\}$ and g: $x \mapsto x^2 + 3$, $\{x \in \mathbb{R}: x > 0\}$.

1. Identify the domain and the range of the functions f(x) and g(x).
2. Find the functions fg(x) and gf(x).
3. Find the inverse function $g^{-1}(x)$, identifying the domain and the range.

Key Words

function
mapping
domain
range
composite function
inverse function

Modulus and Exponential Functions

You must be able to:

- Sketch and use the graph of $y = |ax + b|$ to solve equations or inequalities involving $|ax + b|$
- Sketch the graph of $y = |f(x)|$ and $y = |f(-x)|$ given the graph of $y = f(x)$
- Combine transformations of graphs
- Understand the graph of the exponential function $y = e^{ax+b} + c$.

The Modulus of a Function

- The **modulus** of a number, $|n|$, is the absolute size of n and hence is always positive, e.g. $|5| = 5$, $|-7| = 7$.
- The modulus of a function, $y = |f(x)|$, turns any negative y-values into positive values, e.g. $y = |2x|$ gives $y = 4$ when $x = -2$.
- To sketch the graph of the modulus of a function, sketch the graph of $y = f(x)$ and reflect any parts of the line or curve that are below the x-axis in the x-axis.

Key Point

$y = |f(x)|$:

- $f(x) \geqslant 0$, $|f(x)| = f(x)$
- $f(x) < 0$, $|f(x)| = -f(x)$, a reflection of $f(x)$ in the x-axis.

1) $f(x) = |2x - 3|$

Sketch the graph of $y = f(x)$ and use it to solve the inequality $|2x - 3| \leqslant x + 1$.

To graph $y = f(x)$, draw the line $y = 2x - 3$ and reflect the part that is below the x-axis over the x-axis.

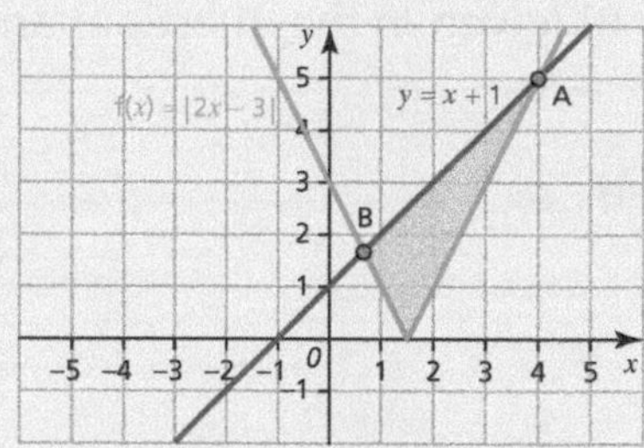

Add the line $y = x + 1$ to the graph and identify the area where $y = |2x - 3|$ is below the line $y = x + 1$.

Point A lies on the intersection of $y = 2x - 3$ and $y = x + 1$.

$2x - 3 = x + 1 \Rightarrow x = 4$

Point B lies on the intersection of $y = -(2x - 3)$ and $y = x + 1$.

$-(2x - 3) = x + 1 \Rightarrow -2x + 3 = x + 1 \Rightarrow x = \frac{2}{3}$

The values of x for which $|2x - 3| \leqslant x + 1$ are $\frac{2}{3} \leqslant x \leqslant 4$.

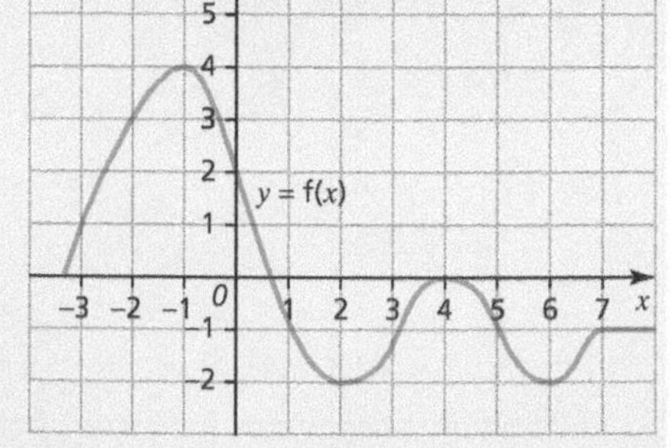

2) Given the graph of $y = f(x)$, sketch the graph of $y = |f(x)|$ and $y = |f(-x)|$.

Reflect the parts of the graph that are below the x-axis above the x-axis.

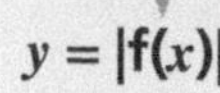

$y = |f(x)|$

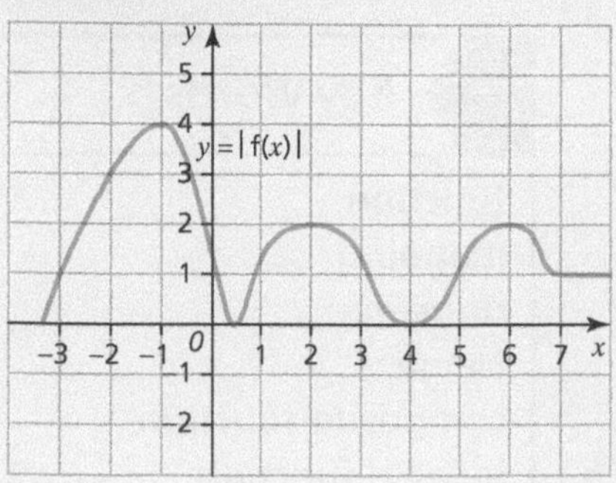

$y = f(-x)$

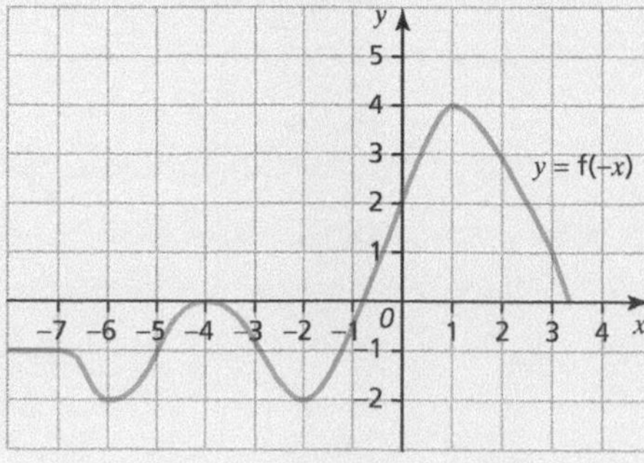

$y = |f(-x)|$

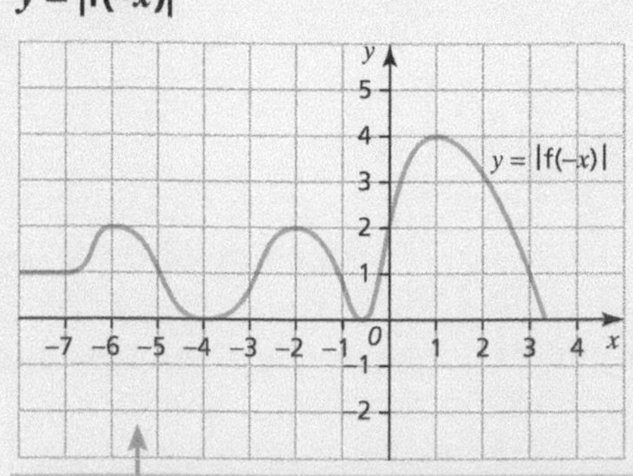

Sketch the graph of $y = f(-x)$ and reflect the parts that are below the x-axis above the x-axis.

Combining Transformations of Graphs

- $y = f(x + a)$ is a translation of $y = f(x)$ by $\begin{pmatrix} -a \\ 0 \end{pmatrix}$.
- $y = f(x) + a$ is a translation of $y = f(x)$ by $\begin{pmatrix} 0 \\ a \end{pmatrix}$.
- $y = f(ax)$ is a horizontal stretch of $y = f(x)$ by a scale factor of $\frac{1}{a}$.
- $y = af(x)$ is a vertical stretch of $y = f(x)$ by a scale factor of a.
- $y = f(-x)$ is a reflection of $y = f(x)$ in the y-axis.
- $y = -f(x)$ is a reflection of $y = f(x)$ in the x-axis.

Key Point

Transformations of $y = f(x)$ can occur in any combination.

Given the graph of $y = f(x)$, sketch the graph of $y = -2f(3x) + 1$.

The transformation is of the form $y = -af(bx) + c$, so it is a horizontal stretch of $y = f(x)$ by a scale factor of $\frac{1}{3}$, a reflection in the x-axis, a vertical stretch by a scale factor of 2 and a translation of 1 unit up. The order is important; be sure to apply the transformation inside the brackets first.

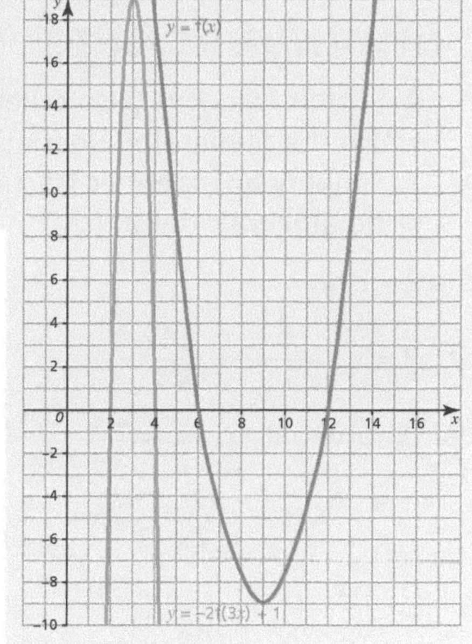

Exponential Graphs of $y = e^{ax+b} + c$

- $y = e^{ax+b} + c$ is a transformation of $y = e^x$.
- To sketch the graph of $y = e^{ax+b} + c$, find the y-intercept and the horizontal **asymptote**.
- To find the horizontal asymptote, look at the limit as $x \to \infty$ and as $x \to -\infty$

Sketch the graph of $y = e^{2x+3} - 1$.

When $x = 0$, $y = e^{(2 \times 0 + 3)} - 1 \approx 19.1$.

When $x \to \infty$, $y \to \infty$.

When $x \to -\infty$, $y \to -1$, so there is an asymptote at $y = -1$.

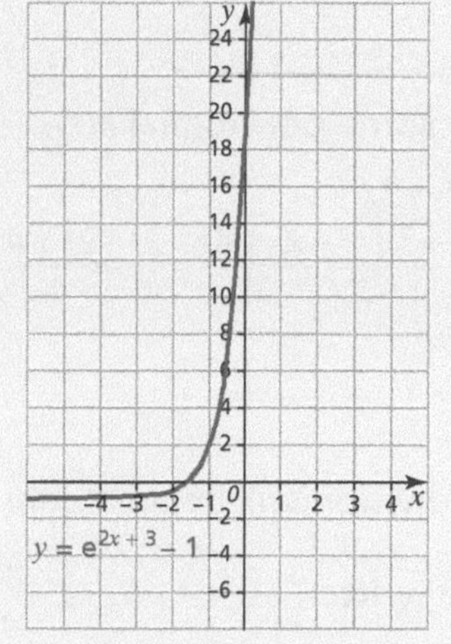

Use the general shape of $y = e^x$.

Quick Test

1. $f(x) = \left|\frac{1}{3}x - 1\right|$
   a) Sketch the graph of $y = f(x)$.
   b) Use the graph to solve the equation $\left|\frac{1}{3}x - 1\right| = 2x + 5$.
2. $f(x) = x^2 - 2x - 3$
   a) Sketch the graph of $y = |f(-x)|$.
   b) On the same grid, sketch the graph of $y = 2\left|f\left(-\frac{1}{3}x\right)\right| + 4$.
3. Sketch the graph of $y = e^{3x-1} + 2$.

Key Words

modulus
transformation
asymptote

Algebraic Fractions

You must be able to:

- Simplify algebraic fractions with linear or quadratic denominators by cancelling or by using algebraic division
- Write rational functions as partial fractions.

Simplifying Algebraic Fractions

- **Algebraic fractions** (also called **rational expressions**) are fractions in which the numerator, denominator, or both are polynomials.
- To simplify an algebraic fraction, factorise and cancel where possible, or do **algebraic division**.

Key Point

Algebraic fractions are just fractions containing polynomials, so they follow the normal rules of operations.

Cancelling

$$f(x) = x - \frac{x}{x+3} + \frac{12}{x^2+2x-3}$$

a) Show that $f(x) = \frac{x^3+x^2-2x+12}{(x+3)(x-1)}$

$$x - \frac{x}{x+3} + \frac{12}{x^2+2x-3} = x - \frac{x}{x+3} + \frac{12}{(x+3)(x-1)}$$

Factorise: $x^2 + 2x - 3 = (x + 3)(x - 1)$

$$= \frac{x(x+3)(x-1) - x(x-1) + 12}{(x+3)(x-1)} = \frac{x^3+x^2-2x+12}{(x+3)(x-1)}$$

Multiply so that each term has the same common denominator of $(x + 3)(x - 1)$ and write as a single fraction.

b) Show also that $f(x) = \frac{x^2-2x+4}{x-1}$

$(-3)^3 + (-3)^2 - (2 \times -3) + 12 = 0$

Use the Factor Theorem to show that $(x + 3)$ is a factor of $x^3 + x^2 - 2x + 12$.

$x^3 + x^2 - 2x + 12 = (x + 3) \times p(x)$ for some polynomial $p(x)$.

$(x^3 + x^2 - 2x + 12) \div (x + 3) = x^2 - 2x + 4$

Use algebraic division.

Therefore, $\frac{x^3+x^2-2x+12}{(x+3)(x-1)} = \frac{(x+3)(x^2-2x+4)}{(x+3)(x-1)} = \frac{x^2-2x+4}{x-1}$

Division

- The remainder from algebraic division will be an algebraic fraction.

Show that $\frac{2x^3+3x^2-x+3}{x-1}$ can be written as $Ax^2+Bx+C+\frac{D}{x-1}$ and find the values of A, B, C and D.

Note that the form $Ax^2+Bx+C+\frac{D}{x-1}$ can also be called a partial fraction.

$(2x^3 + 3x^2 - x + 3) \div (x - 1)$

$$\begin{array}{r l}
 & 2x^2 + 5x + 4 \\
x - 1 \,) & 2x^3 + 3x^2 - x + 3 \\
 & -(2x^3 - 2x^2) \\
 & 5x^2 - x \\
 & -(5x^2 - 5x) \\
 & 4x + 3 \\
 & -(4x - 4) \\
 & 7
\end{array}$$

$$\frac{2x^3+3x^2-x+3}{x-1} = 2x^2 + 5x + 4 + \frac{7}{x-1}$$

Write the remainder as a fraction with the divisor as the denominator.

$A = 2$, $B = 5$, $C = 4$, $D = 7$

Partial Fractions

- Any fraction or **rational function** can be written as the sum of two or more fractions. These are called **partial fractions.**
- A rational function can be decomposed into partial fractions by using substitution or by **equating coefficients.**

Key Point

Look at the denominator of the fraction to see which denominators to use in the partial fractions.

Decompose $\frac{2x-8}{x^2+x-2}$ into partial fractions.

$$\frac{2x-8}{x^2+x-2}=\frac{2x-8}{(x-1)(x+2)}\equiv\frac{A}{x-1}+\frac{B}{x+2}$$

Factorise the denominator and write as partial fractions.

Method 1: Substitution

$$\frac{2x-8}{(x-1)(x+2)}\equiv\frac{A(x+2)+B(x-1)}{(x-1)(x+2)}$$

Multiply so that each fraction has the common denominator $(x-1)(x+2)$, then write as a single fraction.

$2x-8\equiv A(x+2)+B(x-1)$

$2\times 1-8=A(1+2)+B(1-1)$

To find A, substitute $x=1$ to eliminate B.

$-6=3A\Rightarrow A=-2$

$(2\times -2)-8=A(-2+2)+B(-2-1)$

To find B, substitute $x=-2$ to eliminate A.

$-12=-3B\Rightarrow B=4$

$$\therefore\frac{2x-8}{x^2+x-2}\equiv\frac{4}{x+2}-\frac{2}{x-1}$$

Method 2: Equating coefficients

$2x-8\equiv A(x+2)+B(x-1)$

As seen in Method 1.

$\equiv Ax+2A+Bx-B$

$\equiv x(A+B)+(2A-B)$

Expand the brackets. Collect like terms and simplify.

$A+B=2$ **(1)**

$2A-B=-8$ **(2)**

Equate the coefficients. The coefficient of x is 2 and the constant term is -8.

$3A=-6$

$A=-2$

Use simultaneous equations to solve for A and B.

$$-2+B=2\Rightarrow B=4\ \therefore\frac{2x-8}{x^2+x-2}\equiv\frac{4}{x+2}-\frac{2}{x-1}$$

Quick Test

1. Write as a single fraction: $\frac{x+3}{x-1}+\frac{x}{x+3}+1$
2. Simplify $\frac{6x^2+x-1}{x+2}\div\frac{3x+2}{2x^2+5x+2}$.
3. Decompose $\frac{3x^2-14x+10}{(x+1)(x-2)^2}$ into partial fractions.

Key Words

algebraic fraction
rational expression
algebraic division
rational function
partial fraction
equating coefficients

Parametric Equations

You must be able to:

- Convert equations between parametric and Cartesian form
- Identify the domain and range of parametric equations
- Sketch the graph of a parametric equation.

Parametric Equations

- **Parametric equations** are a pair of functions used to define the x- and y-coordinates on a graph. They are often used to describe curves that are not easy to plot.
- For parametric equations $x = \text{p}(t)$, $y = \text{q}(t)$, $t \in \mathbb{R}$, where p(t) and q(t) are functions. The points (x, y) are therefore given in relation to a third variable called the **parameter** (usually t).
- The graph (right) shows a set of parametric equations $x = \text{p}(t)$, $y = \text{q}(t)$ for some functions p(t) and q(t). Each point (x, y) on the curve is found by using the functions so each point $(x, y) = (\text{p}(t), \text{q}(t))$ for some value t.

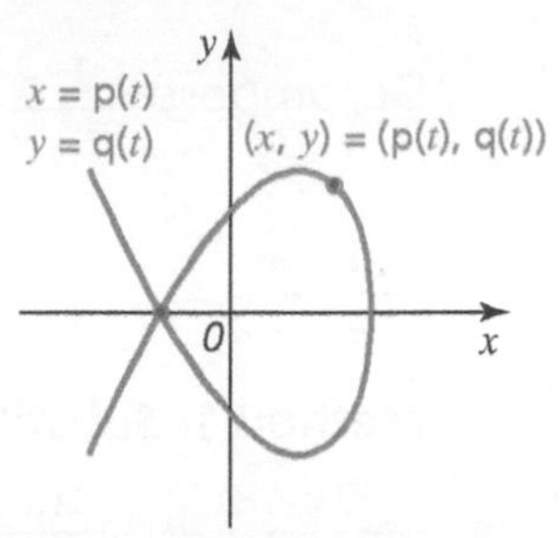

Key Point

For parametric equations $x = \text{p}(t)$, $y = \text{q}(t)$, each point on the curve is $(x, y) = (\text{p}(t), \text{q}(t))$.

Converting to Cartesian Equations

- Parametric equations can sometimes be converted to **Cartesian equations** in order to sketch the curve.
- For parametric equations $x = \text{p}(t)$ and $y = \text{q}(t)$, the domain of the Cartesian equation $y = \text{f}(x)$ is the range of p(t), and the range of f(x) is the range of q(t).
- Use substitution to eliminate the parameter in order to convert a parametric equation into a Cartesian equation.

Key Point

Parametric equations can be converted to Cartesian equations using substitution.

1) A curve has parametric equations $x = \frac{1}{t-1}$, $y = 3t + 1$, $\{t \in \mathbb{R} : t > 1\}$.

a) Write the curve as a Cartesian equation $y = \text{f}(x)$, stating the domain and the range of f(x).

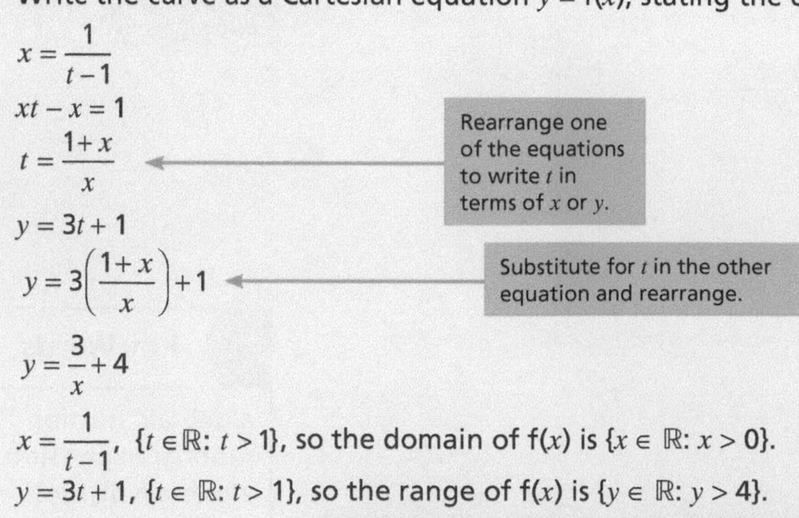

$x = \frac{1}{t-1}$

$xt - x = 1$

$t = \frac{1+x}{x}$ ← Rearrange one of the equations to write t in terms of x or y.

$y = 3t + 1$

$y = 3\left(\frac{1+x}{x}\right) + 1$ ← Substitute for t in the other equation and rearrange.

$y = \frac{3}{x} + 4$

$x = \frac{1}{t-1}$, $\{t \in \mathbb{R}: t > 1\}$, so the domain of f($x$) is $\{x \in \mathbb{R}: x > 0\}$.

$y = 3t + 1$, $\{t \in \mathbb{R}: t > 1\}$, so the range of f($x$) is $\{y \in \mathbb{R}: y > 4\}$.

b) Sketch the curve.

Draw the curve using the domain and the range. The curve is a transformation of $y = \frac{1}{x}$ and has a horizontal asymptote at $y = 4$ (see right).

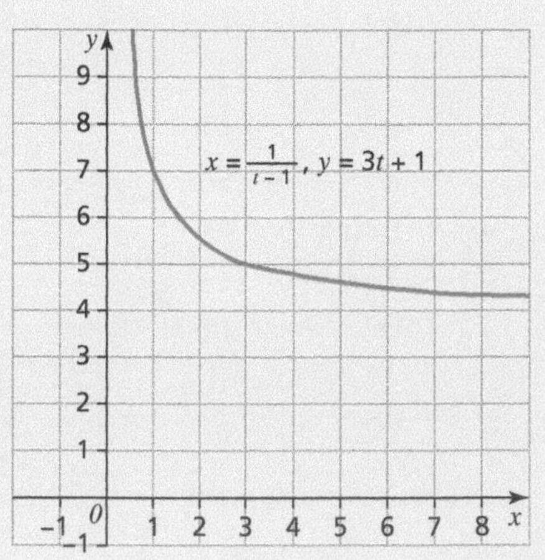

2) A curve has parametric equations $x = 2 + 3\cos(t)$, $y = -1 + 3\sin(t)$, $\{t \in \mathbb{R}\}$. Write the curve as a Cartesian equation $y = f(x)$, stating the domain and the range of $f(x)$.

$x = 2 + 3\cos(t)$

$\cos(t) = \frac{x-2}{3}$ ← Rearrange both equations in the form $t =$

$y = -1 + 3\sin(t)$

$\sin(t) = \frac{y+1}{3}$

$\sin^2(t) + \cos^2(t) \equiv 1$ ← Use the trigonometric identities.

$\left(\frac{y+1}{3}\right)^2 + \left(\frac{x-2}{3}\right)^2 = 1$ ← Substitute and rearrange.

$(y + 1)^2 + (x - 2)^2 = 9$

The curve is a circle with centre (2, −1) and a radius of 3.

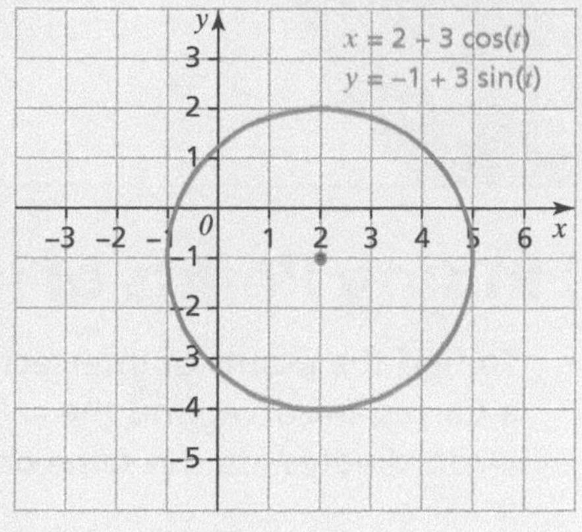

Sketching Parametric Curves

- Parametric curves are often not easily recognisable and can 'double back' on themselves in the x-direction.
- Substitute values of t into the parametric equations and plot the resulting (x, y) points to sketch a parametric curve.

Key Point

To sketch a parametric curve, substitute values of the parameter into each parametric equation and plot the resulting points, connecting the points in order of t.

Sketch the curve defined by the parametric equations $x = t\cos(t)$, $y = t\sin(t)$, $\{t \in \mathbb{R}: 0° \leqslant t \leqslant 3\pi°\}$

t	0	$\frac{\pi}{2}$	π	$\frac{3\pi}{2}$	2π	$\frac{5\pi}{2}$	3π
$x = t\cos(t)$	0	0	−3.14	0	6.28	0	−9.42
$y = t\sin(t)$	0	1.57	0	−4.71	0	7.85	0

Substitute enough values of t to get an idea of the shape of the curve and be sure to cover the whole of the domain of t.

Plot the points in order of t and connect them with a smooth curve.

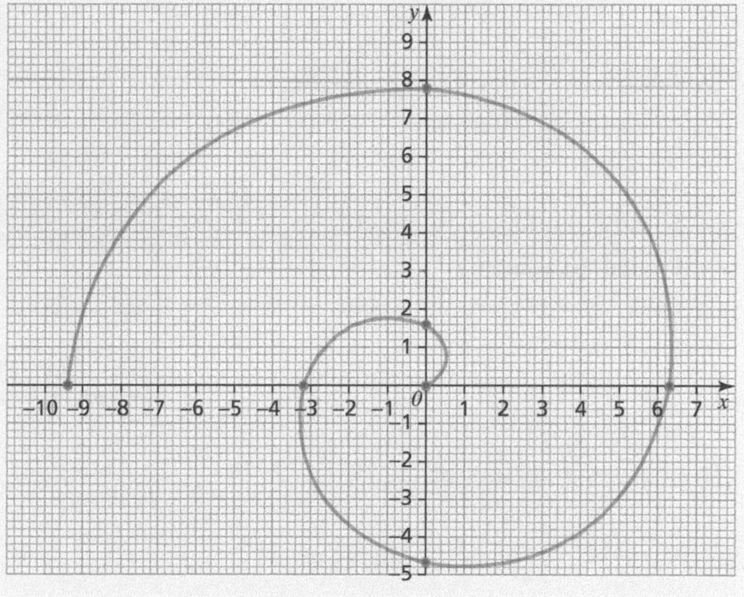

Quick Test

1. $x = 3t$, $y = \frac{5}{t}$, $\{t \in \mathbb{R}: 0 \leqslant t \leqslant 5\}$
   a) Express the parametric equations as a Cartesian equation $y = f(x)$.
   b) Sketch the graph of $y = f(x)$.
2. Sketch the curve given by the parametric equations $x = 2\cos(t) + 1$, $y = 3\sin(t) - 1$, $\{t \in \mathbb{R}: 0 \leqslant t \leqslant 2\pi\}$.

Key Words

parametric equations
parameter
Cartesian equation

Problems Involving Parametric Equations

You must be able to:

- Find points of intersection involving parametric curves
- Use parametric equations to model a variety of real-life contexts.

Finding Points of Intersection

- To find the points of intersection of a parametric curve with a Cartesian curve, find the values at which the x- and y-coordinates are the same on both curves.

Key Point

Use substitution to find the points of intersection of a parametric and a Cartesian curve.

The graph shows a curve with parametric equations

$x = \cos(t),\ y = t + 1 - \sin^2(t),\ \{t \in \mathbb{R}: -\frac{\pi}{2} < t \leq \frac{3\pi}{2}\}$.

Find the coordinates of the point at which the parametric curve intersects the curve $y = x^2$.

Sketch the curve $y = x^2$ on the same graph.

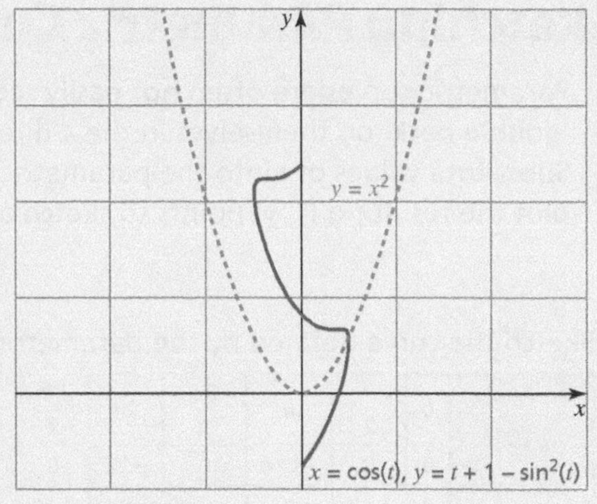

$y = x^2$

$t + 1 - \sin^2(t) = (\cos(t))^2$ ← Substitute the parametric equations for x and y into the Cartesian equation and solve for t.

$t + 1 = \cos^2(t) + \sin^2(t)$

$t + 1 = 1$ ← Remember, $\sin^2(\theta) + \cos^2(\theta) \equiv 1$

$t = 0$

$x = \cos(0) = 1$ ← Substitute the value of t into the parametric equations to find the x- and y- coordinates.

$y = 0 + 1 - 0 = 1$

The curves intersect at (1, 1).

Modelling Movement

- Parametric equations can be used to model the locus or path of the movement of a point.
- **Linear motion** can be modelled by the parametric equations $x = (v\cos(\theta))t + x_0$, $y = (v\sin(\theta))t + y_0$, where v is the constant velocity, θ is the **angle of elevation**, t is time and (x_0, y_0) is the object's initial location when $t = 0$.
- **Parabolic motion** can be modelled by the parametric equations $x = t(v\cos a)$, $y = t(v\sin a) - 4.9t^2 + y_0$, where v is the initial velocity, a is the angle at which the object is initially projected and y_0 is the initial height.

Key Point

Parametric equations are often used to model movement.

1) A plane takes off at a constant speed of $70\,ms^{-1}$. The plane is at a height of 500 m after travelling a horizontal distance of 2000 m.

a) Find the angle of elevation, θ.

$$\tan(\theta)=\frac{500}{2000} \Rightarrow \theta=\tan^{-1}\left(\frac{500}{2000}\right)=14.0° \text{ (to 1 d.p.)}$$

b) Find the parametric equations to model the plane's motion.

$x=(v\cos(\theta))t+x_0,\ y=(v\sin(\theta))t+y_0$

$x=(70\cos(14°))t,\ y=(70\sin(14°))t$

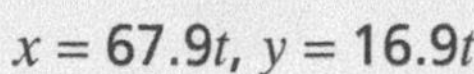

$x=67.9t,\ y=16.9t$

c) Find the height and the horizontal distance of the aeroplane after 10 seconds.

The horizontal distance is the x-coordinate and the height is the y-coordinate.

$x=67.9\times 10=679\text{ m}$ and $y=16.9\times 10=169\text{ m}$

d) Find the time at which the aeroplane reaches a cruising altitude of 11 800 m and hence find the domain of the parametric equations.

$11\,800=16.9t \Rightarrow t=698.2\text{ s (11.63 minutes)}$

Domain is $\{t\in\mathbb{R}: 0<t\leqslant 698.2\}$ where t is the time in seconds after the aeroplane takes off.

2) A cannonball is fired from a height of 2 m above flat ground with an initial velocity of $15\,ms^{-1}$ at an angle of 60° above the horizontal. The path of the ball is modelled by the parametric equations $x=\frac{15}{2}t,\ y=\frac{15\sqrt{3}}{2}t-4.9t^2+2$.

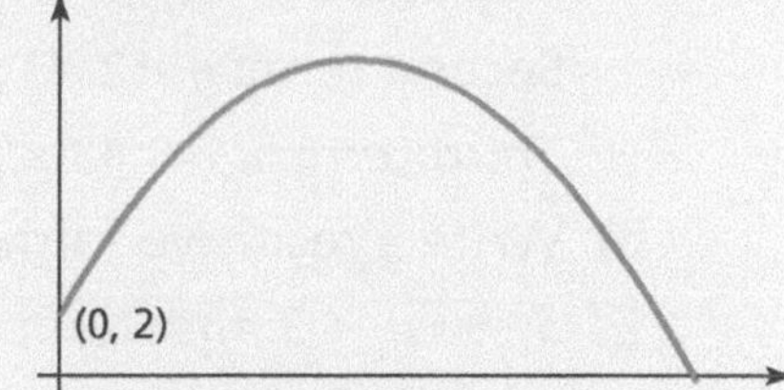

a) Find the time at which the ball hits the ground.

$$0=\frac{15\sqrt{3}}{2}t-4.9t^2+2$$

The ball hits the ground when $y=0$.

$$t=\frac{-\frac{15\sqrt{3}}{2}\pm\sqrt{\left(\frac{15\sqrt{3}}{2}\right)^2-(4\times -4.9\times 2)}}{(2\times -4.9)}=2.797 \text{ seconds (to 3 d.p.)}$$

Use the quadratic formula with $a=-4.9$, $b=\frac{15\sqrt{3}}{2}$ and $c=2$.

b) Find the horizontal distance travelled by the ball before hitting the ground.

$$x=\frac{15}{2}t \Rightarrow x=\frac{15}{2}\times 2.797 \Rightarrow x=20.98\text{ m}$$

The distance travelled is the x-coordinate of the point at which the ball hits the ground, $t=2.797$.

Quick Test

1. Given the parametric equations $x=\frac{1}{(3t+1)^2}$, $y=2t-1$, $\{t\in\mathbb{R}: t>0\}$, find the point of intersection of the curve and the x-axis.
2. A child's ride at a theme park is modelled by the parametric equations $x=2t^2$, $y=2\cos(3t)+2$, $\{t\in\mathbb{R}: 0\leqslant t\leqslant 2\pi\}$, where t is the time in seconds, x is the horizontal distance in metres and y is the height in metres.
   a) Find the initial height of the ride.
   b) Find the horizontal distances at which the ride is at its lowest points.

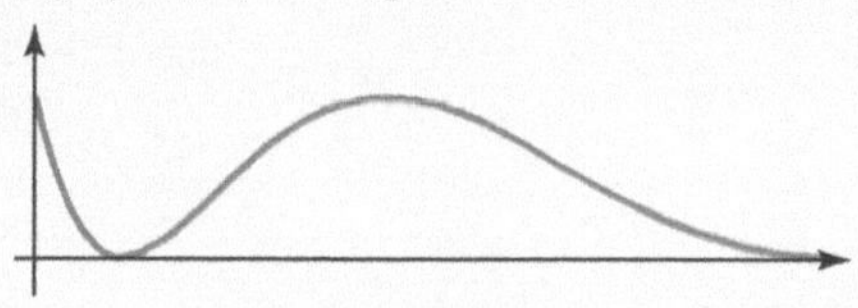

Key Words

linear motion
angle of elevation
parabolic motion

Types of Sequences

You must be able to:

- Understand and use sequences defined by an nth term formula or recursively
- Understand and be able to identify increasing, decreasing and periodic sequences and their formulae
- Use sigma notation for the sum of a series.

Definition of Sequences

- A **sequence** is a set of numbers in a certain order that follow a particular pattern.
- Sequences can be defined using an **nth term formula**, such as $u_n = 3n + 4$.
- Sequences can also be defined using a **recursive formula**, in which the next term is found using the previous terms, e.g. $u_{n+1} = 2u_n$

Key Point

A sequence can be defined using an nth term formula or recursively. The nth term of a sequence is written as u_n.

1) A sequence is given by $u_n = 3n + 4$.

a) Write down the first three terms of the sequence.

First term: $n = 1$, $(3 \times 1) + 4 = 7$

Second term: $n = 2$, $(3 \times 2) + 4 = 10$

Third term: $n = 3$, $(3 \times 3) + 4 = 13$

Substitute the term number for n.

b) Write a recursive formula describing the sequence.

$u_{n+1} = u_n + 3$, first term $u_1 = 7$

To get to each successive term in the sequence, add 3.

2) Write an nth term formula and a recursive formula for the sequence $\frac{1}{10}, \frac{1}{100}, \frac{1}{1000}, \frac{1}{10\,000}, \ldots$

An nth term formula is $u_n = \frac{1}{10^n}$

Notice that the denominators are increasing powers of 10.

A recursive formula is $u_{n+1} = \frac{u_n}{10}$, $u_1 = \frac{1}{10}$

Each term is the previous term divided by 10.

Increasing, Decreasing and Periodic Sequences

- In an increasing sequence, each successive term increases in size.
- In a decreasing sequence, each successive term decreases in size.
- A **periodic sequence** is neither increasing nor decreasing and instead repeats a particular pattern (a period). For a periodic sequence, $u_{k+p} = u_k$ where k is the term number and p is the period.

Identify whether each sequence is increasing, decreasing or periodic.

a) $u_n = \frac{2}{2n-1} + 3$

The sequence defined by $u_n = \frac{2}{2n-1} + 3$ describes a decreasing sequence because as $n \to \infty, \frac{2}{2n-1} \to 0 \Rightarrow u_n \to 3$.

b) $u_n = n^2$

The sequence defined by $u_n = n^2$ describes an increasing sequence because as $n \to \infty$, $u_n \to \infty$.

c) 3, 5, 8, 8, 10, 3, 5, 8, 8, 10,...

The sequence 3, 5, 8, 8, 10, 3, 5, 8, 8, 10,... is a periodic sequence, i.e. the sequence repeats itself so that $u_1 = u_6$, $u_2 = u_7$ and so on. In general, $u_{k+5} = u_k$.

Sigma Notation, $\sum_1^k u_n$

- The sum of the terms in a sequence is called a **series**.
- $\sum_1^k u_n$ means the sum of the first k terms of the series u_n.

Key Point

$\sum_1^k u_n = u_1 + u_2 + u_3 + \ldots + u_k$

$\sum_1^k 1 = k$

1) $u_n = \frac{1}{2n}$

Find $\sum_1^5 u_n$

Find the sum of the first five terms.

$$\sum_1^5 u_n = \frac{1}{2\times1} + \frac{1}{2\times2} + \frac{1}{2\times3} + \frac{1}{2\times4} + \frac{1}{2\times5} = \frac{137}{120}$$

2) Write the series $\frac{1}{5} + \frac{1}{7} + \frac{1}{9} + \frac{1}{11} + \ldots + \frac{1}{41}$ in the form $\sum_1^k u_n$

The nth term formula for the sequence is $u_n = \frac{1}{2n+3}$.

Each successive denominator is increasing by 2; since the first term is $\frac{1}{5}$, the nth term formula is $u_n = \frac{1}{2n+3}$.

The term $\frac{1}{41}$ is the 19th term in the sequence.

Solve $2n + 3 = 41$ for n.

The series can hence be expressed as $\sum_1^{19} \frac{1}{2n+3}$.

Quick Test

1. A sequence is given by the recursive formula $u_{n+1} = 3u_n$ where the first term is 2. Write down the first five terms.
2. Describe the sequences given by the formulae below as increasing, decreasing or periodic.
   a) $u_n = (-1)^n \times 3$
   b) $u_{(n+1)} = \frac{u_n}{2}$
   c) $u_n = 3^n$
3. Write the series 5 + 7 + 9 + 11 + 13 + 15 + 17 using sigma notation and find the sum.

Key Words

sequence
*n*th term formula
recursive formula
periodic sequence
series

Arithmetic Sequences and Series

You must be able to:

- Understand, use and find the nth term formula for an arithmetic sequence
- Understand, use and derive the formula for the sum of the first n terms of an arithmetic series
- Use arithmetic sequences and series in context.

Arithmetic Sequences

- An **arithmetic sequence** is a sequence that increases or decreases by the same amount between each successive term; there is a **common difference**, d, between each term.
- The nth term is given by $u_n = a + d(n - 1)$, where a is the first term, n is the term number and d is the common difference.
- Arithmetic sequences can be defined recursively as $u_{n+1} = u_n + d$.

Key Point

The nth term rule for an arithmetic sequence is $u_n = a + d(n - 1)$.

1) A sequence is defined by $u_n = -3n + 40$.

a) List the first five terms of the sequence.

$n = 1 \Rightarrow u_1 = (-3 \times 1) + 40 = 37$

$n = 2 \Rightarrow u_2 = (-3 \times 2) + 40 = 34$

$n = 3 \Rightarrow u_3 = (-3 \times 3) + 40 = 31$

$n = 4 \Rightarrow u_4 = (-3 \times 4) + 40 = 28$

$n = 5 \Rightarrow u_5 = (-3 \times 5) + 40 = 25$

Notice that each successive term is decreasing by 3:

37 34 31 28 25

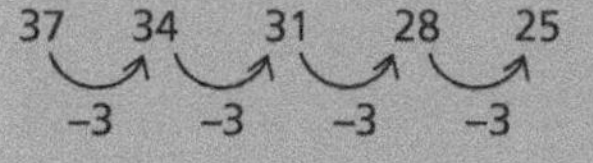

b) Find the first term that is negative.

$-3n + 40 < 0$

$-3n < -40$

$n > \dfrac{-40}{-3}$

$n > 13.33$

The first term that is negative will be the 14th term.

To find the first term that is negative, set up an inequality with the nth term formula.

Don't forget to switch the inequality symbol when multiplying or dividing by a negative number.

2) The 5th term of an arithmetic sequence is 13 and the 12th term is 27. Find the nth term rule and write the first five terms of the sequence.

$u_n = a + d(n - 1)$

$u_5 \Rightarrow 13 = a + 4d$ **(1)**

$u_{12} \Rightarrow 27 = a + 11d$ **(2)**

$27 = a + 11d$ **(2)**

$-(13 = a + 4d)$ **(1)**

$14 = 7d$

$d = 2$

$a = 13 - 8 = 5$

$u_n = 5 + 2(n - 1) = 3 + 2n$

The first five terms are then 5, 7, 9, 11, 13.

Substitute into the general nth term formula and solve the two equations simultaneously.

Subtract equation 1 from equation 2 to eliminate a and solve for d.

Arithmetic Series

- An arithmetic series is the sum of an arithmetic sequence.
- An arithmetic series is also called an **arithmetic progression**.
- The sum of the first n terms of an arithmetic series can be found using the formula $S=\frac{n}{2}(a+l)$ or $S=\frac{1}{2}n(2a+(n-1)d)$, where n is the number of terms, a is the first term, l is the last term and d is the common difference.

Key Point

The sum of the first n terms of an arithmetic series is $S=\frac{n}{2}(a+l)$ or $S=\frac{1}{2}n(2a+(n-1)d)$.

1) Prove that the sum of the first n terms of an arithmetic series is $S=\frac{1}{2}n(2a+(n-1)d)$ and thus, $S_n=\frac{n}{2}(a+l)$.

You must be able to prove the formula for the sum of an arithmetic series.

$$S_n=[a]+[a+d]+\dots\dots\dots\dots+[a+(n-2)d]+[a+(n-1)d]$$
$$S_n=[a+(n-1)d]+[a+(n-2)d]+\dots\dots+[a+d]+[a]$$
$$2S_n=[2a+(n-1)d]+[2a+(n-1)d]+\dots\dots\dots\dots+[2a+(n-1)d]$$

Write S_n in ascending order and in descending order and add.

$$2S_n=n(2a+(n-1)d)$$

There are n equal terms.

$$S_n=\frac{1}{2}n(2a+(n-1)d)=\frac{1}{2}n(a+a+(n-1)d)=\frac{1}{2}n(a+l)$$

$(a+(n-1)d)=l$

2) Find the sum of the first n natural numbers.

You must be able to derive the formula for the first n natural numbers.

$$S_n=1+2+3+\dots\dots\dots\dots+(n-2)+(n-1)+n$$
$$S_n=n+(n-1)+(n-2)\dots\dots\dots\dots+3+2+1$$
$$2S_n=(n+1)+(n+1)+\dots\dots+(n+1)+(n+1)+(n+1)$$

Set up the proof as before to get the sum of n equal terms of $(n+1)$.

$$2S_n=n(n+1)\Rightarrow S_n=\frac{1}{2}n(n+1)$$

3) Sophie saves her pocket money in a piggy bank. She convinces her parents to give her £5 pocket money the first week and to increase the amount by 50p every week.

a) How much pocket money does Sophie receive in the 10th week?

$u_n=a+(n-1)d$

$u_{10}=5+(10-1)0.5=£9.50$

Substitute $n=10$ and $a=5$ into the general formula for the nth term of an arithmetic sequence.

b) Assuming she does not spend any, how much money is in her piggy bank at the end of one year?

$S=\frac{1}{2}n(2a+(n-1)d)$

$S=\frac{1}{2}52(2\times5+(52-1)0.5)$

$=£923.00$

Substitute $a=5$, $d=0.5$ and $n=52$.

Quick Test

1. $u_n=\frac{1}{4}n-2$
   a) Write the first five terms of the sequence.
   b) Find the first term of the sequence that is positive.
   c) Find the sum of the first 100 terms.
2. Find $\sum_{1}^{30}3n-5$.

Key Words

arithmetic sequence
common difference
arithmetic progression

Geometric Sequences and Series

You must be able to:

- Understand, use and find the nth term formula for a geometric sequence
- Understand, use and derive the formula for the sum of the first n terms of a finite geometric series and for the sum to infinity of convergent geometric series
- Use geometric sequences and series in context.

Geometric Sequences

- A **geometric sequence** is a sequence in which each successive term is found by multiplying the previous term by a **common ratio**, r.
- The nth term formula for a geometric sequence is $u_n = a \times r^{n-1}$, where a is the first term, r is the common ratio and n is the number of terms.
- A geometric sequence can be defined recursively as $u_{n+1} = ru_n$.

Key Point

The nth term formula for a geometric sequence is $u_n = a \times r^{n-1}$.

1) A geometric sequence is defined as $u_n = 3 \times 2^{n-1}$.

a) List the first five terms of the sequence.

$n = 1 \Rightarrow u_1 = 3 \times 2^{1-1} = 3 \times 1 = 3$

$n = 2 \Rightarrow u_2 = 3 \times 2^{2-1} = 3 \times 2 = 6$

$n = 3 \Rightarrow u_3 = 3 \times 2^{3-1} = 3 \times 4 = 12$

$n = 4 \Rightarrow u_4 = 3 \times 2^{4-1} = 3 \times 8 = 24$

$n = 5 \Rightarrow u_5 = 3 \times 2^{5-1} = 3 \times 16 = 48$

Notice that each consecutive term is multiplied by 2:

3 → 6 → 12 → 24 → 48 (×2 ×2 ×2 ×2)

b) What is the first term that is greater than 90 000?

$u_n = 3 \times 2^{n-1}$

$3 \times 2^{n-1} > 90000$ — Set up an inequality using the nth term formula.

$\log_{10}(2^{n-1}) > \log_{10}(30000)$ — Use either $\log_{10}$ or ln.

$(n-1)\log_{10}(2) > \log_{10}(30000)$

$$n - 1 > \frac{\log_{10}(30000)}{\log_{10}(2)}$$

$n > 15.87$

The first term that is greater than 90 000 is the 16th term.

2) The 6th term of a geometric sequence is 1600 and the 10th term is 25 600. Given that $r > 0$, write the nth term formula.

$u_{10} = u_6 \times r^4$ — To get to the 10th term from the 6th term, multiply by the common ratio four times, i.e. r^4.

$25600 = 1600 \times r^4$ — Substitute and solve for r.

$16 = r^4$

$r = \sqrt[4]{16}$ $r = 2$, as the question states $r > 0$.

$u_6 = u_1 \times r^5$ — To find the first term, substitute into the nth term formula and solve for u_1.

$1600 = u_1 \times 2^5$

$$\frac{1600}{2^5} = u_1$$

$u_1 = 50$

$u_n = 50 \times (2)^{n-1}$

Finite and Infinite Geometric Series

- A geometric series is the sum of a geometric sequence.
- A geometric series is also called a **geometric progression**.
- The sum of the first n terms of a geometric series can be found using the formula $S_n = \frac{a(r^n - 1)}{(r - 1)}$, $r > 1$, or $S_n = \frac{a(1 - r^n)}{(1 - r)}$, $r < 1$.

Key Point

The sum of the first n terms of a geometric series is $S_n = \frac{a(r^n - 1)}{(r - 1)}$ or $S_n = \frac{a(1 - r^n)}{(1 - r)}$

1) Find the sum of the geometric series $1 + 3 + 9 + 27 + \ldots + 59049$.

$u_2 \div u_1 = 3 \div 1 = 3$ so $r = 3$

$u_1 = a \times 3^0 \Rightarrow a = 1$

$u_n = 1 \times 3^{n-1}$

$59049 = 1 \times 3^{n-1}$

$\log_{10}(59049) = \log_{10}(3^{n-1})$

$\log_{10}(59049) = (n - 1)\log_{10}(3)$

$$n = \frac{\log_{10}(59\,049)}{\log_{10}(3)} + 1 = 11$$

$$S_{11} = \frac{1(3^{11} - 1)}{(3 - 1)} = 88573$$

2) Derive the formula for the sum of the first n terms of a geometric series.

$S_n = a + ar + ar^2 + \ldots\ldots + ar^{n-2} + ar^{n-1}$ **(1)**

$rS_n = ar + ar^2 + ar^3 + \ldots\ldots + ar^{n-1} + ar^n$ **(2)**

Multiply each term in **(1)** by r to get **(2)**.

$rS_n - S_n = (ar - a) + (ar^2 - ar) + (ar^3 - ar^2) + \ldots. + (ar^{n-1} - ar^{n-2}) + (ar^n - ar^{n-1})$

$S_n(r - 1) = -a + ar - ar + ar^2 - ar^2 + \ldots\ldots\ldots - ar^{n-1} + ar^{n-1} - ar^n$

Re-ordering terms and simplifying.

$S_n(r - 1) = ar^n - a = a(r^n - 1)$

$$S_n = \frac{a(r^n - 1)}{(r - 1)}$$

- The sum of an infinite geometric series can **converge** (tend towards a finite limit) or **diverge** (it does not tend towards a finite limit).
- A geometric series will converge if $|r| < 1$, i.e. $-1 < r < 1$, because as $n \to \infty$, $r^n \to 0$.
- The sum of a convergent infinite geometric series can be found using the formula $S_\infty = \frac{a}{1 - r}$.
- The sum of a divergent infinite geometric series is undefined.

Key Point

For a convergent infinite geometric series, $S_\infty = \frac{a}{1 - r}$.

Find the sum of the geometric series $48 + 6 + \frac{3}{4} + \frac{3}{32} + \ldots$

$u_2 \div u_1 = 6 \div 48 = \frac{1}{8}$, so $r = \frac{1}{8}$

$$S_\infty = \frac{48}{1 - \left(\frac{1}{8}\right)} = \frac{384}{7}$$

Quick Test

1. A geometric sequence is defined by $u_n = 400 \times \left(-\frac{1}{4}\right)^{n-1}$
   a) List the first five terms of the sequence.
   b) Find the sum of the first 10 terms of the sequence.
   c) Decide if the series is convergent or divergent and find the sum of the infinite series, if possible.

Key Words

geometric sequence
common ratio
geometric progression
converge
diverge

Binomial Sequences

You must be able to:

- Expand $(1 + bx)^n$ for any rational number, n
- Expand $(a + bx)^n$ for any rational number, n
- Use binomial expansion with partial fractions to expand rational functions.

Expanding $(1 + bx)^n$ for any Rational n

- When n is a fraction or a negative number, the **binomial expansion** is
$$(1+bx)^n = 1 + \frac{nbx}{1!} + \frac{n(n-1)(bx)^2}{2!} + \frac{n(n-1)(n-2)(bx)^3}{3!} + \ldots$$
- The expansion has an infinite number of terms.
- The expansion of $(1 + bx)^n$, when n is a fraction or a negative number, is valid for $|bx| < 1$, or $|x| < \frac{1}{b}$.

Key Point

The binomial expansion is different when n is a fraction or negative number than when n is a positive integer.

a) Find the first four terms of the expansion of $(1+2x)^{\frac{1}{5}}$ in ascending powers of x and state the range of values of x for which the expansion is valid.

$$(1+2x)^{\frac{1}{5}} = 1 + \frac{\frac{1}{5}\times 2x}{1!} + \frac{\frac{1}{5}\left(\frac{1}{5}-1\right)\times(2x)^2}{2!} + \frac{\frac{1}{5}\left(\frac{1}{5}-1\right)\left(\frac{1}{5}-2\right)\times(2x)^3}{3!} + \ldots$$

$$= 1 + \frac{2x}{5} - \frac{8x^2}{25} + \frac{48x^3}{125} + \ldots$$

The expansion is valid for $|2x| < 1 \Rightarrow |x| < \frac{1}{2}$

Replace n with $\frac{1}{5}$ and bx with $2x$ in the binomial expansion of $(1 + bx)^n$.

b) Use the expansion to find an approximate value of $\sqrt[5]{\frac{3}{2}}$

$$\sqrt[5]{\frac{3}{2}} = \left(\frac{3}{2}\right)^{\frac{1}{5}},\ 1+2x = \frac{3}{2} \Rightarrow x = \frac{1}{4}$$

Solve to find an appropriate value of x.

$$(1+2x)^{\frac{1}{5}} = 1 + \frac{2}{5}\times\frac{1}{4} - \frac{8}{25}\times\left(\frac{1}{4}\right)^2 + \frac{48}{125}\times\left(\frac{1}{4}\right)^3 = \frac{543}{500} = 1.086$$

Substitute into the expansion.

$$\sqrt[5]{\frac{3}{2}} \approx 1.086$$

Expanding $(a + bx)^n$ for any Rational n

- To expand $(a + bx)^n$, factor out the a term, then expand $\left(1+\frac{b}{a}x\right)^n$.
- $(a+bx)^n = \left(a\left(1+\frac{b}{a}x\right)\right)^n = a^n\left(1+\frac{b}{a}x\right)^n$
- The expansion of $(a + bx)^n$, when n is a fraction or a negative number, is valid for $\left|\frac{b}{a}x\right| < 1$, or $|x| < \frac{a}{b}$.

Key Point

Factor out the value of a first when expanding $(a + bx)^n$.

Find the first four terms of the expansion of $\frac{1}{(3x+2)^4}$ in ascending powers of x and state the range of values of x for which the expansion is valid.

$$\frac{1}{(3x+2)^4} = (3x+2)^{-4}$$

Rewrite in index form.

$$(3x+2)^{-4} = 2^{-4}\left(1+\tfrac{3}{2}x\right)^{-4}$$

Remember b is the coefficient of x, regardless of whether it is first or second in the brackets.

$$=2^{-4}\left(1+\frac{-4\left(\frac{3}{2}x\right)}{1!}+\frac{-4(-4-1)\left(\frac{3}{2}x\right)^2}{2!}+\frac{-4(-4-1)(-4-2)\left(\frac{3}{2}x\right)^3}{3!}\right)$$

$$=2^{-4}\left(1-6x+\frac{45x^2}{2}-\frac{135x^3}{2}\right)=\frac{1}{16}-\frac{3x}{8}+\frac{45x^2}{32}-\frac{135x^3}{32}$$

The expansion is valid for $\left|\frac{3}{2}x\right|<1 \Rightarrow |x|<\frac{2}{3}$

Binomial Expansion with Partial Fractions

- Using **partial fractions** can simplify the expansion of more complex expressions. The expansion is valid for the range that satisfies the expansion of both the numerator and the denominator.

Find the first four terms of the expansion of $\frac{x-6}{(2-x)(x-3)}$ in ascending powers of x and state the values of x for which the expansion is valid.

$$\frac{x-6}{(2-x)(x-3)} \equiv \frac{A}{2-x}+\frac{B}{x-3}$$

Start by expressing the rational expression as partial fractions (see page 11).

$$\equiv \frac{4}{2-x}+\frac{3}{x-3}=4(2-x)^{-1}+3(x-3)^{-1}$$

Write the fractions in index form.

The expansion of $4(2-x)^{-1}=4\times 2^{-1}\left(1-\frac{x}{2}\right)^{-1}=2\left(1-\frac{x}{2}\right)^{-1}$

Write $(2-x)^{-1}$ in the form $a^n\left(1+\frac{b}{a}x\right)^n$ and simplify. $a=2$, $b=-1$

$$2\left(1-\frac{x}{2}\right)^{-1}=2\left(1+\frac{-1\times\left(-\frac{1}{2}\right)x}{1!}+\frac{-1(-1-1)\left(-\frac{1}{2}x\right)^2}{2!}+\frac{-1(-1-1)(-1-2)\left(-\frac{1}{2}x\right)^3}{3!}+\ldots\right)$$

$$=2\left(1+\frac{x}{2}+\frac{x^2}{4}+\frac{x^3}{8}+\ldots\right)=2+x+\frac{x^2}{2}+\frac{x^3}{4}+\ldots$$

Similarly, the expansion of $3(x-3)^{-1}=-1-\frac{x}{3}-\frac{x^2}{9}-\frac{x^3}{27}-\ldots$

Thus, the expansion of $4(2-x)^{-1}+3(x-3)^{-1}=2+x+\frac{x^2}{2}+\frac{x^3}{4}-1-\frac{x}{3}-\frac{x^2}{9}-\frac{x^3}{27}-\ldots$

$$=1+\frac{2x}{3}+\frac{7x^2}{18}+\frac{23x^3}{108}+\ldots$$

The expansion of $\frac{4}{2-x}$ is valid for $\left|\frac{1}{2}x\right|<1 \Rightarrow |x|<2$.

Find the range of values that satisfy the expansion of both partial fractions.

The expansion of $\frac{3}{x-3}$ is valid for $\left|\frac{1}{3}x\right|<1 \Rightarrow |x|<3$.

The expansion of $\frac{x-6}{(2-x)(x-3)}$ is therefore valid for $|x|<2$.

A number line is a useful way to visualise the range:

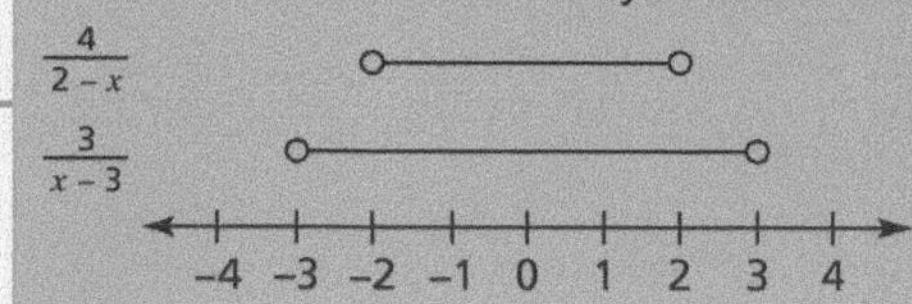

Quick Test

1. Find the first four terms of the expansion of $\frac{1}{\sqrt{1-2x}}$ and state the range of values of x for which the expansion is valid.
2. Find the first four terms of the expansion of $(2x-3)^{-2}$ and state the range of values of x for which the expansion is valid.
3. Find the first four terms of the expansion of $\frac{5x-7}{(x-1)(x-2)}$ and state the range of values of x for which the expansion is valid.

Key Words

binomial expansion

Practice Questions

Composite and Inverse Functions

1 $f(x) = x - 3$

a) Find $f^2(x)$ and $f(x^2)$, stating the domain and the range of each. [4]

b) Find the values for which $f^2(x) = f(x^2)$. [3]

c) Find $f^{-1}(x)$. [1]

2 Given $f(x) = 2\ln(x)$ and $g(x) = e^{3x}$, find $fg(x)$ and $gf(x)$. [4]

Total Marks .... / 12

Modulus and Exponential Functions

1 Given the graph of $y = f(x)$, sketch the graph of $y = |f(x)|$. [2]

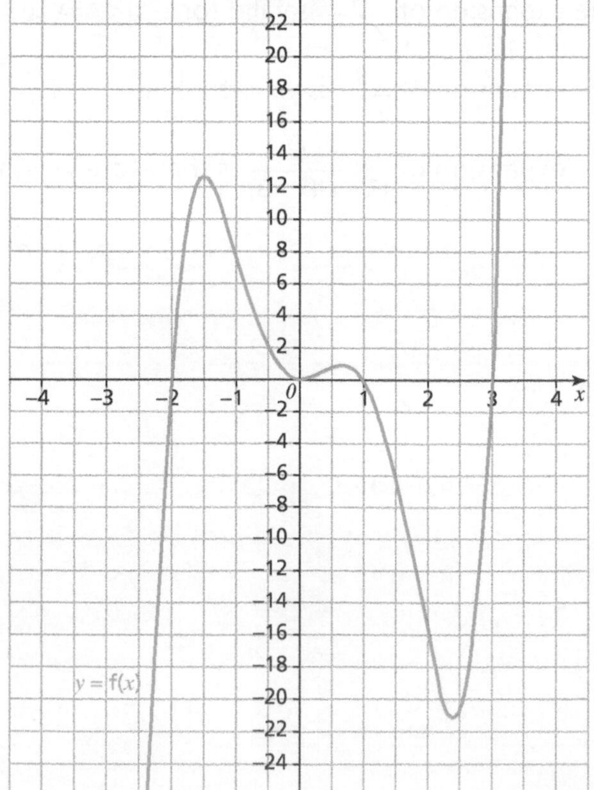

2. Sketch a graph of $y = 2\sin(3x) - 1$ on a separate piece of paper. [4]

3. Sketch a graph of $y = e^{2x} - 1$ on a separate piece of paper. [3]

Total Marks .... / 9

Algebraic Fractions

1. Simplify $\frac{(2x+1)(x^2-2x-3)}{(3x-4)(x+2)} \times \frac{x^2-4}{2x^2-5x-3}$ [5]

2. Using algebraic division, or otherwise, decompose $\frac{3x^4+2x^2-x+2}{x-1}$ into partial fractions. [4]

3. Decompose $\frac{5x^2+3x+13}{(x-1)^2(x+2)}$ into partial fractions. [7]

Total Marks .... / 16

Practice Questions

Parametric Equations

1 A curve has parametric equations $x=\sqrt{t}$, $y=2t+3$, $\{t\in\mathbb{R}: 0\leqslant t\leqslant 9\}$.

a) Express the parametric equations as a Cartesian equation $y=\text{f}(x)$, stating the domain and the range of $\text{f}(x)$. [4]

b) Sketch the curve $y=\text{f}(x)$ on a separate piece of paper. [2]

2 On a separate piece of paper, sketch the curve given by the parametric equations $x=-2+3\cos(t)$, $y=1+3\sin(t)$, $\left\{t\in\mathbb{R}: 0\leqslant t\leqslant \frac{3\pi}{2}\right\}$.

[5]

Total Marks .... / 11

Problems Involving Parametric Equations

1 The path of a figure skater can be modelled by the parametric equations $x=4\cos(t)+2$, $y=3\sin(2t)$, $\{t\in\mathbb{R}: 0\leqslant t\leqslant 2\pi\}$.

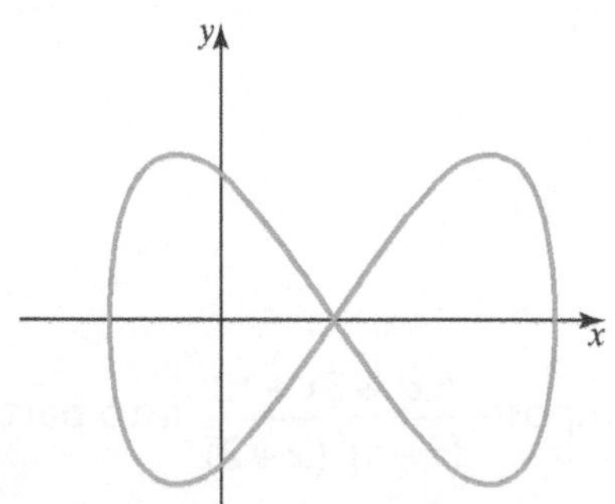

a) Find the coordinates of the point at which the figure skater starts. [2]

b) The figure skater's partner skates onto the ice along the line $y=1$. Find the times at which their paths cross on the ice. [5]

2 The path taken by a flying ant into and out of its nest can be modelled by the parametric equations $x = 2e^{2t}$, $y = e^{4t} - 3e^{2t} + 2$, $\{t \in \mathbb{R}: t \geqslant 0\}$.

Find the time, t, at which the ant flies into and out of its nest. [6]

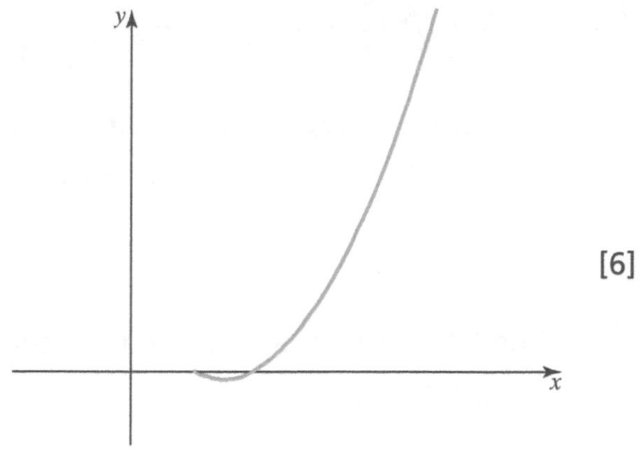

Total Marks .... / 13

Types of Sequences

1 A sequence is given by $u_n = 2n - 1$.

a) Describe the sequence as increasing, decreasing or periodic. [1]

b) Write the first five terms of the sequence. [1]

c) Write a recursive formula to describe the sequence. [1]

d) Find $\sum_{1}^{5} u_n$. [1]

2 Find the sum of the series $\sum_{1}^{5} \left(\frac{1}{2n}\right)^2$. [1]

3 Express the series $1 + \frac{2}{3} + \frac{1}{2} + \frac{2}{5} + \frac{1}{3}$ using sigma notation. [2]

Total Marks .... / 7

Practice Questions

Arithmetic Sequences and Series

1 Mohamed starts a pension with £2500. Each year he deposits £500 more than the previous year. Find the total amount he will have deposited after 30 years. [2]

2 The 10th term of an arithmetic sequence is 152 and the 50th term is –208. Find the formula for the nth term. [5]

3 Cheryl's starting salary is £35 000. She receives a pay rise of £400 per year. Find her salary in her 10th year. [2]

Total Marks .... / 9

Geometric Sequences and Series

1 A geometric sequence is defined by $u_n = \frac{1}{32} \times 4^{n-1}$

a) List the first five terms of the sequence. [1]

b) Find the sum of the first 10 terms of the sequence. [2]

c) Decide if the series is convergent or divergent and find the sum of the infinite series, if possible. [2]

2 Holly opens a savings account with £1250. The account earns 3% interest, compounded annually at the end of each year. She does not deposit any more money.

a) Find the amount of money in the account at the end of five years. [2]

b) Find the number of years after which the account will have £10 000. [3]

3 Seiji starts a new job with a salary of £30 000. His contract states that he will receive a pay rise of 6% of his previous salary each year.

a) How much will he make after five years? [2]

b) After how many years will he make at least £75 000? [2]

Total Marks .... / 14

Binomial Sequences

1 Find the first four terms of the expansion of $(3x - 1)^{-3}$ and state the range of values of x for which the expansion is valid. [4]

2 Find the first four terms of the expansion of $\frac{7x+1}{(x+4)(2x-1)}$ and state the range of values of x for which the expansion is valid. [8]

3 a) Find the first four terms of the expansion of $\frac{2}{x+3}$ and state the range of values of x for which the expansion is valid. [4]

b) Choose a suitable value of x and use the expansion to estimate the value of $\frac{200}{302}$ [3]

Total Marks .... / 19

Radians, Arc Lengths and Areas of Sectors

You must be able to:

- Understand radian measure for angles
- Convert between degrees and radians (and vice versa)
- Know and use the formula for the length of an arc
- Know and use the formula for the area of a sector.

Radian Measure

- An angle is defined to be one **radian** (1^c) if the **arc length** it makes on a circle is equal to the radius.
- Since the circumference of a circle is an arc length of $2\pi r$, the angle at the centre must be $2\pi \times 1$ radian $= 2\pi$ radians.
- Therefore 2π radians $= 360°$.
- So 1 radian $= \frac{360}{2\pi} = \frac{180}{\pi} \approx 57.3°$
- And 1 degree $= \frac{\pi}{180}$ radians.
- To convert degrees to radians, multiply by $\frac{\pi}{180}$.
  For example, $225° = 225 \times \frac{\pi}{180} = \frac{5\pi}{4}$ radians.
- To convert radians to degrees, multiply by $\frac{180}{\pi}$.
  For example, $\frac{2\pi}{5}$ radians $= \frac{2\pi}{5} \times \frac{180}{\pi} = 72°$

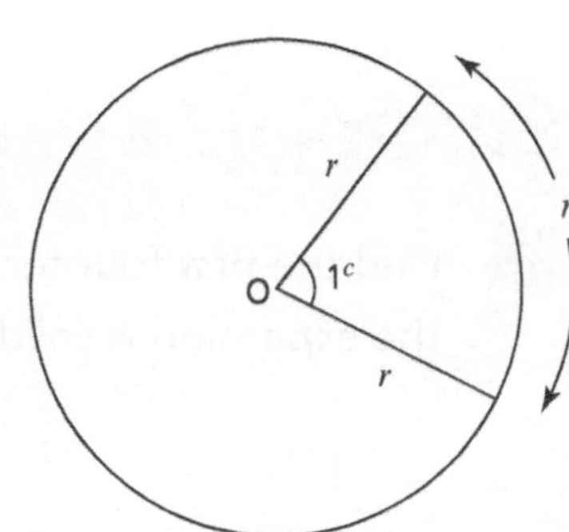

Arc Length

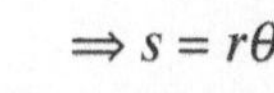

- If s is the arc length of a circle, r the radius, and θ the angle inside the **sector** (in radians), then $s = r\theta$.
- Notice that the following ratios are identical: $\frac{\text{arc length}}{\text{circumference}} = \frac{\theta^c}{2\pi}$
- So $\frac{s}{2\pi r} = \frac{\theta}{2\pi}$

  $\Rightarrow s = \frac{2\pi r\theta}{2\pi}$

  $\Rightarrow s = r\theta$

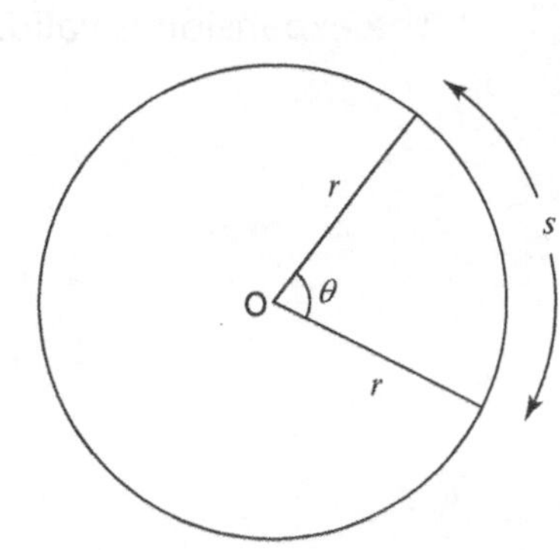

OAB is the sector of a circle. OA = OB = 12 cm and AB = 18 cm.

Find the arc length, s.

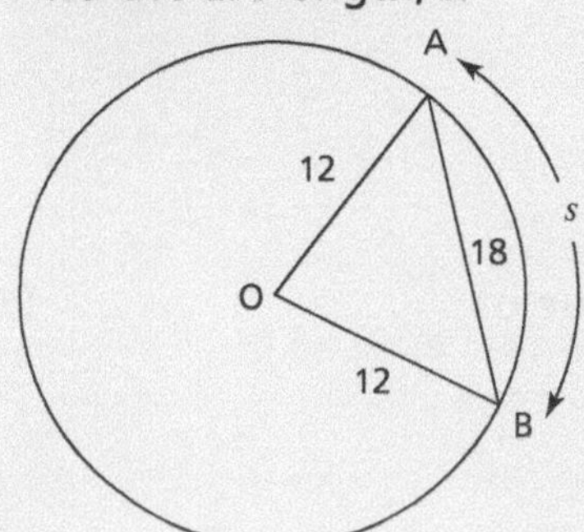

Let M be the midpoint of AB. Then $\sin\alpha = \frac{9}{12}$

So $\alpha = \arcsin\left(\frac{9}{12}\right)$ and $\theta = 2\alpha = 2\arcsin\left(\frac{9}{12}\right) = 1.696^c$

Now applying $s = r\theta$

$s = 12 \times 1.696 = 20.4$ cm

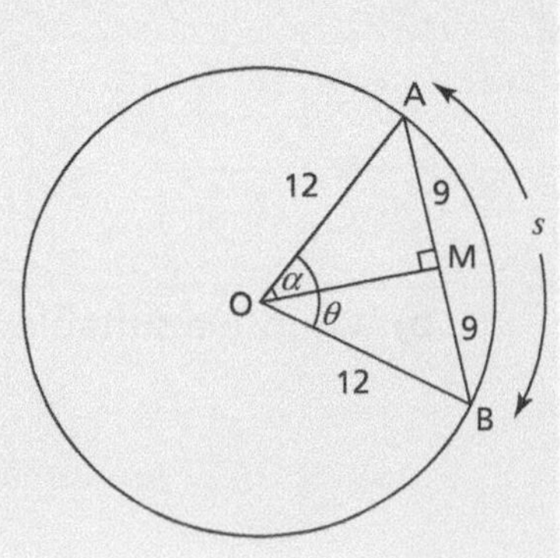

Area of a Sector

- If A is the area of the sector of a circle, r the radius, and θ the angle inside the sector (in radians), then $A = \frac{1}{2}r^2\theta$.
- Notice that the following ratios are identical:
  $\frac{\text{area of minor sector}}{\text{area of circle}} = \frac{\theta}{2\pi^c}$
- So $\frac{A}{\pi r^2} = \frac{\theta}{2\pi}$

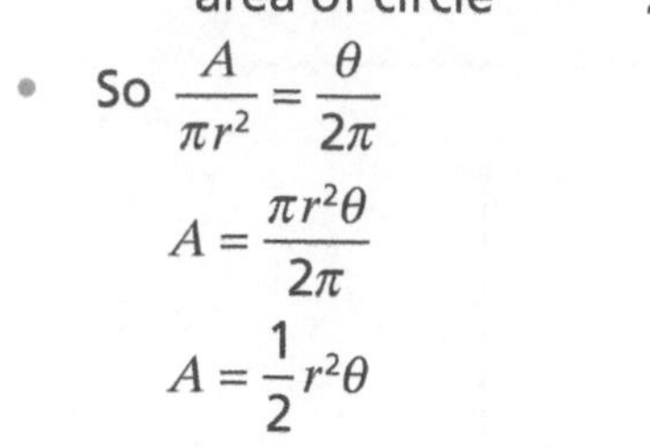

$$A = \frac{\pi r^2\theta}{2\pi}$$

$$A = \frac{1}{2}r^2\theta$$

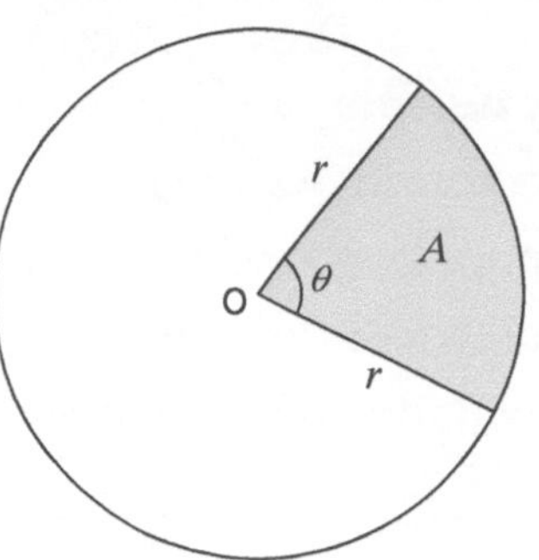

Key Point

Notice that two radii will always split a circle into two sectors. The larger sector is called the major sector, and the smaller one (shaded blue here) is called the minor sector. The area of the major sector here would therefore be equal to $(\pi r^2 - A)$.

The minor sector of a circle has area 100 cm². Given the angle at the centre is 60°, use radians to find the radius of the circle.

First changing the angle into radians: $60° = \frac{\pi}{3}$ radians

Now rearranging $A = \frac{1}{2}r^2\theta$ to make r the subject: $r = \sqrt{\frac{2A}{\theta}}$

Substituting in the given values: $r = \sqrt{\frac{200}{\frac{\pi}{3}}} = 13.8\text{ cm}$

Area of a Segment

- You may also be asked to find the area of a **segment**. Any **chord** of a circle divides the circle into two parts: a major segment and a minor segment (shaded blue here).
- To find the area of the minor segment, subtract the area of the triangle OAB from the area of the sector OAB:
- So area of segment
  $A = \frac{1}{2}r^2\theta - \frac{1}{2}r^2\sin\theta = \frac{1}{2}r^2(\theta - \sin\theta)$
- For example, if the angle is 1 radian and the radius is 5 cm, then: $A = \frac{25}{2}(1 - \sin(1)) = 1.98\text{ cm}^2$

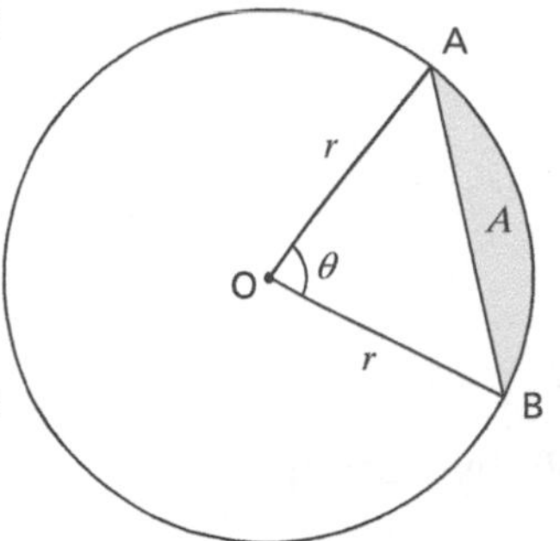

Key Point

The area of the major segment here would therefore be equal to $(\pi r^2 - A)$.

Quick Test

1. Convert $\frac{8\pi}{3}$ radians to degrees.
2. Find the area of a sector with angle 25° and radius 7 cm.
3. Find the perimeter of a circular sector with radius 10 cm and angle $\frac{\pi}{6}$ radians.

Key Words

radian
arc length
sector
segment
chord

Vectors in 3D

You must be able to:

- Use vectors in 3D
- Find magnitudes of vectors in 3D
- Find distances between two points in 3D
- Use the Ratio Theorem in 3D.

Unit Vectors

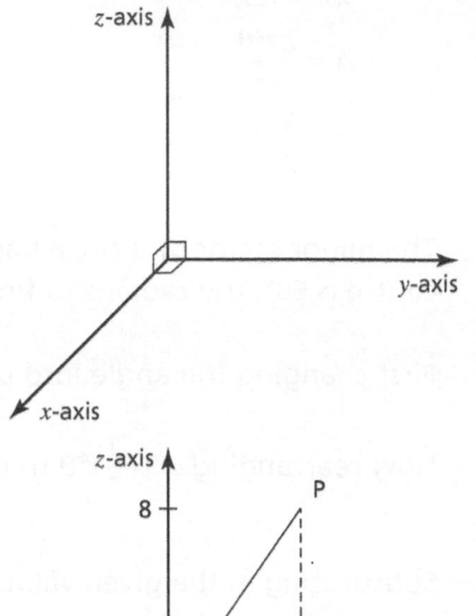

- Coordinate axes in three dimensions are called x-, y- and z-axes. Each axis is perpendicular to the other two.
- From AS-level work, you should know that **i** is a **unit vector** in the direction of the x-axis, and **j** is a unit vector in the direction of the y-axis. Extending this idea, **k** is a unit vector in the direction of the z-axis.
- Now any coordinate in three dimensions may be expressed as a position vector in Cartesian form. For example, if P is the point (3, 5, 8), then $\overrightarrow{OP} = 3\mathbf{i} + 5\mathbf{j} + 8\mathbf{k} = \begin{pmatrix} 3 \\ 5 \\ 8 \end{pmatrix}$.

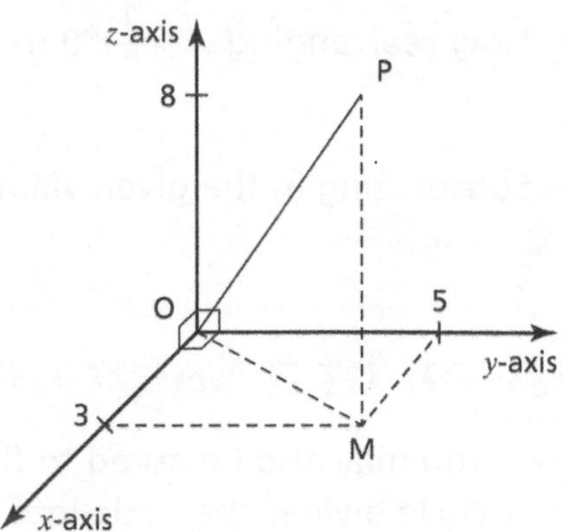

- The **magnitude** of a vector in three dimensions can be obtained using Pythagoras' Theorem.
- In the above case, $OM^2 = 3^2 + 5^2$ and $OP^2 = OM^2 + 8^2 = 3^2 + 5^2 + 8^2$. Therefore $|\overrightarrow{OP}| = \sqrt{3^2 + 5^2 + 8^2} = \sqrt{98} = 7\sqrt{2}$.
- In general, if $\overrightarrow{OP} = x\mathbf{i} + y\mathbf{j} + z\mathbf{k} = \begin{pmatrix} x \\ y \\ z \end{pmatrix}$, then $|\overrightarrow{OP}| = \sqrt{x^2 + y^2 + z^2}$.

Distance Between Two Points

- The above result can be generalised to find the **distance** between any two points in three-dimensional space.
- If $\overrightarrow{OP} = x_1\mathbf{i} + y_1\mathbf{j} + z_1\mathbf{k} = \begin{pmatrix} x_1 \\ y_1 \\ z_1 \end{pmatrix}$ and $\overrightarrow{OQ} = x_2\mathbf{i} + y_2\mathbf{j} + z_2\mathbf{k} = \begin{pmatrix} x_2 \\ y_2 \\ z_2 \end{pmatrix}$, then:

$$|\overrightarrow{PQ}| = \sqrt{(x_2 - x_1)^2 + (y_2 - y_1)^2 + (z_2 - z_1)^2}$$

- The proof is immediate, simply by noting that:

$$|\overrightarrow{PQ}| = |\overrightarrow{OQ} - \overrightarrow{OP}| = \left| \begin{pmatrix} x_2 \\ y_2 \\ z_2 \end{pmatrix} - \begin{pmatrix} x_1 \\ y_1 \\ z_1 \end{pmatrix} \right| = \begin{vmatrix} x_2 - x_1 \\ y_2 - y_1 \\ z_2 - z_1 \end{vmatrix}$$

The points P and Q have position vectors $\overrightarrow{OP} = \begin{pmatrix} 2 \\ -1 \\ -3 \end{pmatrix}$ and $\overrightarrow{OQ} = \begin{pmatrix} 0 \\ 3 \\ 5 \end{pmatrix}$.
Find $|\overrightarrow{PQ}|$.

$$|\overrightarrow{PQ}| = \sqrt{(0-2)^2 + (3-(-1))^2 + (5-(-3))^2} = \sqrt{4+16+64} = \sqrt{84} = 2\sqrt{21}$$

Key Point

$a\mathbf{i} + b\mathbf{j} + c\mathbf{k}$ and $\begin{pmatrix} a \\ b \\ c \end{pmatrix}$ are interchangeable. They are just a different way of writing the same thing.

- You may also be expected to find when the distance between two variable vectors is a minimum.

Points A and B have coordinates $(2t, 1, t)$ and $(1 - t, 3t, t - 2)$ respectively. Find the minimum value of $|\overrightarrow{AB}|$.

Using the distance formula, $|\overrightarrow{AB}|^2 = (1-t-2t)^2 + (3t-1)^2 + (t-2-t)^2$
$= (1 - 3t)^2 + (3t - 1)^2 + (-2)^2$
$= 9t^2 - 6t + 1 + 9t^2 - 6t + 1 + 4$
$= 18t^2 - 12t + 6$

Now differentiate with respect to t and set equal to zero:

$36t - 12 = 0 \Rightarrow t = \frac{1}{3}$

At $t = \frac{1}{3}$, $|\overrightarrow{AB}|^2 = 18\left(\frac{1}{3}\right)^2 - 12\left(\frac{1}{3}\right) + 6 = 4$

Therefore $|\overrightarrow{AB}|_{min} = 2$

$|\overrightarrow{AB}|$ is a minimum when $|\overrightarrow{AB}|^2$ is a minimum, so using this fact it is much easier to differentiate $18t^2 - 12t + 6$ rather than $\sqrt{18t^2 - 12t + 6}$.

Key Point

If you use the formula $\overrightarrow{OC} = \frac{n\mathbf{a} + m\mathbf{b}}{m+n}$, then take careful note of the order that m and n appear in the numerator.

Ratio Theorem

- The **Ratio Theorem** is a quick way of finding a point C when C divides a line AB in a given ratio.
- Suppose $\overrightarrow{OA} = \mathbf{a}$ and $\overrightarrow{OB} = \mathbf{b}$. If C divides AB in the ratio $m : n$, then $\overrightarrow{OC} = \frac{n\mathbf{a} + m\mathbf{b}}{m+n}$
- Proof:
  $\overrightarrow{AB} = -\mathbf{a} + \mathbf{b}$
  So $\overrightarrow{AC} = \frac{m}{m+n}\overrightarrow{AB} = \frac{m}{m+n}(\mathbf{b} - \mathbf{a})$
  So $\overrightarrow{OC} = \overrightarrow{OA} + \overrightarrow{AC} = \mathbf{a} + \frac{m}{m+n}(\mathbf{b} - \mathbf{a}) = \frac{(m+n)\mathbf{a} + m(\mathbf{b} - \mathbf{a})}{m+n} = \frac{n\mathbf{a} + m\mathbf{b}}{m+n}$,
  as required.

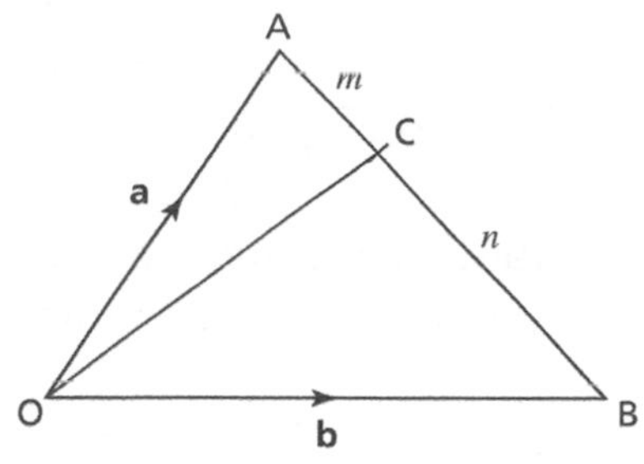

Suppose $\overrightarrow{OA} = \begin{pmatrix} -3 \\ 4 \\ 12 \end{pmatrix}$ and $\overrightarrow{OB} = \begin{pmatrix} 9 \\ -8 \\ 4 \end{pmatrix}$.

Find the coordinates of point C that divides AB in the ratio 2 : 5.

Using $\overrightarrow{OC} = \frac{n\mathbf{a} + m\mathbf{b}}{m+n}$,

$$\overrightarrow{OC} = \frac{5\mathbf{a} + 2\mathbf{b}}{7} = \frac{5\begin{pmatrix} -3 \\ 4 \\ 12 \end{pmatrix} + 2\begin{pmatrix} 9 \\ -8 \\ 4 \end{pmatrix}}{7} = \frac{\begin{pmatrix} -15 \\ 20 \\ 60 \end{pmatrix} + \begin{pmatrix} 18 \\ -16 \\ 8 \end{pmatrix}}{7} = \frac{1}{7}\begin{pmatrix} 3 \\ 4 \\ 68 \end{pmatrix}$$

Quick Test

1. Find the magnitude of the vector $3\mathbf{i} - 3\mathbf{j} + 8\mathbf{k}$.
2. Find the distance between the points $(-1, -1, 5)$ and $(0, 2, -6)$.
3. Find the possible value(s) of t if the magnitude of $\begin{pmatrix} 3 \\ t \\ 4 \end{pmatrix}$ is equal to $\sqrt{33}$.

Key Words

unit vector
magnitude
distance
Ratio Theorem

Trigonometric Ratios

Trigonometry

You must be able to:

- Understand and use small angle approximations
- Know and use exact values for sine, cosine and tangent
- Understand secant, cosecant, cotangent and their graphs
- Understand arcsin, arccos and arctan, and their graphs.

Small Angle Approximations

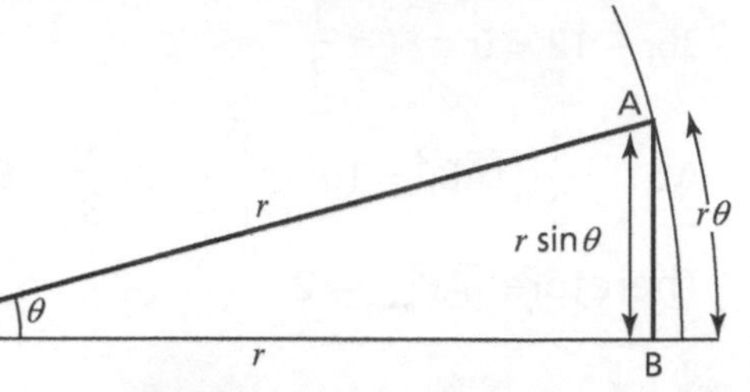

- If you consider a small sector of radius r and angle θ (in radians), the arc length is given as $r\theta$. As θ decreases in size, you can see that the vertical length $r\sin\theta$ becomes a better and better **approximation** to $r\theta$, i.e. $r\sin\theta \approx r\theta$, or $\sin\theta \approx \theta$.
- Similarly, it can be shown that for **small angles** θ (in radians), $\cos\theta \approx 1 - \frac{\theta^2}{2}$ and $\tan\theta \approx \theta$.

Key Point

These approximations are true **only** when θ is measured in radians.

- Using these results, other approximations may be derived. For example:

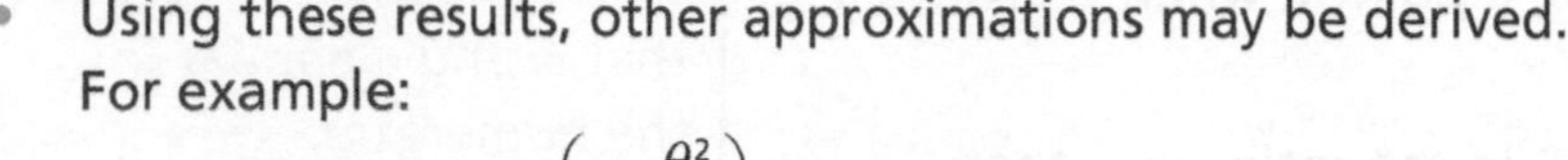

  – $2\sin\theta\cos\theta \approx 2\theta\left(1 - \frac{\theta^2}{2}\right) = 2\theta - \theta^3$

  – $\frac{1}{\cos\theta} \approx \frac{1}{1 - \frac{\theta^2}{2}} = \left(1 - \frac{\theta^2}{2}\right)^{-1} \approx 1 + \frac{\theta^2}{2} + \frac{\theta^4}{4} + \ldots \approx 1 + \frac{\theta^2}{2}$

Using the binomial expansion for $(1 + x)^n$.

- Numerical values can also be determined for some functions of small angles. For example, for small θ,

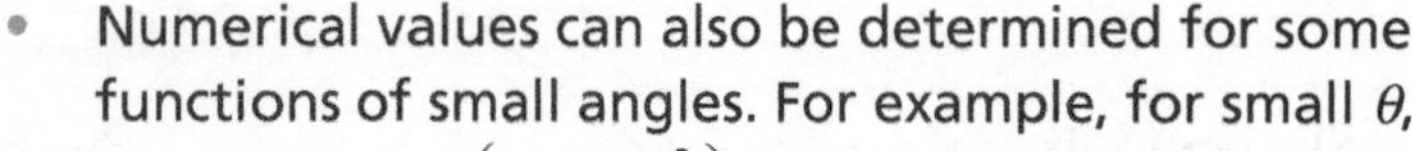

$$\frac{1 - \cos 2\theta}{5\theta\tan\theta} \approx \frac{1 - \left(1 - \frac{(2\theta)^2}{2}\right)}{5\theta^2} \approx \frac{1 - (1 - 2\theta^2)}{5\theta^2} = \frac{2\theta^2}{5\theta^2} = \frac{2}{5}$$

Exact Trigonometric Values

- You must learn some **exact values** for sin, cos and tan (see table right).
- By remembering these values, and knowing in which quadrant the angle lies, it is straightforward to quickly calculate the trigonometric ratios of related angles outside this range. For example:
  – $\sin 210° = \sin(180° + 30°) = -\sin 30° = -\frac{1}{2}$
  – $\tan 315° = \tan(360° - 45°) = -\tan 45° = -1$
  – $\cos 690° = \cos(720° - 30°) = \cos 30° = \frac{\sqrt{3}}{2}$
- You should also be familiar with these ratios when working in radians. For example:
  – $\tan\frac{2\pi}{3} = \tan\left(\pi - \frac{\pi}{3}\right) = -\tan\frac{\pi}{3} = -\sqrt{3}$
  – $\sin\frac{11\pi}{6} = \sin\left(2\pi - \frac{\pi}{6}\right) = -\sin\frac{\pi}{6} = -\frac{1}{2}$
  – $\cos\frac{11\pi}{4} = \cos\left(3\pi - \frac{\pi}{4}\right) = \cos\left(2\pi + \left(\pi - \frac{\pi}{4}\right)\right) = \cos\left(\pi - \frac{\pi}{4}\right) = -\cos\frac{\pi}{4} = -\frac{1}{\sqrt{2}}$

	sin	cos	tan
0°	0	1	0
30°	$\frac{1}{2}$	$\frac{\sqrt{3}}{2}$	$\frac{1}{\sqrt{3}}$
45°	$\frac{1}{\sqrt{2}}$	$\frac{1}{\sqrt{2}}$	1
60°	$\frac{\sqrt{3}}{2}$	$\frac{1}{2}$	$\sqrt{3}$
90°	1	0	–
180°	0	–1	0

Secant, Cosecant and Cotangent

- $\sec\theta \equiv \dfrac{1}{\cos\theta}$ for all values of θ except where $\cos\theta = 0$.
- $\operatorname{cosec}\theta \equiv \dfrac{1}{\sin\theta}$ for all values of θ except where $\sin\theta = 0$.
- $\cot\theta \equiv \dfrac{1}{\tan\theta}$ for all values of θ except where $\tan\theta = 0$.
- You can therefore use these to calculate other exact trigonometric values. For example:
  - $\cot\dfrac{5\pi}{6} = \dfrac{1}{\tan\left(\dfrac{5\pi}{6}\right)} = \dfrac{1}{\tan\left(\pi - \dfrac{\pi}{6}\right)} = -\dfrac{1}{\tan\dfrac{\pi}{6}} = -\dfrac{1}{\dfrac{1}{\sqrt{3}}} = -\sqrt{3}$
  - $\sec\dfrac{11\pi}{3} = \dfrac{1}{\cos\dfrac{11\pi}{3}} = \dfrac{1}{\cos\left(4\pi - \dfrac{\pi}{3}\right)} = \dfrac{1}{\cos\dfrac{\pi}{3}} = \dfrac{1}{\dfrac{1}{2}} = 2$
- You need to know the graphs of each of these functions, as well as their **domains** and **ranges** (see table right).

Inverse Trigonometric Functions

- If you restrict the domain of sin, cos and tan, then they may become one-to-one functions. Their respective inverse functions may then be found by reflecting each function in the line $y = x$.

$y = \arcsin x$ has domain $-1 \leqslant x \leqslant 1$ and range $-\dfrac{\pi}{2} \leqslant y \leqslant \dfrac{\pi}{2}$.

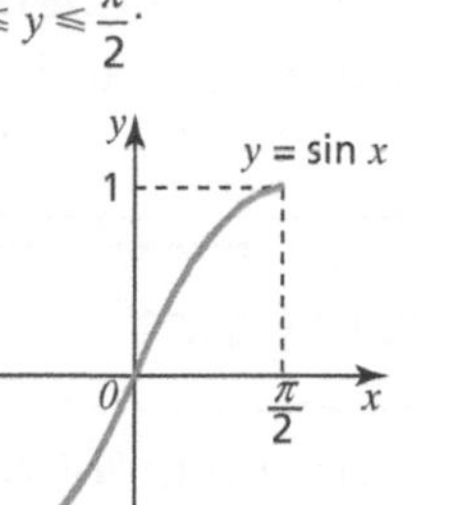

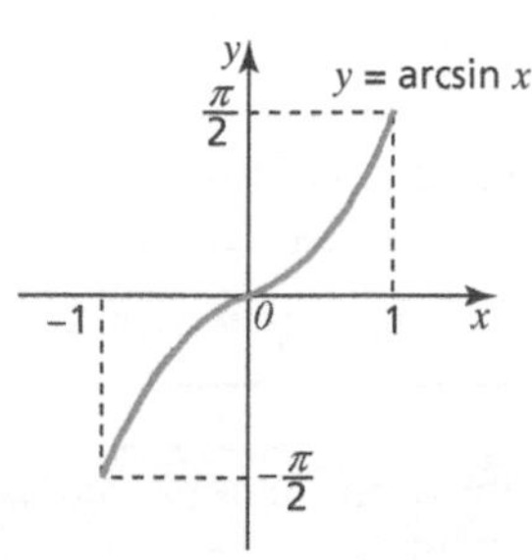

$y = \arccos x$ has domain $-1 \leqslant x \leqslant 1$ and range $0 \leqslant y \leqslant \pi$.

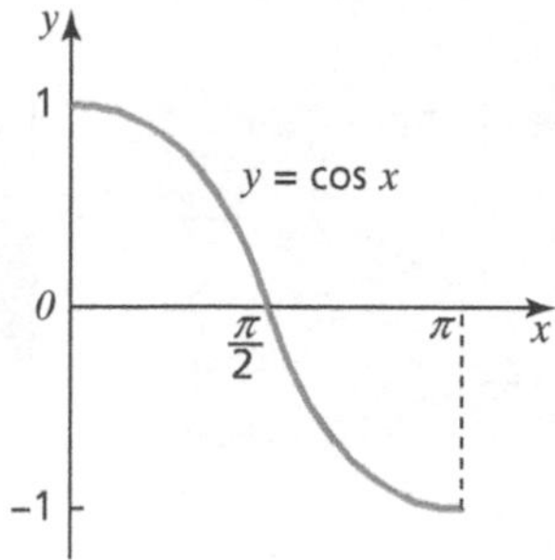

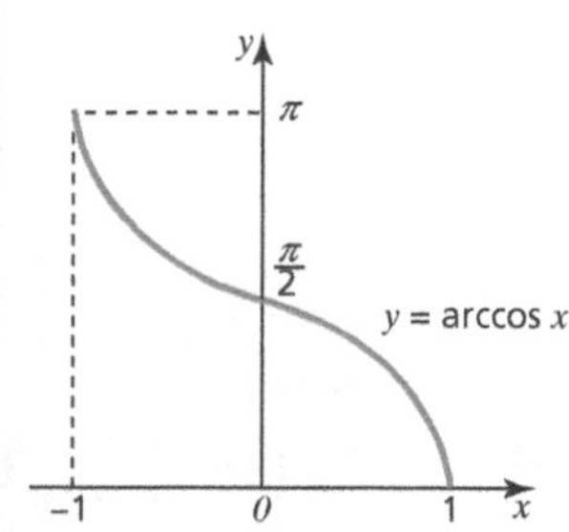

$y = \arctan x$ has domain $x \in \mathbb{R}$ and range $-\dfrac{\pi}{2} \leqslant y \leqslant \dfrac{\pi}{2}$.

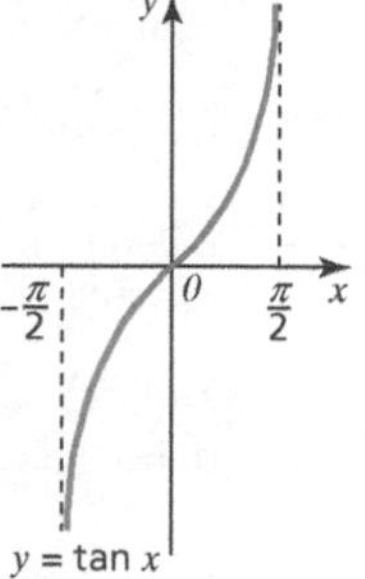

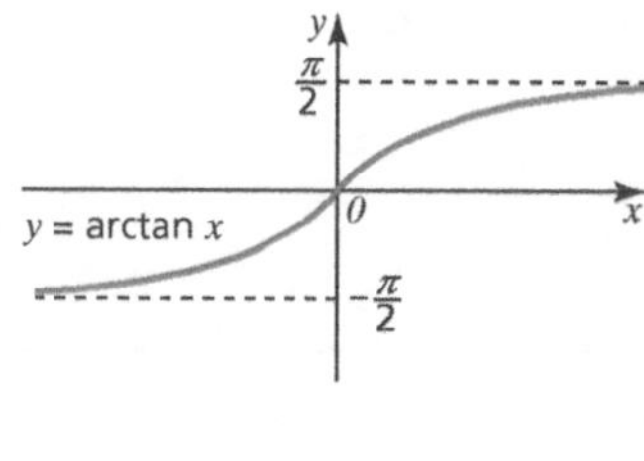

$y = \sec\theta$

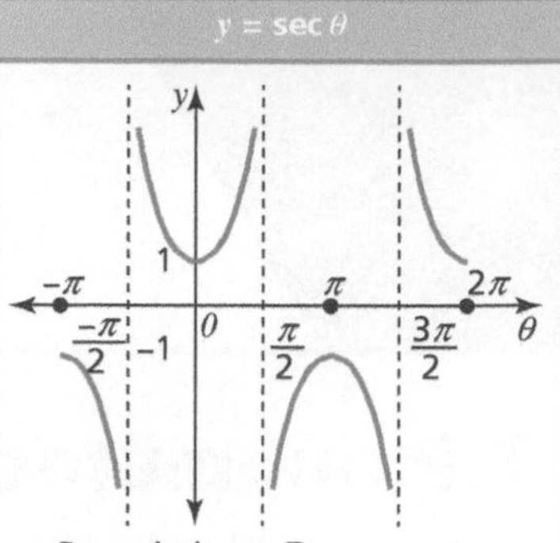

- Domain is $x \in \mathbb{R}$, $x \neq (2n+1)\dfrac{\pi}{2}$
- Range is $y \in \mathbb{R}$, $y \leqslant -1$ or $y \geqslant 1$.
- Graph is symmetrical about the y-axis (i.e. $\sec x$ is an even function).

$y = \operatorname{cosec}\theta$

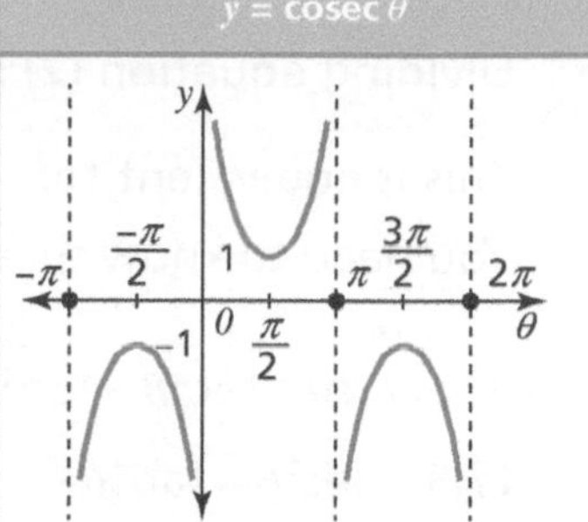

- Domain is $x \in \mathbb{R}$, $x \neq n\pi$.
- Range is $y \in \mathbb{R}$, $y \leqslant -1$ or $y \geqslant 1$.

$y = \cot\theta$

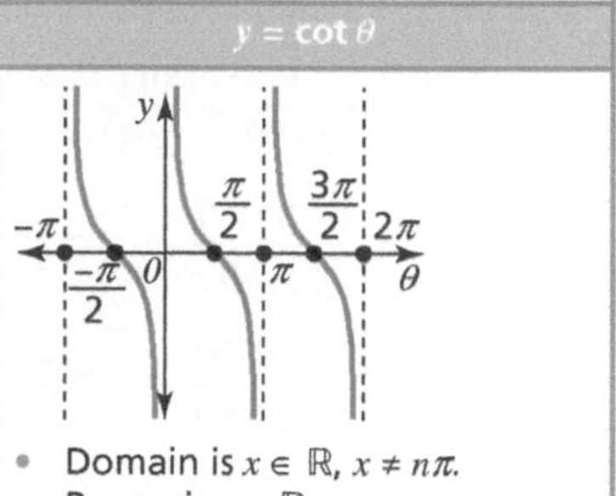

- Domain is $x \in \mathbb{R}$, $x \neq n\pi$.
- Range is $y \in \mathbb{R}$.

Quick Test

1. Without a calculator, show that $\dfrac{1-\sec\theta}{\tan^2\theta} \approx -\dfrac{1}{2}$ for small θ.
2. Without a calculator, find the value of $\operatorname{cosec}\left(\dfrac{2\pi}{3}\right)$.
3. Without a calculator, find the value of $\arctan\left(\tan\dfrac{11\pi}{4}\right)$.

Key Words

approximation
small angles
exact values
domain
range

Trigonometry

Further Trigonometric Equations and Identities

You must be able to:

- Solve equations using $\sec\theta$, $\operatorname{cosec}\theta$ and $\cot\theta$
- Prove identities using $\sec\theta$, $\operatorname{cosec}\theta$ and $\cot\theta$.

Trigonometric Identities

- You should already know the following **identities**:

$\tan\theta = \dfrac{\sin\theta}{\cos\theta}$ **(1)** $\qquad \sin^2\theta + \cos^2\theta = 1$ **(2)**

- Taking equation **(2)**, and dividing by $\sin^2\theta$, gives $\dfrac{\sin^2\theta}{\sin^2\theta} + \dfrac{\cos^2\theta}{\sin^2\theta} = \dfrac{1}{\sin^2\theta}$

This is equivalent to: $1 + \cot^2\theta = \operatorname{cosec}^2\theta$ **(3)**

- Dividing equation **(2)** by $\cos^2\theta$ gives $\dfrac{\sin^2\theta}{\cos^2\theta} + \dfrac{\cos^2\theta}{\cos^2\theta} = \dfrac{1}{\cos^2\theta}$

This is equivalent to: $\tan^2\theta + 1 = \sec^2\theta$ **(4)**

- You need to know these four equations to prove trigonometric identities.

1) Show that $\sec^4\theta - \tan^4\theta \equiv \dfrac{1+\sin^2\theta}{1-\sin^2\theta}$

LHS $= \sec^4\theta - \tan^4\theta$

$= (\sec^2\theta - \tan^2\theta)(\sec^2\theta + \tan^2\theta)$ ← Recognising the difference of two squares.

$= 1(\sec^2\theta + \tan^2\theta)$ ← Using identity **(4)** in first bracket.

$= \dfrac{1}{\cos^2\theta} + \dfrac{\sin^2\theta}{\cos^2\theta}$ ← Using definition of sec θ and identity **(1)**.

$= \dfrac{1+\sin^2\theta}{\cos^2\theta}$ ← Collect over the common denominator.

$= \dfrac{1+\sin^2\theta}{1-\sin^2\theta} =$ RHS ← Using identity **(2)**.

2) Show that $\dfrac{\cos^2\theta - \sin^2\theta}{\cot^2\theta - \tan^2\theta} \equiv \sin^2\theta\cos^2\theta$

LHS $= \dfrac{\cos^2\theta - \sin^2\theta}{\cot^2\theta - \tan^2\theta}$

$= \dfrac{\cos^2\theta - \sin^2\theta}{\dfrac{\cos^2\theta}{\sin^2\theta} - \dfrac{\sin^2\theta}{\cos^2\theta}}$ ← Express the denominator in terms of sin and cos.

$= \dfrac{\cos^2\theta - \sin^2\theta}{\dfrac{\cos^4\theta - \sin^4\theta}{\sin^2\theta\cos^2\theta}}$ ← Express the denominator as a single fraction.

$= \dfrac{(\cos^2\theta - \sin^2\theta)(\sin^2\theta\cos^2\theta)}{\cos^4\theta - \sin^4\theta}$ ← Simplify the expression.

$= \dfrac{(\cos^2\theta - \sin^2\theta)(\sin^2\theta\cos^2\theta)}{(\cos^2\theta + \sin^2\theta)(\cos^2\theta - \sin^2\theta)}$ ← Factorise the new denominator.

$= \sin^2\theta\cos^2\theta =$ RHS ← Use identity **(2)** and cancel.

Key Point

An identity is always true, for any value of θ (or x). An identity can also be referred to as an **equation**, but it is a much stronger condition.

Key Point

When proving identities, always start from one side of the identity and show it is equal to the other side through a sequence of logical mathematical steps.

Key Point

When proving identities (or solving equations), always be on the look-out for using the difference of two squares. This technique is often required when tackling trigonometry questions.

Trigonometric Equations

- The technique used to solve trigonometric equations at A-level remains similar to that used at AS: you can often use identities (if needed) to transform an equation into one involving only one trigonometric function.
- Note that at A-level you may also be expected to express your solutions in **radians**.

Key Point

Remember to factorise where possible, rather than cancelling. For example, $\sin\theta\cot\theta = \cot\theta$ would lead to $\cot\theta(\sin\theta - 1) = 0$. You should not cancel cot to give $\sin\theta = 1$. This is incorrect as it loses possible solutions where $\cot\theta = 0$.

1) Solve $\cot^2\theta - 3\operatorname{cosec}\theta + 3 = 0$ in the range $0 \leqslant \theta \leqslant 2\pi$.

$$\cot^2\theta - 3\operatorname{cosec}\theta + 3 = 0$$

$$(\operatorname{cosec}^2\theta - 1) - 3\operatorname{cosec}\theta + 3 = 0$$ ← Write cot in terms of cosec, using identity **(3)**.

$$\operatorname{cosec}^2\theta - 3\operatorname{cosec}\theta + 2 = 0$$ ← Collect the terms.

$$(\operatorname{cosec}\theta - 2)(\operatorname{cosec}\theta - 1) = 0$$ ← Factorise the quadratic.

$$\operatorname{cosec}\theta = 2$$ ← Solve the first bracket.

$$\sin\theta = \frac{1}{2}$$

$$\Rightarrow \theta = \frac{\pi}{6}, \frac{5\pi}{6}$$ ← Remember there are two solutions in the given range.

Or

$$\operatorname{cosec}\theta = 1$$ ← Solve the second bracket.

$$\sin\theta = 1$$

$$\Rightarrow \theta = \frac{\pi}{2}$$ ← Only one solution here.

2) Solve $\sqrt{3}\sec^2\theta - \left(1+\sqrt{3}\right)\tan\theta + 1 = \sqrt{3}$ in the range $0 \leqslant \theta \leqslant 2\pi$.

$$\sqrt{3}\sec^2\theta - \left(1+\sqrt{3}\right)\tan\theta + 1 = \sqrt{3}$$

$$\sqrt{3}(1+\tan^2\theta) - \left(1+\sqrt{3}\right)\tan\theta + 1 = \sqrt{3}$$ ← Write sec in terms of tan, using identity **(4)**.

$$\sqrt{3} + \sqrt{3}\tan^2\theta - \left(1+\sqrt{3}\right)\tan\theta + 1 = \sqrt{3}$$ ← Multiply out the brackets.

$$\sqrt{3}\tan^2\theta - \left(1+\sqrt{3}\right)\tan\theta + 1 = 0$$ ← Cancel the $\sqrt{3}$.

$$\left(\sqrt{3}\tan\theta - 1\right)(\tan\theta - 1) = 0$$ ← Factorise (or use the quadratic formula).

$$\tan\theta = \frac{1}{\sqrt{3}}$$ ← Solve the first bracket.

$$\Rightarrow \theta = \frac{\pi}{6}, \frac{7\pi}{6}$$ ← Find the two solutions in the given range.

Or

$$\tan\theta = 1$$ ← Solve the second bracket.

$$\Rightarrow \theta = \frac{\pi}{4}, \frac{5\pi}{4}$$ ← Find the two further solutions in the given range.

Quick Test

1. Solve the equation $\cot 2x = \sqrt{3}$ in the range $0 \leqslant \theta \leqslant 2\pi$.
2. Prove the identity $\operatorname{cosec}^4\theta - \cot^4\theta = 1 + 2\cot^2\theta$.
3. Solve the equation $\cot x \sec x + 2\cot x = 0$ in the range $0 \leqslant \theta \leqslant 2\pi$.

Key Words

identity
equation

Using Addition Formulae and Double Angle Formulae

You must be able to:

- Understand and use the addition formulae
- Understand and use the double angle formulae.

Addition Formulae

$$\sin(\alpha+\beta)=\sin\alpha\cos\beta+\cos\alpha\sin\beta \qquad \sin(\alpha-\beta)=\sin\alpha\cos\beta-\cos\alpha\sin\beta$$

$$\cos(\alpha+\beta)=\cos\alpha\cos\beta-\sin\alpha\sin\beta \qquad \cos(\alpha-\beta)=\cos\alpha\cos\beta+\sin\alpha\sin\beta$$

$$\tan(\alpha+\beta)=\frac{\tan\alpha+\tan\beta}{1-\tan\alpha\tan\beta} \qquad \tan(\alpha-\beta)=\frac{\tan\alpha-\tan\beta}{1+\tan\alpha\tan\beta}$$

These six formulae are given in the formula booklet.

- You need to understand the geometrical proofs of these formulae. The diagram (right) offers a way of proving these results.
- The **addition formulae** may be used to find more **exact trigonometric values**, simplify expressions, or prove further identities.

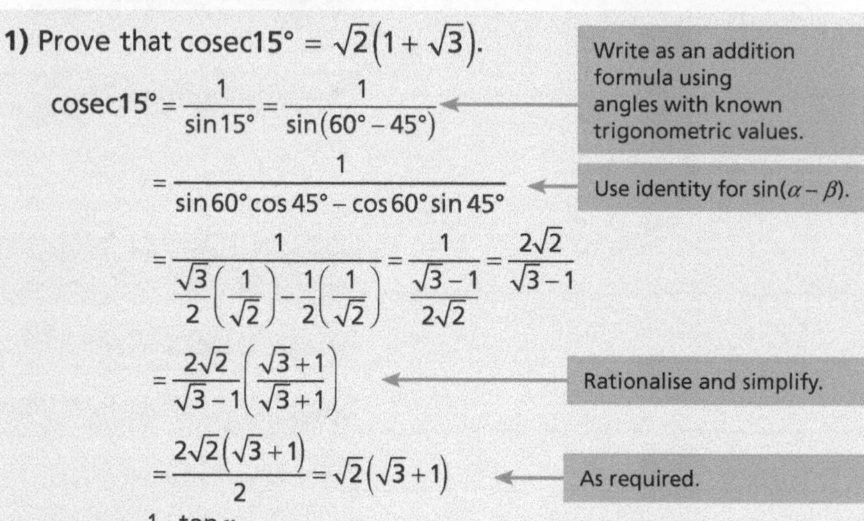

1) Prove that $\text{cosec}15° = \sqrt{2}(1+\sqrt{3})$.

$$\text{cosec}15° = \frac{1}{\sin 15°} = \frac{1}{\sin(60°-45°)}$$

Write as an addition formula using angles with known trigonometric values.

$$= \frac{1}{\sin 60°\cos 45° - \cos 60°\sin 45°}$$

Use identity for $\sin(\alpha-\beta)$.

$$= \frac{1}{\frac{\sqrt{3}}{2}\left(\frac{1}{\sqrt{2}}\right)-\frac{1}{2}\left(\frac{1}{\sqrt{2}}\right)} = \frac{1}{\frac{\sqrt{3}-1}{2\sqrt{2}}} = \frac{2\sqrt{2}}{\sqrt{3}-1}$$

$$= \frac{2\sqrt{2}}{\sqrt{3}-1}\left(\frac{\sqrt{3}+1}{\sqrt{3}+1}\right)$$

Rationalise and simplify.

$$= \frac{2\sqrt{2}(\sqrt{3}+1)}{2} = \sqrt{2}(\sqrt{3}+1)$$

As required.

2) Express $\frac{1-\tan x}{1+\tan x}$ as a single trigonometric expression.

$$\frac{1-\tan x}{1+\tan x} = \frac{\tan\frac{\pi}{4}-\tan x}{1+\tan\frac{\pi}{4}\tan x} = \tan\left(\frac{\pi}{4}-x\right)$$

Recall that $1=\tan\frac{\pi}{4}$ and use the $\tan(\alpha-\beta)$ identity.

3) Given $\sin A = \frac{7}{25}$ and A is obtuse, and $\cos B = \frac{3}{5}$ and B is reflex, find the value of $\cos(A+B)$.

$$\sin A = \frac{7}{25} \Rightarrow \cos A = -\sqrt{1-\sin^2 A} = -\sqrt{1-\left(\frac{7}{25}\right)^2} = -\frac{24}{25}$$

Since A is obtuse, $\cos A < 0$, so take the negative value.

$$\cos B = \frac{3}{5} \Rightarrow \sin B = -\sqrt{1-\cos^2 B} = -\sqrt{1-\left(\frac{3}{5}\right)^2} = -\frac{4}{5}$$

Since B is reflex, $\sin B < 0$, so take the negative value.

Now $\cos(A+B) = \cos A\cos B - \sin A\sin B = -\frac{24}{25}\left(\frac{3}{5}\right)-\frac{7}{25}\left(-\frac{4}{5}\right)$

$$= -\frac{72}{125}+\frac{28}{125} = -\frac{44}{125}$$

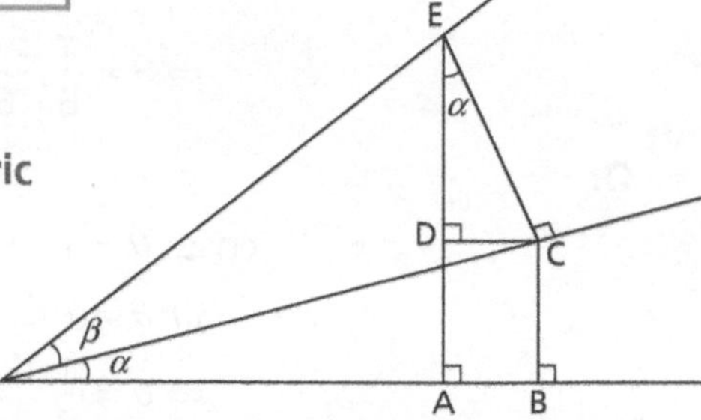

Angle AOE = $\alpha+\beta$. Note that angle DCO = α (alternate angles), so DCE = 90 − α and so DEC = α

- From diagram, $\sin(\alpha+\beta) = \frac{AE}{OE} = \frac{AD+DE}{OE}$

$$= \frac{AD}{OE}+\frac{DE}{OE} = \frac{BC}{OE}+\frac{DE}{OE} = \frac{BC}{OC}\times\frac{OC}{OE}+\frac{DE}{EC}\times\frac{EC}{OE}$$

$$= \sin\alpha\cos\beta+\cos\alpha\sin\beta$$

- Also, replacing β with $(-\beta)$ gives

$$\sin(\alpha-\beta) = \sin(\alpha+(-\beta))$$
$$= \sin\alpha\cos(-\beta)+\cos\alpha\sin(-\beta)$$
$$= \sin\alpha\cos\beta-\cos\alpha\sin\beta$$

- From diagram, $\cos(\alpha+\beta) = \frac{OA}{OE} = \frac{OB-AB}{OE}$

$$= \frac{OB}{OE}-\frac{AB}{OE} = \frac{OB}{OE}-\frac{CD}{OE} = \frac{OB}{OC}\times\frac{OC}{OE}-\frac{CD}{EC}\times\frac{EC}{OE}$$

$$= \cos\alpha\cos\beta-\sin\alpha\sin\beta$$

- Also, replacing β with $(-\beta)$ gives

$$\cos(\alpha-\beta) = \cos(\alpha+(-\beta))$$
$$= \cos\alpha\cos(-\beta)-\sin\alpha\sin(-\beta)$$
$$= \cos\alpha\cos\beta+\sin\alpha\sin\beta$$

- $$\tan(\alpha+\beta) = \frac{\sin(\alpha+\beta)}{\cos(\alpha+\beta)}$$

$$= \frac{\sin\alpha\cos\beta+\cos\alpha\sin\beta}{\cos\alpha\cos\beta-\sin\alpha\sin\beta}$$

$$= \frac{\frac{\sin\alpha\cos\beta}{\cos\alpha\cos\beta}+\frac{\cos\alpha\sin\beta}{\cos\alpha\cos\beta}}{\frac{\cos\alpha\cos\beta}{\cos\alpha\cos\beta}-\frac{\sin\alpha\sin\beta}{\cos\alpha\cos\beta}} = \frac{\tan\alpha+\tan\beta}{1-\tan\alpha\tan\beta}$$

- Finally, replacing β with $(-\beta)$ gives

$$\tan(\alpha+(-\beta)) = \frac{\tan\alpha+\tan(-\beta)}{1-\tan\alpha\tan(-\beta)}$$

$$= \frac{\tan\alpha-\tan\beta}{1+\tan\alpha\tan\beta}$$

Double Angle Formulae

$$\sin 2\alpha = 2\sin\alpha\cos\alpha \qquad \cos 2\alpha = \cos^2\alpha - \sin^2\alpha = 2\cos^2\alpha - 1 = 1 - 2\sin^2\alpha \qquad \tan 2\alpha = \frac{2\tan\alpha}{1-\tan^2\alpha}$$

- You can prove each of these **double angle** formulae by substituting $\beta = \alpha$ into the required addition formula.
- For example, $\sin(\alpha + \beta) = \sin\alpha\cos\beta + \cos\alpha\sin\beta$.
  Putting $\beta = \alpha$ gives $\sin(\alpha + \alpha) = \sin\alpha\cos\alpha + \cos\alpha\sin\alpha$, or $\sin 2\alpha = 2\sin\alpha\cos\alpha$.
- The double angle formulae can be expressed in terms of **half angles:**

$$\sin\alpha = 2\sin\frac{\alpha}{2}\cos\frac{\alpha}{2} \qquad \cos\alpha = \cos^2\frac{\alpha}{2} - \sin^2\frac{\alpha}{2} = 2\cos^2\frac{\alpha}{2} - 1 = 1 - 2\sin^2\frac{\alpha}{2} \qquad \tan\alpha = \frac{2\tan\frac{\alpha}{2}}{1-\tan^2\frac{\alpha}{2}}$$

- The double angle and half angle formulae can be used to find other trigonometric values, solve further equations and prove further identities.

1) Find the value of $\tan\left(\frac{x}{2}\right)$ if $\tan x = 2$ and $0 < x < \frac{\pi}{2}$.

$$\tan x = \frac{2\tan\frac{x}{2}}{1-\tan^2\frac{x}{2}}$$

$$2 = \frac{2t}{1-t^2}$$

$$2 - 2t^2 = 2t$$

$$2t^2 + 2t - 2 = 0, \text{ so } t^2 + t - 1 = 0$$

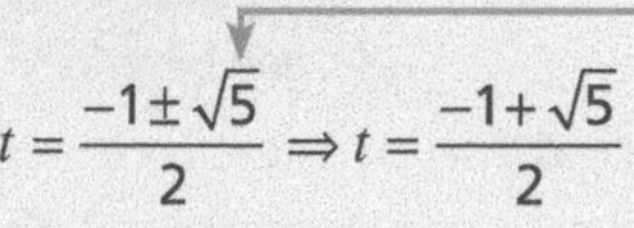

$$t = \frac{-1 \pm \sqrt{5}}{2} \Rightarrow t = \frac{-1+\sqrt{5}}{2}$$

Write $t = \tan\left(\frac{x}{2}\right)$ and substitute in $\tan x = 2$.

Rearrange and simplify the quadratic.

Using the quadratic formula.

Choose the positive root since $0 < x < \frac{\pi}{2}$

2) Given $\tan\theta = \frac{5}{12}$ and θ is acute, find an exact value for $\cos 4\theta$.

(Triangle: hypotenuse 13, opposite 5, adjacent 12, angle θ)

From the adjacent triangle,

$$\tan\theta = \frac{5}{12} \Rightarrow \sin\theta = \frac{5}{13} \text{ and } \cos\theta = \frac{12}{13}$$

$$\cos 2\theta = \cos^2\theta - \sin^2\theta = \left(\frac{12}{13}\right)^2 - \left(\frac{5}{13}\right)^2 = \frac{119}{169}$$

$$\sin 2\theta = 2\sin\theta\cos\theta = 2\left(\frac{5}{13}\right)\left(\frac{12}{13}\right) = \frac{120}{169}$$

Now $\cos 4\theta = \cos^2 2\theta - \sin^2 2\theta = \left(\frac{119}{169}\right)^2 - \left(\frac{120}{169}\right)^2 = -\frac{239}{28561}$

Key Point

The three different forms for $\cos 2\alpha$ are all equally as important, and are found from substituting the identity $\sin^2\alpha + \cos^2\alpha = 1$ into $\cos 2\alpha = \cos^2\alpha - \sin^2\alpha$.

Quick Test

1. Write the expression $\sin\theta\cos\theta - 2\sin^3\theta\cos\theta$ in terms of double angles.
2. Solve the equation $4\sin\theta\cos\theta = 1$ in the range $0 \leq \theta \leq \pi$.
3. Given $\cos\theta = \frac{1}{3}$ and $0 \leq \theta \leq \pi$, find an exact value for $\sin 2\theta$.

Key Words

addition formulae
exact trigonometric values
double angle
half angle

Problem Solving Using Trigonometry

You must be able to:

- Solve equations given in the form $a\cos\theta + b\sin\theta = c$
- Use trigonometric functions to solve problems in context.

Expressions of the Form $a\cos\theta + b\sin\theta$

- Expressions such as $a\cos\theta + b\sin\theta$ can be written in the form:

  $R\sin(x \pm \alpha)$, where $R > 0$ and $0 \leqslant \alpha \leqslant \frac{\pi}{4}$ **or**

  $R\cos(x \pm \alpha)$, where $R > 0$ and $0 \leqslant \alpha \leqslant \frac{\pi}{4}$
- These forms can be useful for **sketching graphs** and **solving equations.**

> **Key Point**
>
> When sketching a curve, try to mark where it crosses the axes and any minimum or maximum points.

1) Express $\sqrt{3}\cos x + \sin x$ in the form $R\cos(x - \alpha)$, where $R > 0$ and $0 \leqslant \alpha \leqslant \frac{\pi}{4}$.

Hence sketch the graph of $y = \sqrt{3}\cos x + \sin x$ for $0 \leqslant x \leqslant 2\pi$.

$\sqrt{3}\cos x + \sin x \equiv R\cos(x - \alpha) = R\cos x\cos\alpha + R\sin x\sin\alpha$ ← Expand the brackets.

$\Rightarrow \sqrt{3} = R\cos\alpha$ ... **(1)** ← **Comparing coefficients** of $\cos x$.

and $1 = R\sin\alpha$ ... **(2)** ← Comparing coefficients of $\sin x$.

$R^2\cos^2\alpha + R^2\sin^2\alpha = R^2(\cos^2\alpha + \sin^2\alpha) = R^2$ ← Square both equations and add.

$= \left(\sqrt{3}\right)^2 + 1^2 = 4$ so $R^2 = 4$ and $R = 2$ ← Since $R > 0$.

$\frac{R\sin\alpha}{R\cos\alpha} = \tan\alpha = \frac{1}{\sqrt{3}} \Rightarrow \alpha = \frac{\pi}{6}$ ← Divide equation **(2)** by equation **(1)** and solve for α.

So $y = \sqrt{3}\cos x + \sin x$ is identical to $y = 2\cos\left(x - \frac{\pi}{6}\right)$. This is a **transformation** of the curve $y = \cos x$, translated $\frac{\pi}{6}$ units in the x-axis direction, and stretched by scale factor 2 in the y-axis direction (see below).

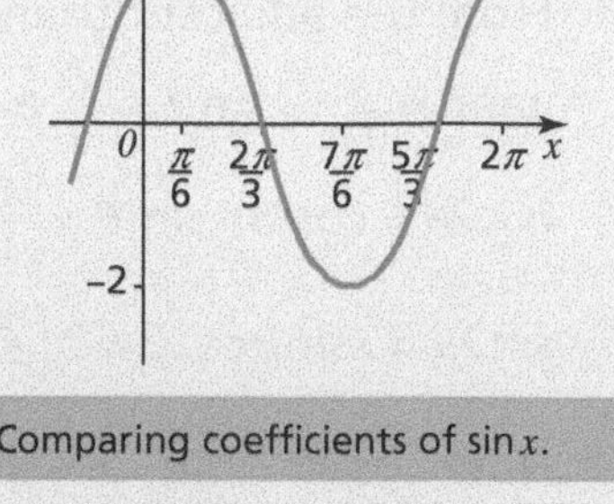

2) Express $5\sin x - 3\cos x$ in the form $R\sin(x - \alpha)$, where $R > 0$ and $0 \leqslant \alpha \leqslant \frac{\pi}{4}$.

Hence solve the equation $5\sin x - 3\cos x = 2$ in the range $0 \leqslant x \leqslant 2\pi$.

$5\sin x - 3\cos x \equiv R\sin(x - \alpha)$

$= R\sin x\cos\alpha - R\cos x\sin\alpha$

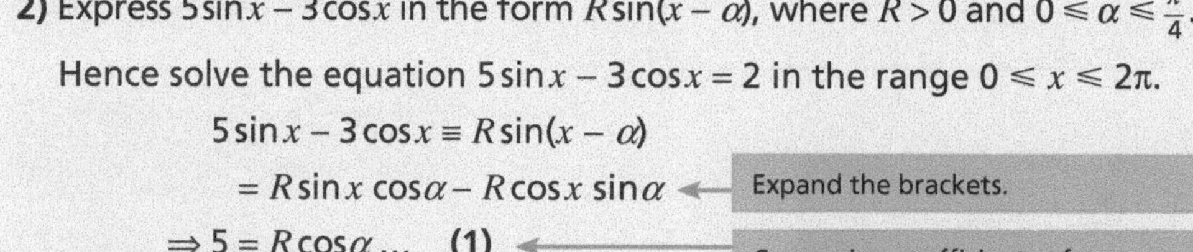

$\Rightarrow 5 = R\cos\alpha$ ... **(1)**

and $3 = R\sin\alpha$ ... **(2)** ← Comparing coefficients of $\sin x$.

$R^2\cos^2\alpha + R^2\sin^2\alpha = 5^2 + 3^2 = 34$

So $R = \sqrt{34}$ ← Since $R > 0$.

$\frac{R\sin\alpha}{R\cos\alpha} = \tan\alpha = \frac{3}{5}$ ← Divide equation **(2)** by equation **(1)**.

$\Rightarrow \alpha = \arctan\left(\frac{3}{5}\right) = 0.5404$

So $5\sin x - 3\cos x \equiv \sqrt{34}\sin(x - 0.5404)$

So the equation to solve is $\sqrt{34}\sin(x - 0.5404) = 2$

or $\sin(x - 0.5404) = \frac{2}{\sqrt{34}}$

so $x - 0.5404 = 0.3501$ or $x - 0.5404 = \pi - 0.3501$

therefore $x = 0.891^c$ or $x = 3.33^c$

- Some real-life applications using this technique are in problems such as those involving wave motion or hours of sunlight.

The depth of water in a harbour can be modelled by the equation $D = 2\cos\left(\frac{\pi t}{6}\right) + 2\sqrt{3}\sin\left(\frac{\pi t}{6}\right) + 15$ where t is the time (in hours) from midnight and D is the depth of the water (in metres).

a) Show that D can be written in the form $A\cos\left(\frac{\pi t}{6} - \alpha\right) + 15$, where A and α are constants to be determined.

$2\cos\left(\frac{\pi t}{6}\right) + 2\sqrt{3}\sin\left(\frac{\pi t}{6}\right) \equiv A\cos\left(\frac{\pi t}{6} - \alpha\right) = A\cos\frac{\pi t}{6}\cos\alpha + A\sin\frac{\pi t}{6}\sin\alpha$ ← Expand the brackets.

$\Rightarrow 2 = A\cos\alpha \ldots$ **(1)** ← Comparing coefficients of $\cos\left(\frac{\pi t}{6}\right)$.

$\Rightarrow 2\sqrt{3} = A\sin\alpha \ldots$ **(2)** ← Comparing coefficients of $\sin\left(\frac{\pi t}{6}\right)$.

$A^2\cos^2\alpha + A^2\sin^2\alpha = 2^2 + \left(2\sqrt{3}\right)^2 = 16$ ← Square both equations and add.

So $A = 4$ ← Since $A > 0$

$\frac{A\sin\alpha}{A\cos\alpha} = \tan\alpha = \frac{2\sqrt{3}}{2} = \sqrt{3} \Rightarrow \alpha = \arctan\sqrt{3} = \frac{\pi}{3}$ ← Divide equation **(2)** by equation **(1)** and then solve for α.

So $2\cos\left(\frac{\pi t}{6}\right) + 2\sqrt{3}\sin\left(\frac{\pi t}{6}\right) \equiv 4\cos\left(\frac{\pi t}{6} - \frac{\pi}{3}\right) \Rightarrow D = 4\cos\left(\frac{\pi t}{6} - \frac{\pi}{3}\right) + 15$

b) Find the maximum depth of the water.

The cosine function's maximum is 1, so the maximum depth of the water is $D = 4 \times 1 + 15 = 19$ m.

c) Find the time between consecutive high tides.

The high tides occur when $\cos\left(\frac{\pi t}{6} - \frac{\pi}{3}\right) = 1$.

So, the first high tide is when $\frac{\pi t}{6} - \frac{\pi}{3} = 0$, i.e. $t = 2$.

The second high tide occurs when $\frac{\pi t}{6} - \frac{\pi}{3} = 2\pi$, i.e. $t = 14$.

So there are 12 hours between consecutive high tides.

d) A boat cannot set sail unless the depth of water is greater than 12 m. Find the time(s) during the first 12 hours that the boat should not set sail.

Solving $4\cos\left(\frac{\pi t}{6} - \frac{\pi}{3}\right) + 15 = 12 \Rightarrow \cos\left(\frac{\pi t}{6} - \frac{\pi}{3}\right) = -\frac{3}{4}$ ← Cosine is negative here, so you are looking for solutions in the second and third quadrants.

So $\frac{\pi t}{6} - \frac{\pi}{3} = \pi - 0.7227$ or $\frac{\pi t}{6} - \frac{\pi}{3} = \pi + 0.7227$

$t = 6.620$ or $t = 9.380$

$0.62 \times 60 \approx 37$ $0.38 \times 60 \approx 23$

So the boat should not set sail between 0637 and 0923.

Quick Test

1. a) Express $f(x) = 8\sin x + 5\cos x$ in the form $R\sin(x + \alpha)$, where $R > 0$ and $0 \leq \alpha \leq \frac{\pi}{4}$.
   b) Hence write down the range of the function $f(x)$.
2. Express $f(x) = \cos x + \sin x$ in the form $R\cos(x - \alpha)$, where $R > 0$ and $0 \leq \alpha \leq \frac{\pi}{4}$.
3. Find the least positive root of the equation $3\sin x - 2\cos x = 1$.

Key Words

sketching graphs
solving equations
comparing coefficients
transformation

Practice Questions

Radians, Arc Lengths and Areas of Sectors

1 The perimeter of a sector of a circle of radius 4 cm is 20 cm. Find the area of the sector. [4]

2 Consider the following diagram:

a) Calculate the angle AOB. [3]

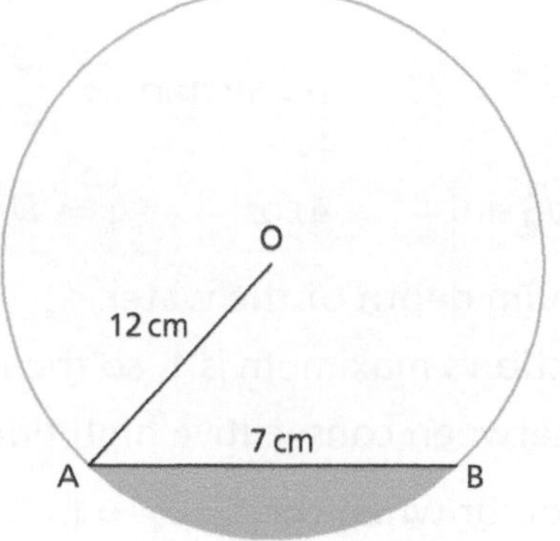

b) Calculate the shaded area. [2]

3 A chord AB splits a circle into a minor segment and a major segment. Given that the area of the minor segment is 50 cm^2 and angle OAB = $\frac{3\pi}{4}$ degrees, find the minor arc length AB. [4]

Total Marks .... / 13

Vectors in 3D

1 Find the distance between the points (2, 1, 3) and (–1, –2, 4). [3]

2 Find the possible value(s) of p if the magnitude of $\begin{pmatrix} -8 \\ p \\ 10 \end{pmatrix}$ is equal to $10\sqrt{2}$. [4]

3 The points A and B have position vectors $\begin{pmatrix} t-1 \\ 3t \\ 2 \end{pmatrix}$ and $\begin{pmatrix} 4t \\ 3 \\ 1-t \end{pmatrix}$ respectively.

a) Find the vector $\overrightarrow{AB}$. [2]

b) Find the value of t that makes $|\overrightarrow{AB}|$ a minimum. [6]

Total Marks .... / 15

Trigonometric Ratios

1 Find the exact values of the following, without using a calculator.

a) $\sin\frac{11\pi}{6}$ [2]

b) $\tan\frac{13\pi}{4}$ [2]

c) $\cos 4\pi$ [1]

2 Find the exact values of the following, without using a calculator.

a) $\arcsin\left(\sin\frac{9\pi}{4}\right)$ [2]

b) $\arccos\left(\sin\left(-\frac{5\pi}{6}\right)\right)$ [2]

c) $\arctan\left(\tan\frac{17\pi}{6}\right)$ [2]

3 Find an approximation for the expression $\frac{x\sin 2x}{1-\cos x}$ if x is small and in radians. [2]

Total Marks .... / 13

Further Trigonometric Equations and Identities

1 Prove the identity $\cos^4 x - \sin^4 x \equiv 2\cos^2 x - 1$. [3]

2 Solve the equation $\sec^2 x - 6\tan x + 4 = 0$ in the range $0 \leqslant x < 2\pi$. [7]

3 Prove the identity $\tan x + \cot x \equiv \sec x \operatorname{cosec} x$. [3]

4 Solve the equation $\cot\left(x - \frac{\pi}{4}\right) = 1$ in the range $0 \leqslant x < 2\pi$. [5]

Total Marks .... / 18

Using Addition Formulae and Double Angle Formulae

1 Given $\sin A = \frac{3}{5}$ (where $0 < A < \frac{\pi}{4}$) and $\sin B = -\frac{2}{5}$ (where $\pi < B < \frac{3\pi}{2}$), find an exact value for $\sin(A - B)$. [6]

2 Given $\tan x = -\frac{7}{24}$ and x is an obtuse angle, find an exact value for $\sin 2x$. [5]

3 Solve the equation $\sin 2x = 3\cos^2 x$ in the range $0 \leqslant x < 2\pi$. [6]

Total Marks .... / 17

Problem Solving Using Trigonometry

1 Express $5\cos x + 12\sin x$ in the form $R\cos(x - \alpha)$ where $R > 0$ and $0 \leqslant \alpha \leqslant \frac{\pi}{4}$. [6]

2 Express $7\cos x + 24\sin x$ in the form $R\cos(x - \alpha)$ where $R > 0$ and $0 \leqslant \alpha \leqslant \frac{\pi}{4}$.

Hence solve the equation $7\cos x + 24\sin x = 5$ in the range $0 \leqslant x < 2\pi$. [11]

3 For the function $f(x) = 8 + 7\cos x + 24\sin x$, find the minimum and the maximum values of $f(x)$ and the smallest positive value of x for when the maximum occurs. [5]

Total Marks .... / 22

Review Questions

Composite and Inverse Functions

1 $f(x) = x^2 - 4$, $\{x \in \mathbb{R}: x \geqslant 0\}$

a) Find $f^{-1}(x)$, stating the domain and the range. [2]

b) Sketch the graph of $y = f^{-1}(x)$ on a separate piece of paper. [3]

c) Find the values for which $f^{-1}(x) = f(x)$. [3]

2 Given $f: x \mapsto \dfrac{2x+5}{3}$ and $g: x \mapsto 2x - 4$

a) Find fg(11). [2]

b) Solve the equation $fg(x) = gf(x)$. [3]

Total Marks .... / 13

Modulus and Exponential Functions

1 Sketch a graph on a separate piece of paper and use it to solve the equation $\left|\frac{1}{2}x+1\right| = 2x-3$. [4]

2 Sketch a graph of $y = e^{(4x+1)} - 2$ on a separate piece of paper. [3]

3 Given the graph of $y = f(x)$, sketch the graph of $y = 2\left|f\left(\frac{1}{2}x+1\right)\right| + 2$. [6]

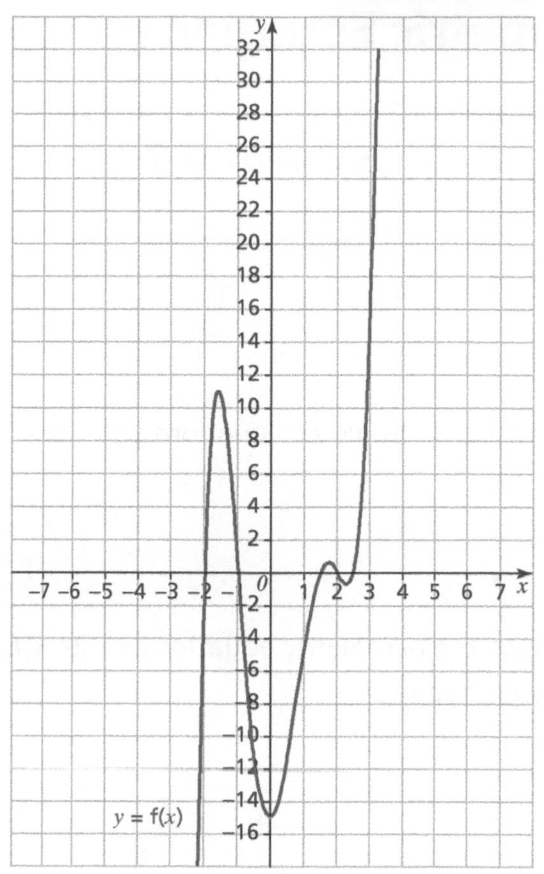

Total Marks .... / 13

Algebraic Fractions

1 Simplify $\dfrac{2x^2+x-3}{3x^2-13x+4} \div \dfrac{2x^2-x-6}{x^2-5x+4}$ [4]

2 Decompose $\dfrac{7x^2+20x+5}{(x+3)(x^2-1)}$ into partial fractions. [6]

3 Show that $\dfrac{2x^4+x^3+x^2-2x+3}{x+1} = Ax^3+Bx^2+Cx+D+\dfrac{E}{x+1}$ and hence find the values of A, B, C and D. [5]

Total Marks .... / 15

Review Questions

Parametric Equations

1 $x = 5t + 1,\ y = \frac{2}{t} + 3t - 1,\ \{t \in \mathbb{R} : t > 0\}$. Express the parametric equations as a Cartesian equation $y = f(x)$, stating the domain of $f(x)$. [4]

2 Sketch the parametric equations $x = t + 3,\ y = 2t^2 - 1,\ \{t \in \mathbb{R}: -2 \leqslant t \leqslant 2\}$ on a separate piece of paper. [6]

3 Sketch the parametric equations $x = 2\cos(t),\ y = 3\sin(t), \left\{t \in \mathbb{R} : -\frac{\pi}{2} \leqslant t \leqslant \frac{\pi}{2}\right\}$ on a separate piece of paper. [5]

Total Marks .... / 15

Problems Involving Parametric Equations

1 A water skier jumps off a ramp from 2 m high at a speed of 15 ms^{-1} at an angle of $\theta = 30°$.

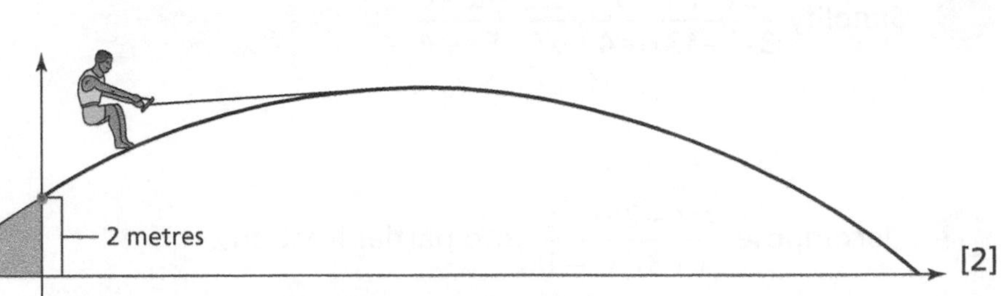

a) Write parametric equations to model the skier's jump. [2]

b) Find the number of seconds the skier is in the air. [3]

c) Find the horizontal distance the skier has travelled when he lands. [1]

2 The path of a comet orbiting the Earth is given by the parametric equations $x = t^3 - at$, $y = t^2 - 4$.
Given the curve intersects the y-axis at (0, 4), find a. [3]

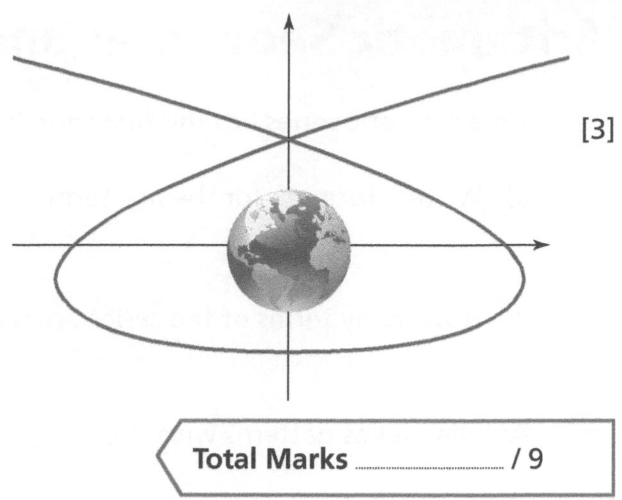

Total Marks .... / 9

Types of Sequences

1 A sequence is given by $u_{n+1} = (u_n)^2 - 4$, $u_1 = 2$.

a) Describe the sequence as increasing, decreasing or periodic. [1]

b) Write the first six terms of the sequence. [1]

c) Find $\sum_{1}^{6} u_n$. [1]

2 Find the value of $\sum_{1}^{4} 3n^2 - 2n$. [1]

3 Express the series 2 + 9 + 28 + 65 + 126 + 217 using sigma notation. [2]

Total Marks .... / 6

Review Questions

Arithmetic Sequences and Series

1 An arithmetic series has the first term 23 and a common difference of 7.

a) Write a formula for the nth term. [1]

b) How many terms of the series are required to give a sum of at least 400? [3]

2 Amelia makes patterns with sticks: 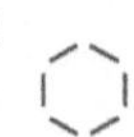  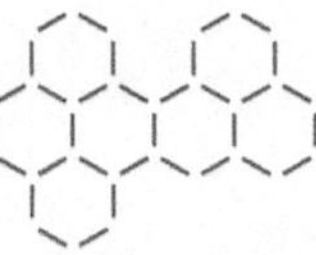

a) Explain whether or not the number of sticks used in each patterns is an arithmetic sequence. [1]

b) Find the number of sticks in the 20th pattern. [2]

c) Find the total number of sticks used to make 20 patterns. [1]

3 Show that the sum of the first 100 even numbers is 10 100. [3]

Total Marks .... / 11

Geometric Sequences and Series

1 A geometric sequence is defined by $u_n = 5\times\left(\frac{1}{5}\right)^{n-1}$

a) List the first five terms of the sequence. [1]

b) Find the sum of the first 10 terms of the sequence. [1]

c) Decide if the series is convergent or divergent and find the sum of the infinite series, if possible. [2]

2 A geometric sequence has the first term 100 and the fourth term $\frac{25}{4}$.

a) Write the nth term formula for the sequence. [2]

b) Find the first term that is less than $\frac{1}{4}$. [3]

c) Find the sum of the first three terms of the series. [2]

d) Find the sum of the infinite series. [1]

3 The population of an insect is decreasing by 8% per year. In the first year, there were 180 000 and in the last recorded year there were 85 000.

a) For how many years was the population recorded? Give your answer to the nearest year. [3]

b) After how many years will the population be extinct? [3]

Total Marks ________ / 18

Binomial Sequences

1 Find the first four terms of the expansion of $\frac{1}{\sqrt{x+4}}$ and state the range of values of x for which the expansion is valid. [4]

2 Given the coefficient of x^3 in the expansion of $\frac{1}{kx-5}$ is $-\frac{8}{625}$, find the value of k. [3]

3 Find the first four terms of the expansion of $\sqrt[4]{1+3x}$ and state the range of values of x for which the expansion is valid. [3]

Total Marks ________ / 10

Differentiating Exponentials, Logarithms and Trigonometric Ratios

You must be able to:

- Differentiate e^{kx}, $\sin kx$, $\cos kx$, $\tan kx$ and related sums, differences and constant multiples
- Understand and use the derivative of $\ln x$
- Understand the condition $f''(x) > 0$ implies a minimum and $f''(x) < 0$ implies a maximum for points where $f'(x) = 0$
- Differentiate using the chain rule.

The Chain Rule

- The **chain rule** is used to differentiate composite functions.
- If $y = f(u)$ and $u = g(x)$, then $\frac{dy}{dx} = \frac{dy}{du} \times \frac{du}{dx}$.
- $\frac{dy}{dx} = \frac{1}{\left(\frac{dx}{dy}\right)}$

Key Point

The chain rule can also be written in the form:

If $y = [f(x)]^n$, then $\frac{dy}{dx} = n[f(x)]^{n-1}f'(x)$

$y = (5x + 4)^5$. Find $\frac{dy}{dx}$.

$u = 5x + 4,\ y = u^5$

$\frac{dy}{du} = 5u^4,\ \frac{du}{dx} = 5$

$\frac{dy}{dx} = 5 \times 5u^4 = 25(5x+4)^4$

Differentiating Logarithms and Exponentials

- If $y = \ln x$, $\frac{dy}{dx} = \frac{1}{x}$. You must learn this.
- If $y = e^x$, $\frac{dy}{dx} = e^x$. You must learn this.

Key Point

Learn the following results:

If $y = \ln x$, $\frac{dy}{dx} = \frac{1}{x}$.

If $y = e^x$, $\frac{dy}{dx} = e^x$.

If $y = \cos x$, $\frac{dy}{dx} = -\sin x$.

If $y = \sin x$, $\frac{dy}{dx} = \cos x$.

1) If $y = 3\ln x$, find $\frac{dy}{dx}$.

$\frac{dy}{dx} = 3 \times \frac{1}{x} = \frac{3}{x}$

2) If $y = 4e^{2x}$, find $\frac{dy}{dx}$.

$\frac{dy}{dx} = 4 \times e^{2x} \times 2 = 8e^{2x}$ ← Remember to apply the chain rule.

Differentiating Trigonometric Ratios from First Principles

- $f'(x) = \lim_{\delta x \to 0}\left[\frac{f(x+\delta x) - f(x)}{\delta x}\right]$

Let $f(x) = \sin x$, $f'(x) = \lim_{\delta x \to 0}\left[\frac{\sin(x+\delta x) - \sin(x)}{\delta x}\right]$

$f'(x) = \lim_{\delta x \to 0}\left[\frac{\sin x \cos \delta x + \cos x \sin \delta x - \sin x}{\delta x}\right]$ ← Use $\sin(A + B)$.

As $\delta x \to 0$, $\cos \delta x \to 1$ and $\sin \delta x \to \delta x$

$f'(x) = \lim_{\delta x \to 0}\left[\frac{\sin x + \cos x \times \delta x - \sin x}{\delta x}\right] = \cos x$

Differentiating Trigonometric Ratios

- If $y = \cos x$, $\frac{dy}{dx} = -\sin x$. You should learn this result.
- If $y = \sin x$, $\frac{dy}{dx} = \cos x$. You should learn this result.
- If $y = \tan x$, $\frac{dy}{dx} = \sec^2 x$.

1) If $y = \tan^3 x$, find $\frac{dy}{dx}$.

$\frac{dy}{dx} = 3(\tan^2 x) \times \sec^2 x = 3\tan^2 x \sec^2 x$

2) If $y = \cos(4x - 3)$, find $\frac{dy}{dx}$.

$\frac{dy}{dx} = -\sin(4x - 3) \times 4 = -4\sin(4x - 3)$

3) If $y = \sin 3x$, find $\frac{dy}{dx}$.

$\frac{dy}{dx} = (\cos 3x) \times 3 = 3\cos 3x$

Key Point

Always remember to apply the chain rule.

Determining the Nature of the Turning Points

- If $\frac{dy}{dx} = 0$ and $\frac{d^2y}{dx^2} < 0$, maximum.
- If $\frac{dy}{dx} = 0$ and $\frac{d^2y}{dx^2} > 0$, minimum.
- At an **inflection point**, $f''(x)$ changes sign.
- If $\frac{dy}{dx} = 0$ and $\frac{d^2y}{dx^2} = 0$, but $\frac{d^3y}{dx^3} \neq 0$, point of inflection.

Key Point

$$\frac{dy}{dx} = \frac{1}{\left(\frac{dx}{dy}\right)}$$

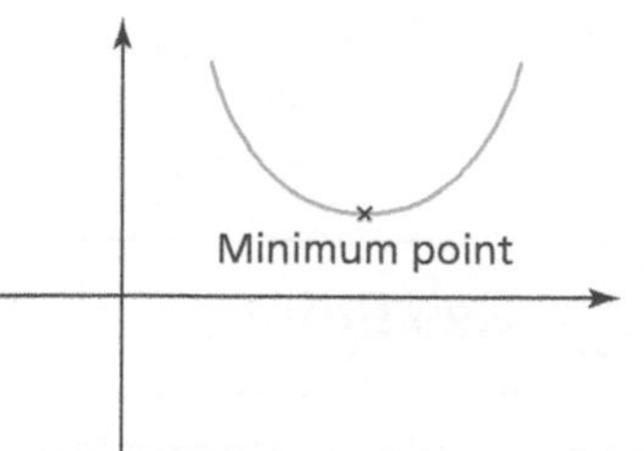

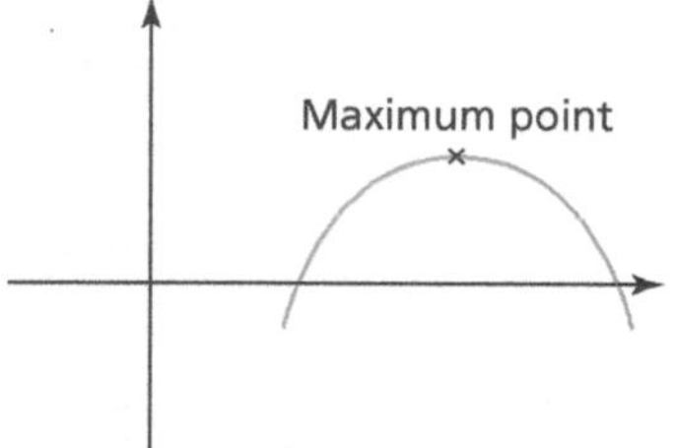

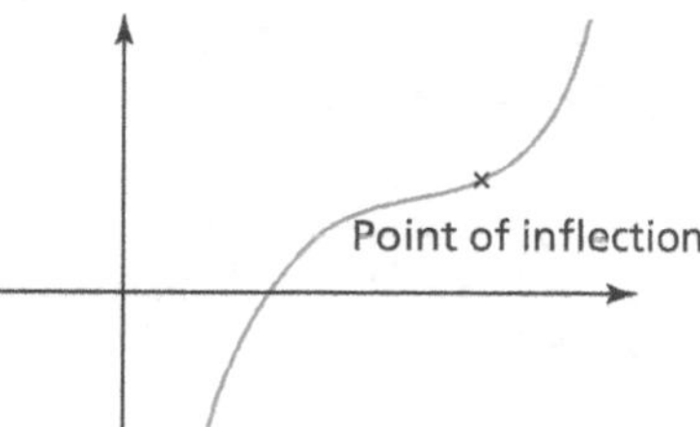

Find and determine the nature of the stationary point on the curve with equation $y = x^3 - 5$.

$\frac{dy}{dx} = 3x^2$ and when $3x^2 = 0$, $x = 0$.

Therefore a stationary point at (0, –5).

$\frac{d^2y}{dx^2} = 6x$.

When $x = 0$, $\frac{d^2y}{dx^2} = 0$ and $\frac{d^3y}{dx^3}(= 6) \neq 0$, therefore point of inflection.

Quick Test

1. If $y = \cos(6x - 2)$, find $\frac{dy}{dx}$.
2. If $y = 5e^{4x}$, find $\frac{dy}{dx}$.
3. If $y = 6\ln(4x)$, find $\frac{dy}{dx}$.
4. If $y = (5 - 2x)^3$, find the value of $\frac{dy}{dx}$ at the point (1, 27).

Key Words

chain rule
point of inflection

Product Rule, Quotient Rule and Implicit Differentiation

You must be able to:

- Differentiate using the product rule
- Differentiate using the quotient rule
- Differentiate implicitly
- Differentiate inverse trigonometric functions.

The Product Rule

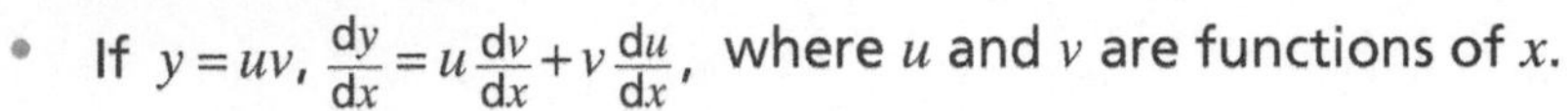

- If $y = uv$, $\frac{dy}{dx} = u\frac{dv}{dx} + v\frac{du}{dx}$, where u and v are functions of x.

Given that $f(x) = x^2(1 + 3x)^5$, find $f'(x)$.

Let $u = x^2$ and $v = (1 + 3x)^5$

$\frac{du}{dx} = 2x \quad \frac{dv}{dx} = 15(1+3x)^4$

$\frac{dy}{dx} = (x^2)15(1+3x)^4 + (2x)(1+3x)^5 = 2x(1+3x)^5 + 15x^2(1+3x)^4$

$\frac{dy}{dx} = (1+3x)^4(2x(1+3x) + 15x^2) = (1+3x)^4(21x^2 + 2x)$

$\frac{dy}{dx} = x(21x+2)(1+3x)^4$

Key Point

Remember, you might have to apply the chain rule within the product and quotient rule.

Apply the chain rule to differentiate v.

Factorise and simplify when possible.

The Quotient Rule

- If $y = \frac{u(x)}{v(x)}$, $\frac{dy}{dx} = \frac{v\frac{du}{dx} - u\frac{dv}{dx}}{v^2}$.

Find the value of $\frac{dy}{dx}$ at the point when $x = 1$ on the curve with equation $y = \frac{2x}{3x + 1}$.

$u = 2x$, $v = 3x + 1$ and $\frac{du}{dx} = 2$, $\frac{dv}{dx} = 3$

$\frac{dy}{dx} = \frac{(3x+1)(2) - 2x(3)}{(3x+1)^2}$

When $x = 1$, $\frac{dy}{dx} = \frac{(3\times1+1)\times2 - 2\times1\times3}{(3\times1+1)^2} = \frac{1}{8}$

Key Point

You can also differentiate a quotient by rearranging the function and using the product rule instead, i.e. $\frac{2x}{3x + 1} = 2x(3x+1)^{-1}$.

Implicit Differentiation

- By the chain rule, $\frac{d}{dx}(y^n) = ny^{n-1}\frac{dy}{dx}$.
- By the product rule, $\frac{d(xy)}{dx} = x\frac{dy}{dx} + y \times 1$.

Find $\frac{dy}{dx}$ in terms of x and y where $x^2 + 6y^3 + 5xy = 7$.

$2x + 18y^2\frac{dy}{dx} + 5x\frac{dy}{dx} + 5y = 0$

Use the product rule when differentiating $5xy$.

Key Point

Always remember to look for the product and quotient rule when using **implicit differentiation**.

Differentiating $y = a^{kx}$

- If $y = a^{kx}$, then $\ln y = \ln(a^{kx})$ and $\ln y = kx\ln a$

$$\frac{1}{y}\frac{dy}{dx} = k(\ln a)$$

Differentiate both sides.

Remember, differentiate $\ln y$ using implicit differentiation.

$$\frac{dy}{dx} = k(\ln a) \times y$$

$$\frac{dy}{dx} = k(\ln a)a^{kx}$$

You should learn this result.

Find the equation of the tangent to the curve with equation $y = 3^{2x} + 3^{-x}$ at the point $(1, \frac{28}{3})$.

$$\frac{dy}{dx} = 2(\ln 3)3^{2x} - (\ln 3)3^{-x}$$

For 3^{-x}, $k = -1$.

When $x = 1$, $\frac{dy}{dx} = \frac{53}{3}(\ln 3)$

$$y - \frac{28}{3} = \frac{53}{3}(\ln 3)(x-1)$$

Using $y - y_1 = m(x - x_1)$.

$3y - 28 = 53(\ln 3)x - 53(\ln 3)$

$3y - 53(\ln 3)x + 53(\ln 3) - 28 = 0$

Key Point

When differentiating implicitly, remember to differentiate both the left and right-hand side of the equation.

Inverse Trigonometric Functions

1) If $y = \arctan(x)$, find $\frac{dy}{dx}$.

$\tan y = x$

Apply the tan function to both sides of the equation.

$$\sec^2 y\frac{dy}{dx} = 1$$

Use implicit differentiation.

$$\frac{dy}{dx} = \frac{1}{\sec^2 y}$$

$\sec^2 y = 1 + \tan^2 y$, $\tan y = x$

$$\frac{dy}{dx} = \frac{1}{1+x^2}$$

2) If $y = \arcsin x$, find $\frac{dy}{dx}$.

$\sin y = x$

$$\cos y\frac{dy}{dx} = 1, \text{ so } \frac{dy}{dx} = \frac{1}{\cos y}$$

$$\frac{dy}{dx} = \frac{1}{\sqrt{1-x^2}}$$

$1 - \sin^2 x = \cos^2 x$

Key Point

When differentiating inverse trigonometric functions, remember to use the trigonometric identities.

Quick Test

1. Given that $5x^2 - 7xy + 6y^2 = 4$, find $\frac{dy}{dx}$.
2. If $y = 2^{3x} + 2^x$, find the equation of the normal at the point when $x = 1$.
3. If $y = 6x(x^2 + 2)^2$, find $\frac{dy}{dx}$.

implicit differentiation

Differentiating Parametric Equations and Differential Equations

You must be able to:

- Differentiate simple functions and relations defined parametrically
- Set up a differential equation using given information.

Differentiating Parametric Equations

- When an equation is defined **parametrically**, you can differentiate using a rearranged version of the chain rule.
- $\frac{dy}{dx} = \frac{dy}{dt} \div \frac{dx}{dt}$ or $\frac{dy}{dx} = \frac{dy}{dt} \times \frac{dt}{dx}$

1) Find the gradient at the point P where $t = 3$ on the curve given parametrically by $x = t^3 + 5t$, $y = t^2 + 3$.

$\frac{dy}{dt} = 2t$ and $\frac{dx}{dt} = 3t^2 + 5$

$\frac{dy}{dx} = \frac{dy}{dt} \div \frac{dx}{dt}$

$\frac{dy}{dx} = \frac{2t}{3t^2 + 5}$

When $t = 3$, $\frac{dy}{dx} = \frac{6}{3 \times 9 + 5} = \frac{6}{32} = \frac{3}{16}$

Key Point

You might have to use the chain rule, product rule and quotient rule when differentiating parametric functions.

2) Find the equation of the normal to the curve with parametric equations $x = 2t - 3\sin t$ and $y = t^2 + t\cos t$ at the point P when $t = \frac{\pi}{2}$.

$\frac{dx}{dt} = 2 - 3\cos t$ $\quad \frac{dy}{dt} = 2t - t\sin t + \cos t$

$\frac{dy}{dx} = \frac{2t - t\sin t + \cos t}{2 - 3\cos t}$

When $t = \frac{\pi}{2}$, $\frac{dy}{dx} = \frac{\pi}{4}$

Gradient of the normal $= \frac{-4}{\pi}$

When $t = \frac{\pi}{2}$, $x = \pi - 3$, $y = \frac{\pi^2}{4}$

Equation of the normal: $y - \frac{\pi^2}{4} = \frac{-4}{\pi}(x - (\pi - 3))$

$$4(\pi)y - \pi^3 = -16x + 16(\pi) - 48$$

Use the formula for the equation of a straight line: $y - y_1 = m(x - x_1)$.

Key Point

You need to substitute in the value of the third variable, i.e. t, to find the x and y-coordinates of the point on the tangent or normal.

Differential Equations

- You can set up differential equations using information from a real context. They are used in many problems in science and economics.
- $\frac{dy}{dx} - 6 - 5x$ and $\frac{dn}{dt} = 7n$ are examples of differential equations.

1) A population grows at a rate proportional to the population at a given time. Write an equation for the rate of growth of the population.

Let P represent the size of the population and t represent time.

The rate of change of the population is $\frac{dP}{dt}$.

The population grows at a rate proportional to the size of the population $\frac{dP}{dt} = kP$

As the population is growing, k is a positive constant.

2) Liquid is pouring into a container at a rate of $40\,cm^3s^{-1}$.

At time t seconds, liquid is leaking from the container at a rate of $\frac{1}{16}V\,cm^3s^{-1}$, where V is the volume of the container at any given time.

Show that $16\frac{dV}{dt} = 640 - V$.

Liquid flowing in $= 40\,cm^3s^{-1}$

Liquid flowing out $= \frac{1}{16}V\,cm^3s^{-1}$

Rate of change of liquid $= 40 - \frac{1}{16}V$

$$\frac{dV}{dt} = 40 - \frac{1}{16}V$$

Multiply all terms by 16.

$$16\frac{dV}{dt} = 640 - V$$, therefore shown.

3) A cube has side length L cm and surface area $A\,cm^2$. The length is expanding at a rate of $3\,cm\,s^{-1}$.

Show that $\frac{dA}{dt} = 18L$.

$$A = 6L^2$$

$$\frac{dL}{dt} = 3$$

$$\frac{dA}{dL} = 12L$$

$$\frac{dA}{dt} = \frac{dA}{dL} \times \frac{dL}{dt}$$

$$\frac{dA}{dt} = 12L \times 3 = 36L$$

Key Point

Interpret the information carefully in the question to make sure the constant of proportionality has the correct sign. If the population size is decreasing, the constant of proportionality will be negative.

Key Point

To find the total rate of change, you may have to combine values given in the question.

Key Point

The chain rule may need to be used for connected rates of change.

Quick Test

1. A curve has the parametric equation $x = t^2$, $y = 2t^3$. Find the equation of the tangent at the point (1, 2).
2. A circular patch of paint has a radius of r cm and the radius is increasing at a rate which is inversely proportional to the radius. Write a differential equation relating r and t, where t is time.

Standard Integrals and Definite Integrals

You must be able to:

- Evaluate the area of a region bounded by a curve and given straight lines, or between two curves
- Integrate e^{kx}, $\frac{1}{x}$, $\sin kx$, $\cos kx$ and related sums, differences and constant multiples
- Use trigonometric identities to integrate.

Finding the Area Between a Curve and a Line and Between Two Curves

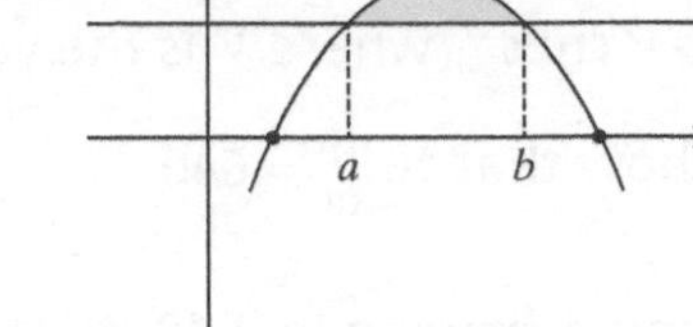

- To find the area between a curve and a line, or between two curves, you can calculate $\int_a^b (y_2 - y_1)\mathrm{d}x$ or $\int_a^b y_2\mathrm{d}x - \int_a^b y_1\mathrm{d}x$.
- It can be easier to work out the area under the curve using integration and the area under the line by recognising the polygon formed.

The diagram shows a curve with the equation $y = x^2 + 3$ and the line with equation $y = 7$. The line intersects the curve at the points A and B. Find the area of the finite region bounded by the curve and the line.

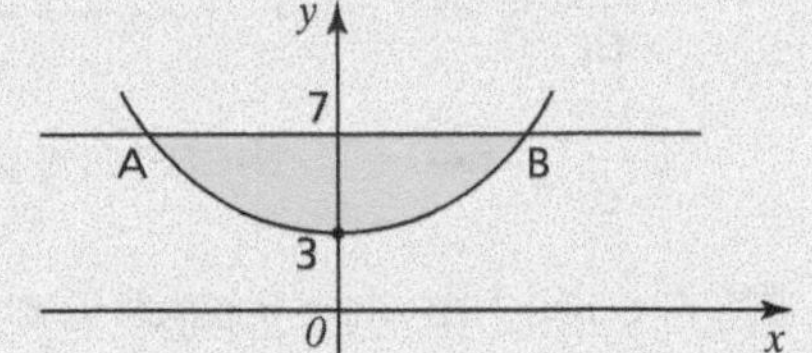

$x^2 + 3 = 7$

$x^2 = 4$

$x = 2$ (B) and -2 (A)

Find the x-coordinates of A and B by solving simultaneously.

There are two possible methods to find the area:

Method 1

$$\int_{-2}^{2} (\text{line} - \text{curve})\mathrm{d}x$$

$$\int_{-2}^{2} 7 - (x^2 + 3)\mathrm{d}x$$

$$\int_{-2}^{2} (-x^2 + 4)\mathrm{d}x$$

$$\left[\frac{-x^3}{3} + 4x\right]_{-2}^{2}$$

$$\left(\frac{-(2)^3}{3} + 4\times 2\right) - \left(\frac{-(-2)^3}{3} + 4\times -2\right)$$

$$= \frac{32}{3}$$

Method 2

Area of rectangle – Area under curve

Area of rectangle $= 4 \times 7 = 28$

The height of the rectangle is 7 and the width is the length A to B.

$$\int_{-2}^{2} (x^2 + 3)\mathrm{d}x$$

$$\left[\frac{x^3}{3} + 3x\right]_{-2}^{2}$$

$$\left(\frac{(2)^3}{3} + 3\times 2\right) - \left(\frac{(-2)^3}{3} + 3\times -2\right) = \frac{52}{3}$$

$$28 - \frac{52}{3} = \frac{32}{3}$$

Integrating e^{kx}, $\frac{1}{x}$, $\sin kx$ and $\cos kx$

- These standard integrals are given in the formula booklet for the exam:

$$\int e^x \, dx = e^x + c \qquad \int \frac{1}{x} dx = \ln|x| + c$$
$$\int \sin x \, dx = -\cos x + c \qquad \int \cos x \, dx = \sin x + c$$

- You may have to use the chain rule in reverse to integrate multiples of standard integrals.

Find the following integrals:

a) $\int (e^x + \cos 2x) dx$

$$\int (e^x + \cos 2x) dx = e^x + \frac{1}{2}\sin 2x + c$$

b) $\int \left(3\sin 6x + \frac{1}{x}\right) dx$

$$\int \left(3\sin 6x + \frac{1}{x}\right) dx = 3 \times \frac{-1}{6}\cos 6x + \ln|x| + c$$
$$= -\frac{1}{2}\cos 6x + \ln|x| + c$$

Key Point

You should also be familiar with the following integrals, although you are not expected to learn them:

$\int \sec^2 x \, dx = \tan x + c$

$\int \operatorname{cosec} x \cot x \, dx = -\operatorname{cosec} x + c$

$\int \operatorname{cosec}^2 x \, dx = -\cot x + c$

$\int \sec x \tan x = \sec x + c$

You must reverse the chain rule; $\sin 2x$ differentiates to $2\cos 2x$ so $\frac{1}{2}$ is needed.

Key Point

You should always check your integration is correct by differentiating.

Integrating Using Trigonometric Identities

- In order to use the list of standard integrals, you may have to rearrange your function using a trigonometric identity.

1) Find $\int \tan^2 x \, dx$.

$$\int \tan^2 x dx = \int (\sec^2 x - 1) dx$$
$$\int (\sec^2 x - 1) dx = \tan x - x + c$$

Use the identity $1 + \tan^2 x = \sec^2 x$.

2) Find $\int \sin^2 x \, dx$.

$$\cos 2x = 1 - 2\sin^2 x$$
$$\sin^2 x = \frac{1}{2} - \frac{1}{2}\cos 2x$$
$$\int \sin^2 x dx = \int \left(\frac{1}{2} - \frac{1}{2}\cos 2x\right) dx$$
$$= \frac{1}{2}x - \frac{1}{4}\sin 2x + c$$

Quick Test

1. Find $\int \frac{1}{3x} dx$.
2. Find $\int (e^{3x} + \cos 5x) dx$.
3. Find $\int (1 + 4x)^7 dx$.
4. Find $\int (\sin 4x \cos 4x) dx$.

Further Integration 1

You must be able to:
- Use substitution as a method of integration
- Use integration by parts to integrate some expressions.

Integration by Substitution

- You can sometimes simplify an integral by changing the variable:

1. Find $\frac{du}{dx}$ and make dx the subject by writing in the form $dx = f(x)du$.
2. Substitute in your expression for dx and simplify if possible.
3. Substitute every x for an expression in u, leaving an integral involving u and du only.
4. Integrate with respect to u.
5. Either substitute back in your expression for u in terms of x to find an indefinite integral, or change limits in terms of u to find a definite integral.

1) Use the substitution $u = e^{2x}$ to find $\int \frac{e^{2x}}{(3+e^{2x})^3} dx$.

1. $\frac{du}{dx} = 2e^{2x}$, $dx = \frac{1}{2e^{2x}} du$
2. $\int \frac{e^{2x}}{(3+e^{2x})^3} dx = \int \frac{e^{2x}}{(3+e^{2x})^3} \times \frac{1}{2e^{2x}} du = \int \frac{1}{2(3+e^{2x})^3} du$
3. $\int \frac{1}{2(3+e^{2x})^3} du = \int \frac{1}{2(3+u)^3} du$
4. $\int \frac{1}{2(3+u)^3} du = \frac{1}{2}\int (3+u)^{-3} du = -\frac{1}{4}(3+u)^{-2}$
5. $-\frac{1}{4}(3+u)^{-2} = -\frac{1}{4(3+e^{2x})^2} + c$

Therefore $\int \frac{e^{2x}}{(3+e^{2x})^3} dx = -\frac{1}{4(3+e^{2x})^2} + c$

Key Point

Always find $\frac{du}{dx}$ first and substitute it in.

2) Using the substitution $u = 3x^2 + 4$, evaluate $\int_0^1 x(3x^2+4)^5 dx$.

1. $\frac{du}{dx} = 6x$, $dx = \frac{1}{6x} du$
2. $\int_0^1 x(3x^2+4)^5 dx = \int_0^1 x(3x^2+4)^5 \times \frac{1}{6x} du = \int_0^1 \frac{1}{6}(3x^2+4)^5 du$
3. $\int_0^1 \frac{1}{6}(3x^2+4)^5 du$

   When $x = 1$, $u = 7$ and when $x = 0$, $u = 4$, therefore $\int_4^7 \frac{1}{6}u^5 du$.
4. $\int_4^7 \frac{1}{6}u^5 du = \left[\frac{1}{36}u^6\right]_4^7$
5. $\left(\frac{1}{36}\times 7^6\right) - \left(\frac{1}{36}\times 4^6\right) = 3154.25$

Key Point

Remember to change the limits to be in terms of u if necessary.

Substitute the x-values into the expression for u.

Integration by Parts

- Some functions written as products can be integrated using integration by parts.
- You should remember the product rule for differentiation: $\dfrac{d(uv)}{dx} = v\dfrac{du}{dx} + u\dfrac{dv}{dx}$
- Integrate both sides with respect to x to achieve: $uv = \int v\dfrac{du}{dx}dx + \int u\dfrac{dv}{dx}dx$
- This rearranges to (given in the formula booklet): $\int u\dfrac{dv}{dx}dx = uv - \int v\dfrac{du}{dx}dx$

1) Find $\int x\sin x\,dx$.

Let $u = x$ and $\dfrac{dv}{dx} = \sin x$.

$\dfrac{du}{dx} = 1$ and $v = -\cos x$

You need to differentiate u to find $\frac{du}{dx}$ and integrate $\frac{dv}{dx}$ to find v.

$\int x\sin x\,dx = -x\cos x - \int -\cos x \times 1\,dx$

$= -x\cos x + \sin x + c$

Substitute into the rule given above.

2) Find $\int x^2e^{3x}dx$.

Let $u = x^2$ and $\dfrac{dv}{dx} = e^{3x}$.

$\dfrac{du}{dx} = 2x,\ v = \dfrac{1}{3}e^{3x}$

$\int x^2e^{3x}\,dx = \dfrac{1}{3}x^2e^{3x} - \int \dfrac{2}{3}xe^{3x}\,dx$

You will have to use integration by parts again.

$\int \dfrac{2}{3}xe^{3x}\,dx$, let $u = \dfrac{2}{3}x$ and $\dfrac{dv}{dx} = e^{3x},\ \dfrac{du}{dx} = \dfrac{2}{3},\ v = \dfrac{1}{3}e^{3x}$

$\int \dfrac{2}{3}xe^{3x}dx = \dfrac{2}{9}xe^{3x} - \int \dfrac{2}{9}e^{3x}dx = \dfrac{2}{9}xe^{3x} - \dfrac{2}{27}e^{3x}$

Therefore $\int x^2e^{3x}\,dx = \dfrac{1}{3}x^2e^{3x} - \left(\dfrac{2}{9}xe^{3x} - \dfrac{2}{27}e^{3x}\right)$

$= \dfrac{1}{3}x^2e^{3x} - \dfrac{2}{9}xe^{3x} + \dfrac{2}{27}e^{3x} + c$

3) Find $\int \ln(x)dx$.

Let $\int \ln(x)dx = \int 1 \times \ln(x)dx$, and let $u = \ln(x)$ and $\dfrac{dv}{dx} = 1$.

$\dfrac{du}{dx} = \dfrac{1}{x}$ and $v = x$

$\int \ln(x)dx = x\ln(x) - \int x \times \dfrac{1}{x}dx$, so $\int \ln(x)dx = x\ln(x) - x + c$.

Key Point

Integration by parts is used to integrate the product of two functions. One part of the product is u and the other is $\frac{dv}{dx}$.

Key Point

In most cases when using integration by parts, set the x term as u. The exception to this is if one part of your product is a function of $\ln(x)$, in which case set $\ln(x)$ as u.

Key Point

You may have to use integration by parts twice within an integration.

Key Point

Take care with minus signs.

Key Point

A trick worth remembering: when integrating $\ln(x)$, write it as $1 \times \ln(x)$ and then use integration by parts.

Quick Test

1. Use the substitution $u = x^2 + 3$ to find $\int x(x^2+3)^3\,dx$.
2. Use the substitution $x = 4\sin^2 u$ to evaluate $\int_0^2 \sqrt{x(4-x)}dx$.
3. Find $\int x^2\ln(x)dx$.

Further Integration 2

You must be able to:

- Integrate by inspection by reversing the chain rule
- Integrate using partial fractions.

Integration by Inspection

- It is possible to integrate expressions which are the product of a function and its differential by **inspection**. They will be in the form $f'(x)[f(x)]^n$ or $\frac{f'(x)}{f(x)}$
- You can guess the **form** of the integral, differentiate and compare with the original and adjust the constant.
- This is the reverse process of the chain rule.

1) $\int 2x(x^2+5)^9\,dx$

$f'(x)=2x$ $\quad f(x)=x^2+5$

By inspection: $\int 2x(x^2+5)^9\,dx = \frac{1}{10}(x^2+5)^{10}+c$

Check by differentiating using the chain rule.

2) $\int \frac{12x^2}{4x^3+6}\,dx$

$f'(x)=12x^2$

$f(x)=4x^3+6$

By inspection: $\int \frac{12x^2}{4x^3+6}\,dx = \int (12x^2)(4x^3+6)^{-1} = \ln|4x^3+6|+c$

Key Point

When integrating by inspection, you can only adjust your guess by adjusting constants.

- You might need to adjust the constant.

$\int x^3\sqrt{4-3x^4}\,dx = \int x^3(4-3x^4)^{\frac{1}{2}}\,dx$

This is in the form $f'(x)[f(x)]^n$.

Try $y=(4-3x^4)^{\frac{3}{2}}$

$\frac{dy}{dx} = \frac{3}{2}(4-3x^4)^{\frac{1}{2}} \times 12x^3 = 18x^3(4-3x^4)^{\frac{1}{2}}$

You can now compare with the original and adjust your first guess.

$\int x^3\sqrt{4-3x^4}\,dx = \frac{1}{18}(4-3x^4)^{\frac{3}{2}}+c$

- Some are easier to spot than others!
- A commonly missed example of this is $\int 3\tan x\sec^5 x\,dx$.

$\int 3\tan x\sec^5 x\,dx = \int 3\tan x\sec x\sec^4 x\,dx$

$\sec x\tan x$ is the derivative of $\sec x$.

Try $y=\sec^5 x$

$\frac{dy}{dx} = 5\times\sec^4 x\times\sec x\tan x = 5\tan x\sec^5 x$

Therefore $\int 3\tan x\sec^5 x\,dx = \frac{3}{5}\sec^5 x + c$.

Integration Using Partial Fractions

- You can separate an expression into partial fractions to allow you to integrate. Here are two examples:

1) $\int \frac{6x-2}{(x-3)(x+1)}\,dx = \int \frac{4}{x-3}+\frac{2}{x+1}\,dx = 4\ln|x-3|+2\ln|x+1|+c$

2) $\int \frac{3x^2-2x-4}{x^2-3x+2}\,dx = \int 3+\frac{3}{x-1}+\frac{4}{x-2}\,dx = 3x+3\ln|x-1|+4\ln|x-2|+c$

Key Point

Always check if your fraction is top heavy and, if so, divide before separating into partial fractions.

Deciding which Method to Use

- Once you have mastered all the methods, the challenge in the examination is to decide which method is the best to use. The flow chart below should help you to decide.

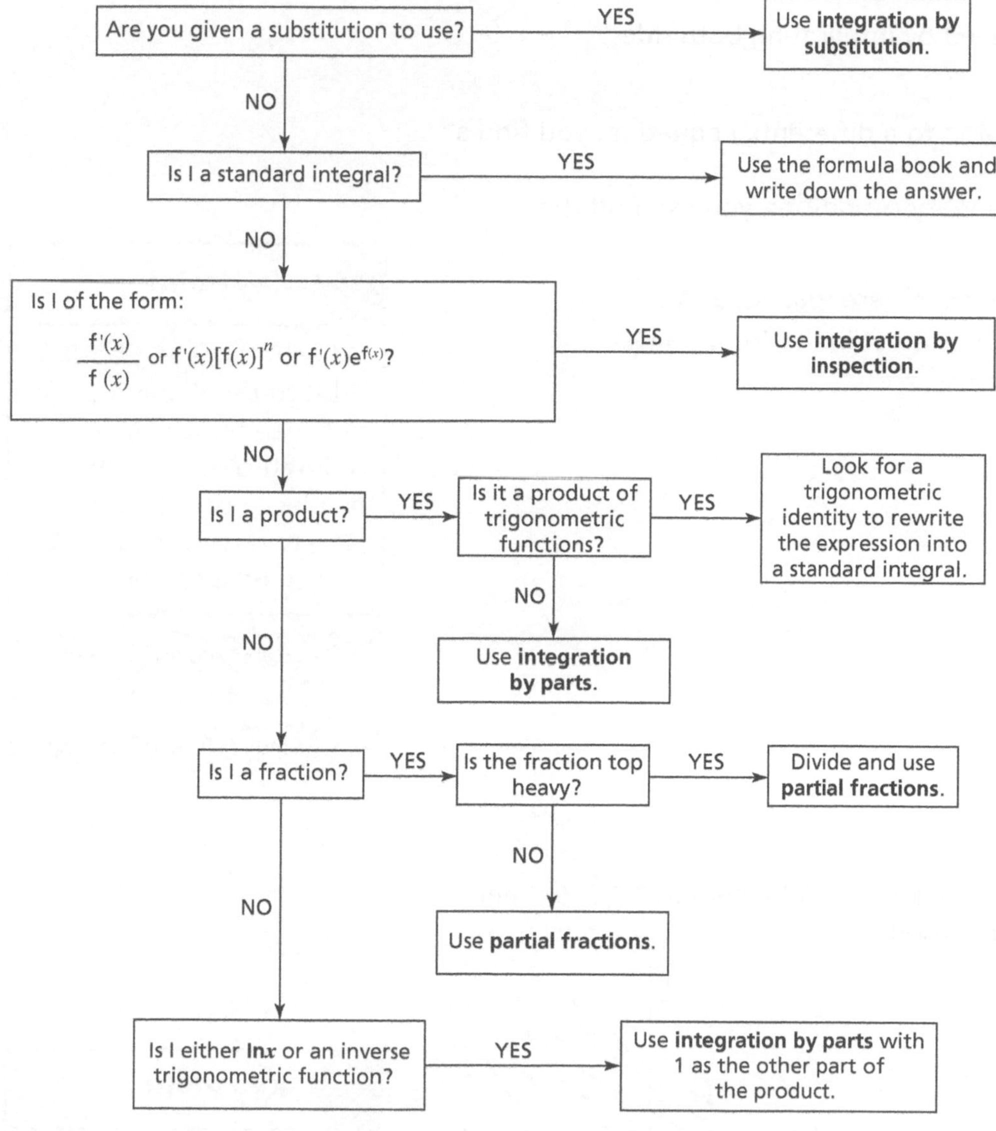

Quick Test

1. Find $\int (2x+6)^7\,dx$.
2. Find $\int \frac{6x^2}{2x^3-6}\,dx$.
3. Find $\int \frac{5x+3}{(2x-3)(x+3)}\,dx$.

Key Words

integration by inspection

Differential Equations

You must be able to:

- Evaluate the analytical solution of simple first order differential equations with separable variables, including finding particular solutions
- Interpret the solution of a differential equation in the context of solving a problem.

Separating the Variables

- When you have an equation in the form $\frac{\mathrm{d}y}{\mathrm{d}x} = \mathrm{f}(x)\mathrm{g}(y)$, it can be rearranged into the form $\frac{1}{\mathrm{g}(y)}\mathrm{d}y = \mathrm{f}(x)\mathrm{d}x$. This is called separating the variables.
- This equation can now be solved by integrating both sides: $\int \frac{1}{\mathrm{g}(y)}\mathrm{d}y = \int \mathrm{f}(x)\mathrm{d}x$
- When finding a **general solution** to a differential equation, you find a family of solution curves.
- When you are also given **boundary conditions**, you can find the **particular solution.**

1) Find the general solution to the differential equation $\frac{\mathrm{d}y}{\mathrm{d}x} = (1+y)(1-2x)$. Leave your answer in the form $y = \mathrm{f}(x)$.

$$\frac{\mathrm{d}y}{\mathrm{d}x} = (1+y)(1-2x)$$

$$\frac{1}{1+y}\mathrm{d}y = (1-2x)\mathrm{d}x$$

$$\int \frac{1}{1+y}\mathrm{d}y = \int (1-2x)\mathrm{d}x$$

$$\ln|1+y| = x - x^2 + c$$

$$1 + y = \mathrm{e}^{x-x^2+c}$$

To make y the subject, use e to inverse ln.

$$1 + y = \frac{A\mathrm{e}^x}{\mathrm{e}^{x^2}}$$

Use the indices rules to rearrange.

$$y = \frac{A\mathrm{e}^x}{\mathrm{e}^{x^2}} - 1$$

Key Point

When integrating both sides of the equation, you only need one constant of integration, which can be placed on either side. (The two constants will be combined into one.)

2) Find the general solution to the differential equation $3\frac{\mathrm{d}y}{\mathrm{d}x} = 5x^2$ and sketch the family of solution curves.

$$3\frac{\mathrm{d}y}{\mathrm{d}x} = 5x^2$$

$$3\mathrm{d}y = 5x^2\mathrm{d}x$$

$$\int 3\mathrm{d}y = \int 5x^2\,\mathrm{d}x$$

$$3y = \frac{5}{3}x^3 + c$$

$$y = \frac{5}{9}x^3 + c$$

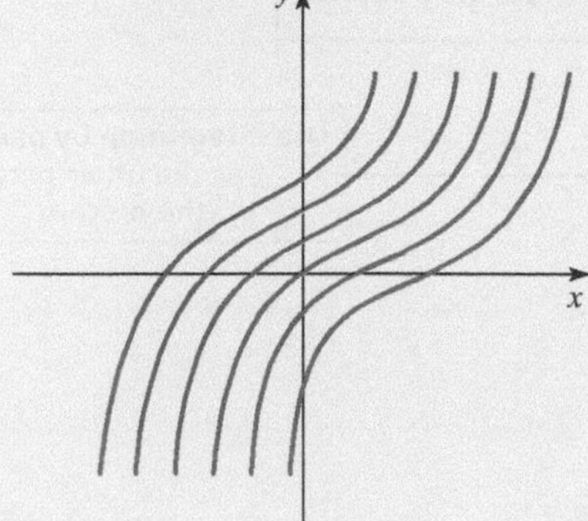

Key Point

When there is no x or y term in the original equation, remember to integrate constants.

3) Find the particular solution to the differential equation $\frac{1}{x}\frac{dy}{dx} = \frac{1}{x^2 + 1}$ given that when $y = 0$, $x = 0$.

$$\frac{1}{x}\frac{dy}{dx} = \frac{1}{x^2+1}$$

$$1dy = \frac{x}{x^2+1}dx$$

$$\int 1dy = \int \frac{x}{x^2+1}dx$$

$$y = \frac{1}{2}\ln|x^2+1| + c$$

When $x = 0$, $y = 0$

$0 = \frac{1}{2}\ln|1| + c$, so $c = 0$

Therefore, the particular solution is $y = \frac{1}{2}\ln|x^2+1|$.

Differential Equations in Context

- Sometimes a differential equation will arise out of a contextual problem and you will need to construct the equation from the given information.

The rate of change of the radius (r) of a circular spot of ink, t seconds after it is first measured, is given by $\frac{dr}{dt} = \frac{7}{1+t}$.

Given that when $t = 0$, $r = \frac{1}{2}$, find the time taken for the radius to double its initial value.

$$1dr = \frac{7}{1+t}dt$$

$$\int 1dr = \int \frac{7}{1+t}dt$$

$$r = 7\ln(1+t) + c$$

$$t = 0,\ r = \frac{1}{2}$$

$\frac{1}{2} = 7\ln(1) + c$, so $c = \frac{1}{2}$

$$r = 7\ln(1+t) + \frac{1}{2}$$

When $r = 1$ ← r was initially $\frac{1}{2}$

$$1 = 7\ln(1+t) + \frac{1}{2}$$

$\frac{1}{2} = 7\ln(1+t)$, so $\frac{1}{14} = \ln(1+t)$

$e^{\frac{1}{14}} - 1 = t$, so $t = 0.0740$

Key Point

Initial values are usually measured from $t = 0$.

Key Point

You may need to use the rules of logs to arrange the equation into the required form.

Quick Test

1. Solve the differential equation $\frac{dy}{dx} = 2e^{x-y}$. Give your answer in the form $y = f(x)$.
2. Find the general solution of the differential equation $\frac{dy}{dx} = x\cos^2 y$.
3. Find the particular solution to the differential equation with the given boundary conditions:
   $\frac{dy}{dx} = \sec^2 x \sin^2 y$, $y = \frac{\pi}{3}$, $x = \frac{\pi}{4}$

Key Words

general solution
boundary conditions
particular solution

Practice Questions

Differentiating Exponentials, Logarithms and Trigonometric Ratios

1 Differentiate the following:

a) $\sin 2x$ [1]

b) $(1 + 6x^2)^3$ [2]

c) $5\ln(2x)$ [2]

2 Find and determine the nature of the turning points on the graph with equation $y = x^3 + 2x^2 + x$. [6]

Total Marks .... / 11

Product Rule, Quotient Rule and Implicit Differentiation

1 Differentiate the following:

a) $x^2(\ln x)$ [2]

b) $\frac{(1+2x)^2}{x^3}$ [2]

c) $\arcsin(x)$ [3]

2 Given that $y = 3^t$, show that $\frac{dy}{dt} = 3^t\ln 3$. [3]

3 a) Differentiate $e^{3x}\sin x$ with respect to x. [3]

The curve C has equation $y = e^{3x}\sin x$.

b) Show that the turning points on C occur when $\tan x = -3$. [4]

c) Find the equation of the normal to C at the point where $x = \frac{\pi}{2}$. [3]

4 Find $\frac{dy}{dx}$ in terms of x and y where $6x^3 + 3xy + x^2 - 6 = 2x$. [4]

Total Marks .... / 24

Differentiating Parametric Equations and Differential Equations

1 The graph shows part of a curve with parametric equations $x = (1 + t)^2$, $y = t^3 + 8$.

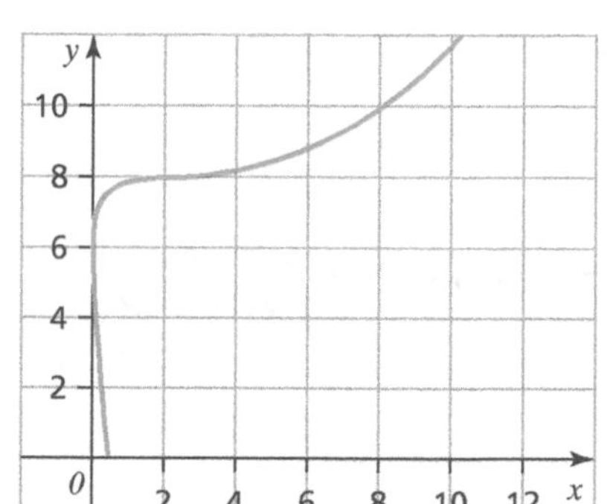

The curve crosses the x-axis at the point A.

a) Find the x-coordinate of point A. [2]

b) Find $\frac{dy}{dx}$ in terms of t. [3]

c) Find the equation of the tangent at point A. [3]

2 Given that $x = y^3\ln(2y)$

a) Find $\frac{dx}{dy}$. [2]

b) Use your answer to part **a)** to find the value of the gradient at the point when $y = e^2$. [3]

Total Marks .... / 13

Standard Integrals and Definite Integrals

1 Calculate the following integrals:

a) $\int_3^6 x(x^2+7)^3 dx$ [3]

b) $\int_0^{\pi} \sin^2 x dx$ [4]

c) $\int_2^5 \frac{1}{2x+1} dx$ [3]

d) $\int_{\frac{\pi}{6}}^{\frac{\pi}{3}} \sec^2 x dx$ [3]

Total Marks .... / 13

Further Integration 1 & 2

1 a) Show that $\frac{5x^2-8x+1}{2x(x-1)^2} = \frac{A}{x} + \frac{B}{x-1} + \frac{C}{(x-1)^2}$ where A, B and C are constants to be found. [3]

b) Hence find $\int \frac{5x^2-8x+1}{2x(x-1)^2} dx$. [3]

2 By using the substitution $u = 16 - x^2$, or otherwise, find the exact value of $\int_0^3 \frac{x}{\sqrt{(16-x^2)}} dx$. [5]

3 Shown here is the graph with equation $y = xe^x(3 - x)$.

Find the area bounded by the curve and the positive x-axis. [6]

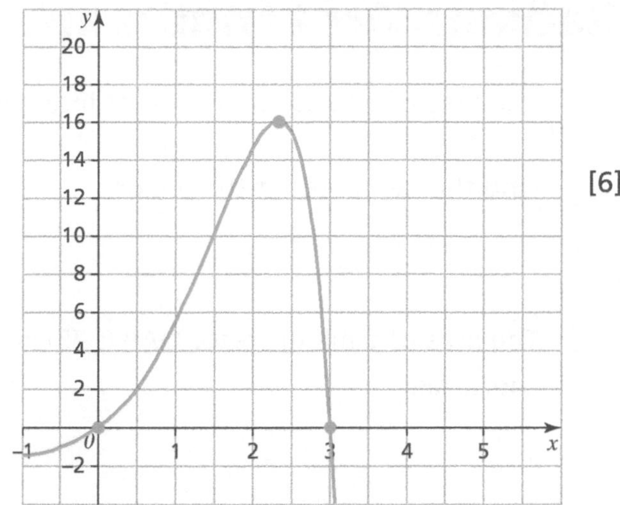

4 Show that:

a) $\int_0^{\pi} x \sin x \, dx = \pi$ [4]

b) $\int_0^{\pi} x^2 \cos x \, dx = -2\pi$ [4]

Total Marks .... / 25

Differential Equations

1 a) Find $\int x e^{-2x} dx$. [4]

b) Solve the differential equation $e^{2x} \frac{dy}{dx} = x \cos^2 y$, given that $y = \frac{\pi}{4}$ when $x = 0$. [4]

Total Marks .... / 8

Review Questions

Radians, Arc Lengths and Areas of Sectors

1. The area of a sector of a circle of internal angle $\frac{2\pi}{3}$ is $100\,\text{cm}^2$.

   Find the perimeter of the sector. [4]

2. The area of a minor sector OAB is $75\,\text{cm}^2$. Given angle $AOB = \frac{\pi}{6}$, find the length of the chord AB. [4]

3. The perimeter of the sector of a circle OAB is numerically equal to twice its area.

   Given angle $AOB = \frac{5\pi}{6}$, find the perimeter of the sector. [5]

Total Marks .... / 13

Vectors in 3D

1. Find the distance between the points (4, 10, 6) and (2, 4, 8). [3]

2. Given $\mathbf{a} = 2\mathbf{i} + 3\mathbf{j} - 8\mathbf{k}$ and $\mathbf{b} = 4\mathbf{i} + 2\mathbf{k}$, find $|3\mathbf{a} - 2\mathbf{b}|$. [4]

3. Given that $\overrightarrow{OP} = \begin{pmatrix} 4 \\ 16 \\ 20 \end{pmatrix}$ and $\overrightarrow{OQ} = \begin{pmatrix} -4 \\ -8 \\ -12 \end{pmatrix}$, find the coordinates of the point R that divides PQ in the ratio 3 : 5. [4]

Total Marks .... / 11

Trigonometric Ratios

1 Find the exact values of the following, without using a calculator.

a) $\cot\frac{11\pi}{6}$ [2]

b) $\sec\frac{11\pi}{4}$ [2]

c) $\text{cosec}\frac{5\pi}{3}$ [2]

2 Find an approximation for the expression $\frac{x - x\cos 2x}{\sin^3 x}$ if x is small and in radians. [3]

3 State the domain and the range of each of the following functions.

a) $y = \cot 2x$ [3]

b) $y = \arcsin\left(\frac{x}{2}\right)$ [2]

c) $y = \sec x$ [4]

Total Marks .... / 18

Review Questions

Further Trigonometric Equations and Identities

1 Prove the identity $\frac{(\sin x + \cos x)^2}{\cos x} \equiv \sec x + 2\sin x$. [4]

2 Solve the equation $\text{cosec}\,x \cot x - 2\cot x = 0$ in the range $0 \leqslant x < 2\pi$. [5]

3 Prove the identity $(\cot x + \text{cosec}\, x)^2 = \frac{1 + \cos x}{1 - \cos x}$. [6]

4 Solve the equation $\sec 2x = -3$ in the range $0 \leqslant x < 2\pi$. [6]

Total Marks .... / 21

Using Addition Formulae and Double Angle Formulae

1 Use a half-angle identity to show that the exact value of cos 105° is $-\frac{\sqrt{2-\sqrt{3}}}{2}$. [6]

2 Given $\cos x = -\frac{1}{3}$ and $\pi < x < \frac{3\pi}{2}$, find exact values for:

a) $\sin\left(\frac{x}{2}\right)$ [3]

b) $\cos\left(\frac{x}{2}\right)$ [3]

3 Solve the equation $\sin\left(\frac{x}{2}\right) - \sin x = 0$ in the range $0 \leqslant x < 2\pi$. [6]

Total Marks .... / 18

Problem Solving Using Trigonometry

1 Express $6\sin x - 4\cos x$ in the form $R\sin(x - \alpha)$, where $R > 0$ and $0 \leqslant \alpha \leqslant \frac{\pi}{4}$. [6]

2 Express $\cos x - 2\sin x$ in the form $R\cos(x + \alpha)$, where $R > 0$ and $0 \leqslant \alpha \leqslant \frac{\pi}{4}$.
Hence solve the equation $\cos x - 2\sin x = \sqrt{2}$ in the range $0 \leqslant x < 2\pi$. [11]

3 For the function $f(x) = 3 - \cos x + 2\sin x$, find the minimum and the maximum values of $f(x)$ and the smallest positive value of x for when the maximum occurs. [5]

Total Marks .... / 22

Iterative Methods and Finding Roots

You must be able to:

- Locate roots of $f(x) = 0$ by considering changes of sign
- Understand how and why the 'changes of sign' method can fail
- Solve equations approximately using the iteration $x_{n+1} = f(x_n)$
- Draw cobweb and staircase diagrams illustrating such iterations.

Locating Roots of $f(x) = 0$

- Sometimes you cannot find a root of an equation in an exact form.
- For example, consider the function $f(x) = \ln x - \sin x$: it is impossible to find an exact value $x = \alpha$ such that $f(\alpha) = 0$. However, $f(2) = \ln 2 - \sin 2 = -0.216 < 0$ and $f(2.5) = \ln 2.5 - \sin 2.5 = 0.318 > 0$.
- Since $f(2) < 0$ and $f(2.5) > 0$, the graph crosses the x-axis somewhere between $x = 2$ and $x = 2.5$. Therefore you can say there exists a **root** α such that $2 < \alpha < 2.5$ (see right).

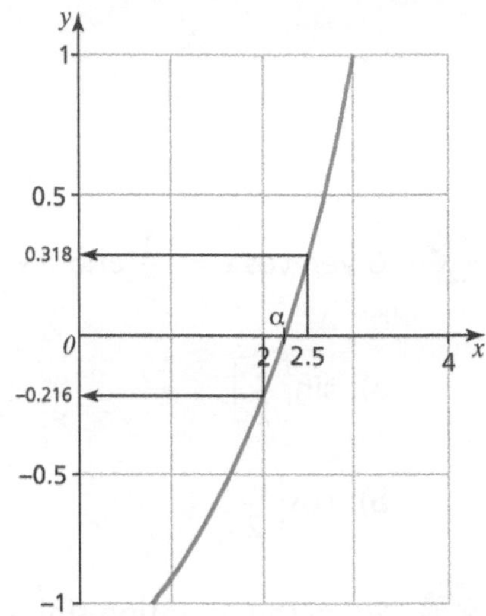

- In general, if there is an **interval** in which $f(x)$ changes sign, then that interval will contain a root α such that $f(\alpha) = 0$. However, this **only** holds true if the function is **continuous**. For example, consider the function $f(x) = \tan x$: $f(1.5) = 14.1 > 0$, $f(2) = -2.19 < 0$
- There is again a sign change, but in this case you **cannot** say there is a root in the interval $1.5 < x < 2$, because $\tan x$ is not continuous in this interval. In fact, there is an asymptote at $x = \frac{\pi}{2}$. In this context, $x = \frac{\pi}{2}$ could also be termed a 'point of discontinuity'.
- You also need to be careful when considering the converse to this method of locating roots. For example, if you consider an interval $a < x < b$ and find that $f(a) > 0$ and $f(b) > 0$ (i.e. there is no sign change), this does **not** necessarily mean that there is **no** root in this interval. There may be two, or four, or six, or any even number of roots.
- Consider the function $f(x) = (\cos x)(\ln x)$:
  $f(0.5) = (\cos 0.5)(\ln 0.5) = -0.61 < 0$, $f(2) = (\cos 2)(\ln 2) = -0.29 < 0$
- Here, there is no sign change, but this does **not** mean you can say with certainty that there are no roots in the interval $0.5 < x < 2$. In fact, there are two roots, as the graph (right) of the function makes clear, where $\alpha_1 = 1$ and $\alpha_2 = \frac{\pi}{2}$.

Key Point

You can think of 'continuity' as meaning you are able to draw the graph without taking your pen from the paper.

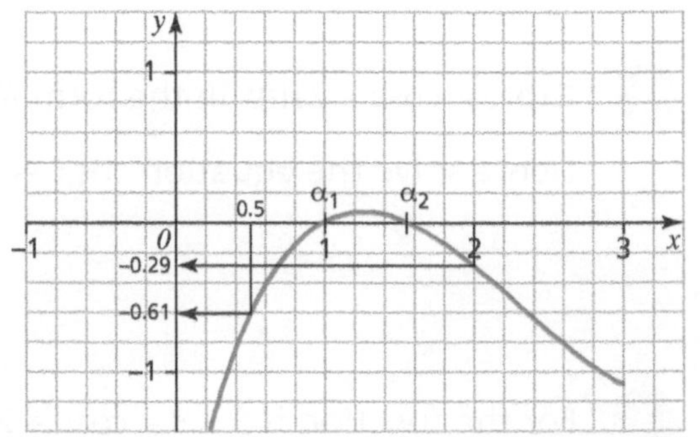

Solving Equations Approximately Using the Iteration $x_{n+1} = f(x_n)$

- To solve an equation $f(x) = 0$ using this method, rearrange $f(x) = 0$ into a form $x = g(x)$, then use the iterative formula $x_{n+1} = g(x_n)$.
- There are three possible situations:
  - The iterative sequence **converges** to a root α, as shown by a **staircase diagram.**
  - The iterative sequence converges to a root α, as shown by a **cobweb diagram.**
  - The iterative sequence may not converge at all.

By rearranging the equation $f(x) = 0.1x^3 - x + 0.1 = 0$ into a suitable form, use an **iteration formula** to find an approximation for a root α to 2 decimal places. Start with $x_0 = 2$.

The equation can be rearranged as $x^3 = \frac{x-0.1}{0.1}$, or $x = \sqrt[3]{\frac{x-0.1}{0.1}}$

This is now in the form $x = g(x)$ so you can use the iteration $x_{n+1} = \sqrt[3]{\frac{x_n - 0.1}{0.1}}$

$x_0 = 2$

$x_1 = \sqrt[3]{\frac{2-0.1}{0.1}} = 2.6684$ $\quad x_2 = \sqrt[3]{\frac{2.6684-0.1}{0.1}} = 2.9504$

$x_3 = \sqrt[3]{\frac{2.9504-0.1}{0.1}} = 3.0547$ $\quad x_4 = \sqrt[3]{\frac{3.0547-0.1}{0.1}} = 3.0915$

$x_5 = \sqrt[3]{\frac{3.0915-0.1}{0.1}} = 3.1043$ $\quad x_6 = \sqrt[3]{\frac{3.1043-0.1}{0.1}} = 3.1087$

This is an increasing sequence, and can be partly illustrated by the 'staircase' diagram on the right.

The numerical results appear to show that the sequence is converging to a value of 3.11.

You can ensure this is the correct value by checking for a sign change in the interval 3.11 ± 0.005

$f(3.105) = 0.1(3.105)^3 - 3.105 + 0.1 = -0.0115 < 0$

$f(3.115) = 0.1(3.115)^3 - 3.115 + 0.1 = 0.00755 > 0$

Since there is a sign change, and $f(x)$ is continuous in this interval, there is a root α such that $3.105 < \alpha < 3.115$. So $\alpha = 3.11$ to 2 d.p.

- Consider the equation $f(x) = x^5 - x - 2 = 0$. By looking for a sign change, you can see a root α exists such that $1 < \alpha < 2$.
  - Rearrange the equation $x^5 - x - 2 = 0$ into $x = x^5 - 2$.
  - You could try the iteration $x_{n+1} = x_n^5 - 2$ with a starting value of $x_0 = 1.5$. This gives the sequence $x_1 = 1.5^5 - 2 = 5.594$, $x_2 = 5.594^5 - 2 = 5475$
  - This sequence is clearly **not** going to converge to a root. However, rearranging the equation into the form $x = \sqrt[5]{2+x}$ and using the iteration $x_{n+1} = \sqrt[5]{2+x_n}$ with $x_0 = 1.5$, **does** deliver a converging sequence: $x_1 = 1.285$, $x_2 = 1.269$, $x_3 = 1.267$
- So the success (or not) of this method **may** depend on the choice of iteration formula. It **may** also depend on the starting value x_0.

Consider the equation $f(x) = 2\sin\left(\frac{x}{2}\right) + 2 - x = 0$. By using a suitable iteration formula and a starting value of $x_0 = 3$, find a root α (to 2 decimal places) such that $f(\alpha) = 0$.

$x_0 = 3$, so $x_1 = 2\sin\left(\frac{3}{2}\right) + 2 = 3.995$

You can use the iteration $x_{n+1} = 2\sin\left(\frac{x_n}{2}\right) + 2$

$x_2 = 3.821$, $x_3 = 3.886$, $x_4 = 3.863$, $x_5 = 3.871$

The sequence is converging to a value of $\alpha = 3.87$ to 2 d.p.

Key Point

The values shown in the example to the left have all been rounded to 4 d.p. When calculating a sequence of iterations, you should always use your calculator's memory at each step to ensure accuracy is not lost.

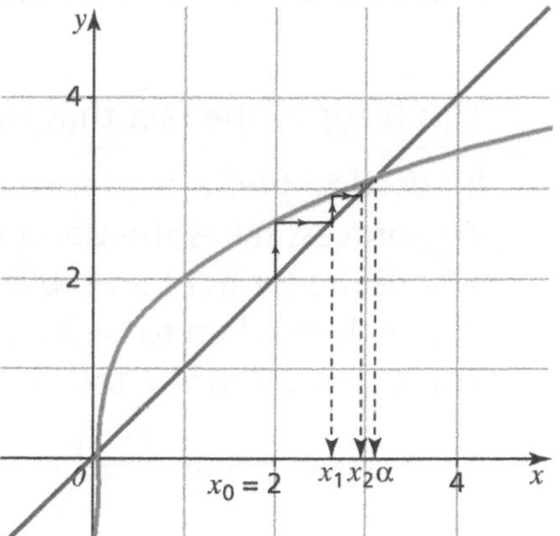

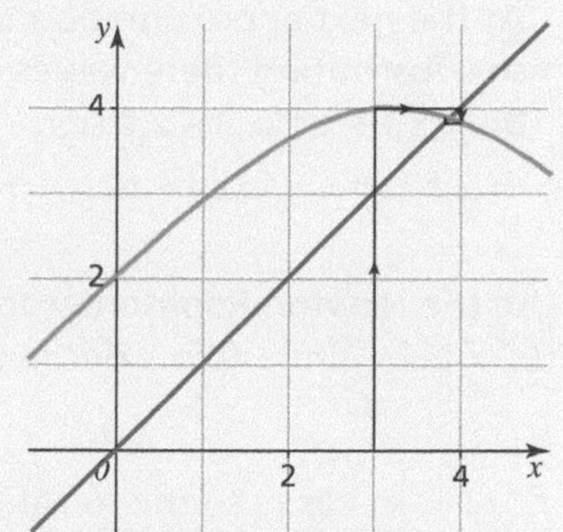

Notice how the sequence 'oscillates' around the root α. This type of sequence can be shown by a 'cobweb diagram'.

Quick Test

1. By rearranging and using a suitable iteration formula, find the root of the equation $xe^x = 1$ to 3 decimal places.
2. $f(x) = \frac{1}{2}x^5 - 3x^3 + \frac{1}{2}x^2 + 1$ has three real roots. By looking for sign changes, find three intervals in which these roots occur, expressing them in the form $a < x < b$, where a and b are consecutive integers.
3. By using the iteration formula $x_{n+1} = \cos(0.5x_n)$, and taking $x_0 = 0.5$, find a root of the equation $x - \cos(0.5x) = 0$ to 2 decimal places.

Key Words

root
interval
continuous (function)
converge
staircase diagram
cobweb diagram
iteration formula

Numerical Integration and Solving Problems in Context

You must be able to:

- Solve equations using the Newton-Raphson method
- Understand why the Newton-Raphson method can sometimes fail
- Use the trapezium rule in order to estimate some definite integrals.

Newton-Raphson

- If f(x) is a **continuous function**, and x_0 is taken to be a first **approximation** to a root α, i.e. where f(α) = 0, then a better approximation x_1 can be found by using the formula $x_1 = x_0 - \frac{f(x_0)}{f'(x_0)}$.
- This leads to the fact that the iteration formula $x_{n+1} = x_n - \frac{f(x_n)}{f'(x_n)}$ may be used to find a sequence of values that converge to α.
- The process takes the point on the curve where $x = x_0$, and considers where the **tangent** line to the curve at this point intersects the x-axis. This is then taken to be x_1, the better approximation to α (above right).
- The equation of the tangent at the point (x_0, y_0) on a general curve y = f(x) is given by $\frac{y - f(x_0)}{x - x_0} = f'(x_0)$. This can be rearranged to give $y = f(x_0) + (x - x_0)f'(x_0)$.
- At the next approximation to the root, $x = x_1$ and $y = 0$, so substituting these values into the equation gives:
  $0 = f(x_0) + x_1 f'(x_0) - x_0 f'(x_0)$
  $\Rightarrow x_1 f'(x_0) = x_0 f'(x_0) - f(x_0) \Rightarrow x_1 = x_0 - \frac{f(x_0)}{f'(x_0)}$

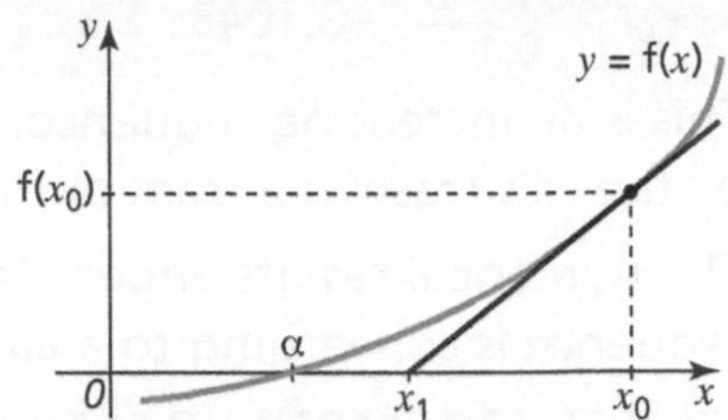

Use the Newton-Raphson process to find a root to the equation f(x) = ln$x$ + sin$x$ = 0 to 3 decimal places, taking a starting value of $x_0 = 1$.

$f'(x) = \frac{1}{x} + \cos x$, so the iteration to be used is $x_{n+1} = x_n - \left(\frac{\ln x_n + \sin x_n}{\frac{1}{x_n} + \cos x_n}\right)$

$$x_1 = 1 - \left(\frac{\ln 1 + \sin 1}{1 + \cos 1}\right) = 0.4537$$

$$x_2 = 0.4537 - \left(\frac{\ln 0.4537 + \sin 0.4537}{\frac{1}{0.4537} + \cos 0.4537}\right) = 0.5671$$

$$x_3 = 0.5671 - \left(\frac{\ln 0.5671 + \sin 0.5671}{\frac{1}{0.5671} + \cos 0.5671}\right) = 0.5786$$

$$x_4 = 0.5786 - \left(\frac{\ln 0.5786 + \sin 0.5786}{\frac{1}{0.5786} + \cos 0.5786}\right) = 0.5787$$

- In the example below left, the sequence is converging fairly quickly to a value of $\alpha = 0.579$. Generally, this will always hold true. If an initial value of x_0 is chosen close to the root, the sequence obtained via Newton-Raphson should converge quite quickly. However, there are instances where this will not be the case.
- In the example below left, if $x_0 = 0$ were chosen, the Newton-Raphson iteration could not be used, as both lnx_0 and $\frac{1}{x_0}$ are undefined at this point.
- Unlikely though it may be, if you were to choose an initial value x_0 such that $\frac{1}{x_0} + \cos x_0 = 0$ (point A on the graph below), then again Newton-Raphson could not be used, since division by zero is undefined.

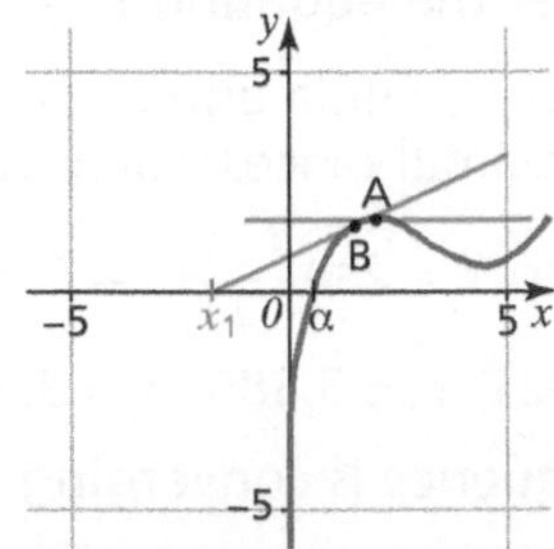

- Geometrically, this corresponds to a **stationary point** on the curve. So any tangent line at such a point would never intersect the x-axis again.
- Finally, if an x-value corresponding to a point of shallow gradient were chosen (e.g. point B on the graph above), any tangent line drawn would intersect the x-axis at x_1 (in this case, some distance from the required root α). In this case, the point of intersection would not be part of the domain of the function, so could not be used.
- With other functions, the sequence obtained may eventually converge, but to a different root than that required.

The Trapezium Rule

- The trapezium rule is a way of estimating the area under a curve, used particularly with functions that cannot be integrated using algebra:

$$\int_a^b y\,dx \approx \frac{h}{2}\left[y_0 + 2(y_1 + y_2 + \ldots + y_{n-1}) + y_n\right] \text{ where } h = \frac{b-a}{n}$$

where n is the number of strips and h is the 'height' of each strip.

Key Point

The greater the number of strips used, the better the approximation will be. You will normally be told in the examination how many strips to use.

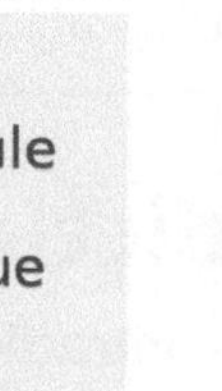

Estimate the value of the integral $\int_0^{\frac{\pi}{2}} \sqrt{\cos x}\,dx$ using the trapezium rule with four strips. By sketching this function, suggest whether the value you obtain is an underestimate or an overestimate compared to the actual value.

$$h = \frac{\frac{\pi}{2} - 0}{4} = \frac{\pi}{8}$$

x	0	$\frac{\pi}{8}$	$\frac{\pi}{4}$	$\frac{3\pi}{8}$	$\frac{\pi}{2}$
$\sqrt{\cos x}$	1	0.9612	0.8409	0.6186	0

$$\int_0^{\frac{\pi}{2}} \sqrt{\cos x}\,dx \approx \frac{\left(\frac{\pi}{8}\right)}{2}[1 + 2(0.9612 + 0.8409 + 0.6186) + 0] = 1.15 \text{ units}^2$$

As can be seen from the (blue) shaded areas (right), there are evident 'gaps' between the four trapezia and the curve. The estimate obtained is therefore an underestimate in this case.

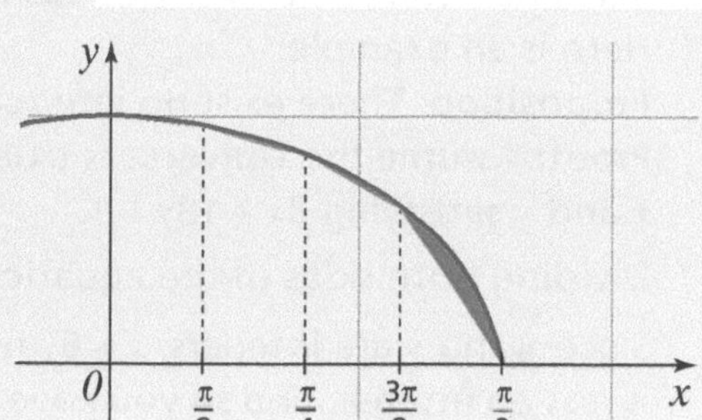

Solving Problems in Context

- Many real-world examples produce equations with no exact solutions. The best you can do is use an iteration procedure to find the root to a required degree of accuracy. You will be told which method to use.

A plant's height, h cm, after a time, t days, is given by the equation $h = 0.9t + 0.08\sin(2\pi t)$. By using a first approximation of $t_0 = 2$ days, use the Newton-Raphson process once to find a better approximation to the time when the plant's height will be 2 cm.

The equation to be solved is $0.9t + 0.08\sin(2\pi t) = 2$, or $0.9t + 0.08\sin(2\pi t) - 2 = 0$.

The Newton-Raphson iteration is therefore $t_1 = t_0 - \left(\frac{0.9t + 0.08\sin 2\pi t - 2}{0.9 + 0.16\pi\cos 2\pi t}\right)$

So $t_1 = 2 - \left(\frac{1.8 + 0.08\sin 4\pi - 2}{0.9 + 0.16\pi\cos 4\pi}\right) = 2 - \left(-\frac{0.2}{1.40265}\right) = 2.14$ days

Key Point

Remember not to round your calculated y-values prematurely. Use your calculator's memory.

Quick Test

1. Use the Newton-Raphson process once to estimate a solution to $2x^3 - 7x^2 + 1 = 0$ to 2 d.p. Take $x_0 = 3.5$.
2. The Newton-Raphson process is used to estimate a root to the equation $x^4 - 2 = 0$. Suggest why a first approximation of $x_0 = 0$ should **not** be used.
3. Use the trapezium rule with four strips to estimate a value for $\int_1^5 \ln x\,dx$. Suggest if this will be an underestimate or an overestimate.

Key Words

approximation
tangent
stationary point

Proof

Proof by Contradiction and Further Trigonometric Identities

You must be able to:

- Apply the method of 'proof by contradiction' to both familiar and unfamiliar proofs
- Prove further trigonometric identities involving addition formulae, double angles, and half-angles.

Proof by Contradiction

- A mathematical proof by **contradiction** begins by stating an opposite proposition is true, and then attempts to show that such an **assumption** leads to a contradiction, thus falsifying the opposite **proposition**.
- Here is an example:

  Proposition: There exist **no** integers x and y such that $3x + 18y = 1$.

  Proof: Assume the **converse** is true, i.e. that there **are** integer values of x and y satisfying $3x + 18y = 1$.

  Dividing both sides of the equation by 3 gives $x+6y=\frac{1}{3}$

  Since x and y are integers, $x + 6y$ must also be an integer. But clearly $\frac{1}{3}$ is not an integer, and so you have a contradiction. Therefore the original assumption must be incorrect. So the only conclusion possible is that there are **no** integers x and y satisfying $3x + 18y = 1$.
- There are two important proofs you need to know:

Key Points

Recall that a prime number is a number having only two whole number factors: 1 and itself.

Key Points

Recall that a rational number is one that can be written as a fraction $\frac{p}{q}$, where p and q are integers.

1. There exists an infinite number of prime numbers	2. $\sqrt{2}$ is irrational
The following is the most common proof, attributed to Euclid, and using proof by contradiction: Assume the converse is true, i.e. that there is a **finite** number of prime numbers. Let these prime numbers be $p_1, \dots p_n$ Now consider the number $N = p_1 \times p_2 \times \dots \times p_n + 1$ If the number N was prime, it would be on the original list of primes, and it is not, since N is larger than any of $p_1, \dots p_n$. Therefore N is not prime. This means there is some prime number p that divides into N exactly. However, this prime p must also be one of the primes on the original list. Therefore p divides into $p_1 \times p_2 \times \dots \times p_n$ exactly. Now you have that $N = p_1 \times p_2 \times \dots \times p_n + 1$, p divides N, and p divides $p_1 \times p_2 \times \dots \times p_n$. This means p must also divide 1. But $p > 1$, so you have a contradiction. Therefore the original assumption is false, i.e. there is **not** a finite number of primes. So there must be an **infinite** number of primes.	You start by assuming the converse is true, i.e. assume $\sqrt{2}$ is **rational**. Therefore you can write $\sqrt{2} = \frac{p}{q}$ where p and q have no common factors (other than 1). Squaring both sides gives $2 = \frac{p^2}{q^2}$ or $p^2 = 2q^2$. Now $2q^2$ is clearly an even number, therefore p^2 must be even, and so p must be even. So let $p = 2r$, say. Therefore $(2r)^2 = 2q^2$, so $4r^2 = 2q^2$, or $q^2 = 2r^2$ Now $2r^2$ is clearly an even number, so q^2 must be even and q must be even. Now you have the position that p and q are both even, so have a common factor of 2. But this contradicts the original assumption that p and q have no common factors. Therefore the original assumption must be false, i.e. $\sqrt{2}$ is **not** rational. So $\sqrt{2}$ must be irrational.

Further Trigonometric Identities

- You may need to prove trigonometric identities that include the addition formulae or double angle formulae. For this reason alone, it is worth trying to memorise each of these formulae. This will help you to recognise when certain formulae could be applied.
- Always start with one side of the identity that has to be proven, and aim to reach the other side in a series of logical steps. You do not have to explain what you are doing in each step, though it should be clear.

Key Point

You should normally start with the more 'complicated looking' side of the identity. This should allow you to make progress at the start. Note that example **2)** on page 79 is an exception to this.

1) Prove the identity $\sin(x+y)\sin(x-y) \equiv \sin^2 x \sin^2 y$.

$\sin(x+y)\sin(x-y)$

$= (\sin x\cos y + \cos x\sin y)(\sin x\cos y - \cos x\sin y)$ ← Use addition formulae to expand.

$= \sin^2 x\cos^2 y - \cos^2 x\sin^2 y + \sin x\sin y\cos x\cos y - \sin x\sin y\cos x\cos y$ ← Expand brackets.

$= \sin^2 x\cos^2 y - \cos^2 x\sin^2 y$ ← Cancel final terms.

$= \sin^2 x(1-\sin^2 y) - (1-\sin^2 x)\sin^2 y$ ← Use identity for $\cos^2$.

$= \sin^2 x - \sin^2 x\sin^2 y - \sin^2 y + \sin^2 x\sin^2 y$ ← Multiply brackets.

$= \sin^2 x \sin^2 y$ ← Cancel terms.

2) Prove the identity $\cos 3\alpha \equiv 4\cos^3\alpha - 3\cos\alpha$.

$\cos 3\alpha = \cos(2\alpha + \alpha)$ ← Use addition formula for cos.

$= \cos 2\alpha\cos\alpha - \sin 2\alpha\sin\alpha$

$= (2\cos^2\alpha - 1)\cos\alpha - 2\sin\alpha\cos\alpha\sin\alpha$ ← Use double angle formulae for both sin and cos.

$= 2\cos^3\alpha - \cos\alpha - 2\sin^2\alpha\cos\alpha$

$= 2\cos^3\alpha - \cos\alpha - 2(1-\cos^2\alpha)\cos\alpha$ ← Express everything in terms of cos.

$= 2\cos^3\alpha - \cos\alpha - 2\cos\alpha + 2\cos^3\alpha$

$= 4\cos^3\alpha - 3\cos\alpha$ ← Expand and collect like terms.

3) Prove the identity $\sin 2x \equiv \dfrac{2\tan x}{1+\tan^2 x}$.

$$\frac{2\tan x}{1+\tan^2 x} = \frac{2\left(\frac{\sin x}{\cos x}\right)}{1+\frac{\sin^2 x}{\cos^2 x}}$$ ← Use $\tan x = \frac{\sin x}{\cos x}$

$$= \left(\frac{\cos^2 x}{\cos^2 x}\right)\left(\frac{2\left(\frac{\sin x}{\cos x}\right)}{1+\frac{\sin^2 x}{\cos^2 x}}\right) = \frac{2\sin x\cos x}{\cos^2 x+\sin^2 x} = 2\sin x\cos x = \sin 2x$$

← Multiply the top and bottom of the fraction by $\cos^2 x$ and simplify.

← Use the identity $\cos^2 x + \sin^2 x = 1$.

4) Prove the identity $\cot\dfrac{\theta}{2} \equiv \dfrac{\sin\theta}{1-\cos\theta}$. ← Use half-angle identities for sine and cosine.

$$\frac{\sin\theta}{1-\cos\theta} = \frac{2\sin\frac{\theta}{2}\cos\frac{\theta}{2}}{1-\left(1-2\sin^2\frac{\theta}{2}\right)} = \frac{2\sin\frac{\theta}{2}\cos\frac{\theta}{2}}{2\sin^2\frac{\theta}{2}}$$ ← Simplify.

$$= \frac{\cos\frac{\theta}{2}}{\sin\frac{\theta}{2}} = \cot\frac{\theta}{2}$$ ← Cancel $2\sin\frac{\theta}{2}$ from the top and bottom of the fraction.

Quick Test

1. Prove the identity $\operatorname{cosec}\left(\theta+\frac{\pi}{2}\right) \equiv \sec\theta$.
2. Prove the identity $\sin 2\theta = 2\cot\theta\sin^2\theta$.
3. Prove by contradiction that, if p^2 is even, where p is an integer, then p is even.

Key Words

contradiction
assumption
proposition
converse
finite
infinite
rational

Practice Questions

Iterative Methods and Finding Roots

1 Show that the equation $e^{\sin x} - \ln x = 0$ has a root in the interval $3.0 < \alpha < 3.1$. [5]

2 Show that the equation $f(x) = x^4 - 7x + 3 = 0$ has a root α such that $1.5 < \alpha < 2$.

By using the iteration $x_{n+1} = \sqrt[4]{7x - 3}$ with $x_0 = 2$, write down x_1, x_2, x_3 and x_4, and hence find α to 2 decimal places. [8]

3 Show that the equation $f(x) = x^7 - 7x + 1 = 0$ has a root α such that $1 < \alpha < 1.5$.

By using the iteration $x_{n+1} = \sqrt[7]{7x - 1}$ with $x_0 = 1$, find α to 3 decimal places. [7]

Total Marks .... / 20

Numerical Integration and Solving Problems in Context

1 Use the trapezium rule with four strips to estimate the value for $\int_0^{\frac{\pi}{2}} e^x \sin x \, dx$. [6]

2 Consider the equation $f(x) = x^2 - 3$. By taking $x_0 = 1.7$ as an initial approximation, apply the Newton-Raphson process twice in order to find $\sqrt{3}$ to 3 decimal places. [5]

3 Use the trapezium rule with six strips to estimate the value for $\int_2^8 (\ln x)^2 \, dx$. [6]

Total Marks .... / 17

Proof by Contradiction and Further Trigonometric Identities

1 Prove the trigonometric identity $\sin\left(x+\frac{\pi}{6}\right)\cos\left(x+\frac{\pi}{6}\right) \equiv \frac{\sin 2x + \sqrt{3}\cos 2x}{4}$. [6]

2 Prove the trigonometric identity $\sin x \sin 2x = 2\cos x - 2\cos^3 x$. [4]

3 Prove by contradiction that $a^2 + b^2 \geqslant 2ab$. [4]

Total Marks .... / 14

Review Questions

Differentiating Exponentials, Logarithms and Trigonometric Ratios

1 $y = x\ln(x)$

a) Find $\frac{dy}{dx}$. [3]

b) Find the coordinates of the turning point. [3]

c) Determine the nature of this turning point. [2]

Total Marks .... / 8

Product Rule, Quotient Rule and Implicit Differentiation

1 Differentiate with respect to x, giving your answer in its simplest form:

a) $x^2\ln(3x)$ [4]

b) $\frac{\sin 4x}{x^3}$ [4]

2 The curve C has equation $y = \frac{3 + \sin 2x}{2 + \cos 2x}$.

a) Show that $\frac{dy}{dx} = \frac{6\sin 2x + 4\cos 2x + 2}{(2 + \cos 2x)^2}$. [4]

b) Find an equation of the tangent to C at the point on C where $x = \frac{\pi}{2}$.

Write your answer in the form $y = ax + b$, where a and b are exact constants. [4]

3 a) Given that $y = \frac{\ln(x^2+1)}{x}$, find $\frac{dy}{dx}$. [4]

b) Given that $x = \tan y$, show that $\frac{dy}{dx} = \frac{1}{1+x^2}$. [5]

Total Marks .... / 25

Differentiating Parametric Equations and Differential Equations

1 Foxes were introduced to an island. The number of foxes, F, t years after they were introduced is modelled by the equation:

$$F = 30e^{\frac{t}{5}},\ t \in \mathbb{R},\ t \geqslant 0$$

a) Write down the number of foxes that were introduced to the island. [1]

b) Find the number of years it would take for the number of foxes to first exceed 500. [2]

c) Find $\frac{dF}{dt}$. [2]

d) Find F when $\frac{dF}{dt} = 50$. [3]

Total Marks .... / 8

Review Questions

Standard Integrals and Definite Integrals

1 Find the following integrals:

a) $\int e^{-9x-3}\,dx$ [2]

b) $\int \frac{2x}{x^2+1}\,dx$ [2]

c) $\int \frac{\sec^2 3x}{\tan 3x}\,dx$ [2]

2 The graph shows the region R enclosed by the curve $y = x + \sin x$, the x-axis and the line $x = \pi$.

Find the exact value of the area of the region R. [5]

Total Marks .... / 11

Further Integration 1 & 2

1 a) Find $\int \frac{1}{(x+2)(2x-1)}\,dx$ [4]

b) Find $\int x\,\text{cosec}^2 x\,dx$ [4]

2 a) Express $\frac{2x}{(3+x)^2}$ in the form $\frac{A}{3+x}+\frac{B}{(3+x)^2}$ where A and B are constants to be determined. [3]

b) Show that $\int_0^3 \frac{2x}{(3+x)^2}\,dx = p + \ln q$, where p and q are integers to be determined. [3]

3 Using the substitution $u = \cos x$, find $\int e^{\cos x} \sin x\, dx$. [6]

4 Find $\int \frac{-x^2-2}{x(x+1)^2}\,dx$. [5]

Total Marks .... / 25

Differential Equations

1 $\frac{dy}{dx} = e^{2x} y^{-1}$

a) Find the general solution. [4]

b) Find the particular solution for $x = 0$, $y = 3$. [2]

2 Solve $\frac{dy}{dx} = 3y(y+2)$. [4]

Total Marks .... / 10

Set Notation and Conditional Probability

You must be able to:

- Use set notation to describe events
- Understand and use the addition and multiplication rules of probability
- Understand and use the conditional probability formula
- Understand and use the rules associated with independent events.

Using Set Notation

- P(A) is the notation to describe the probability of event A occurring.
- You can use set notation to identify different areas on a Venn diagram:

Key Point

Remember, if two events are **mutually exclusive** $P(A \cap B) = 0$.

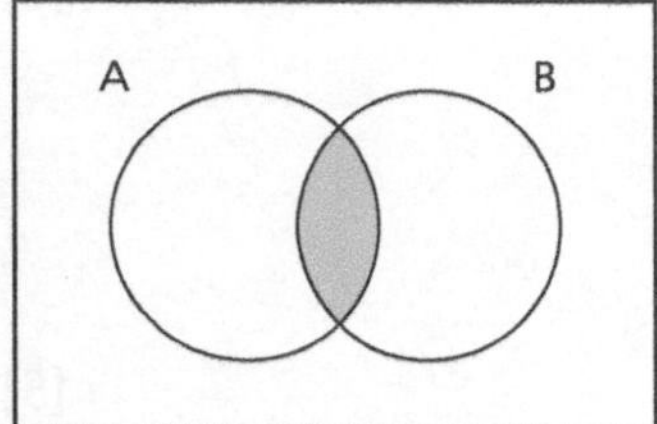

$P(A \cap B)$ = probability that both A and B occur

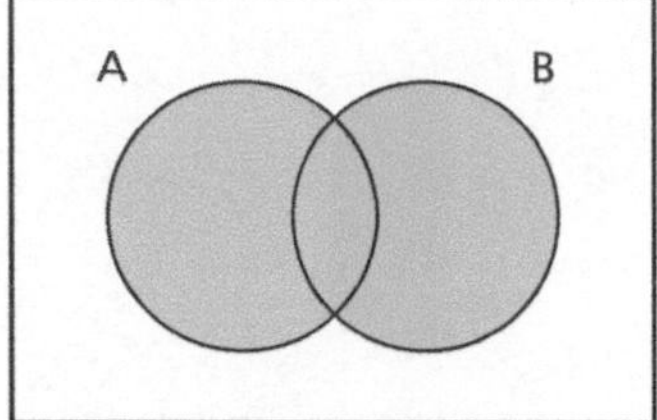

$P(A \cup B)$ = probability that either A or B occurs, or both occur

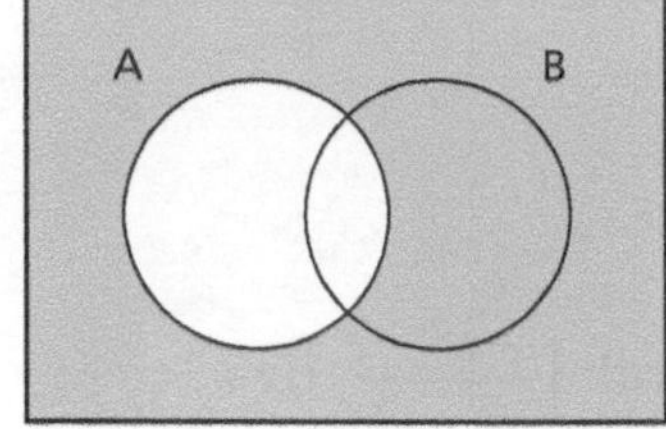

$P(A') = 1 - P(A)$ = probability that A does not occur

Conditional Probability

- The probability of an event occurring may be different if a dependent event has already occurred. This is called **conditional probability.**
- The second branch of a probability tree may represent a conditional probability, if its probability is dependent on that of the first branch.
- $P(B \mid A)$ = the probability of event B occurring **given that** event A has already occurred.

Seventy students attend a sixth-form college. Of these, 27 students study chemistry, 55 study mathematics and 16 study both chemistry and mathematics. Find the probability that a student studies chemistry given that they study mathematics.

There are 55 students who study mathematics and, of those 55, 16 also study chemistry.

Let A be the event that a student studies mathematics.

Let B be the event that a student studies chemistry.

$P(B \mid A) = \frac{16}{55}$

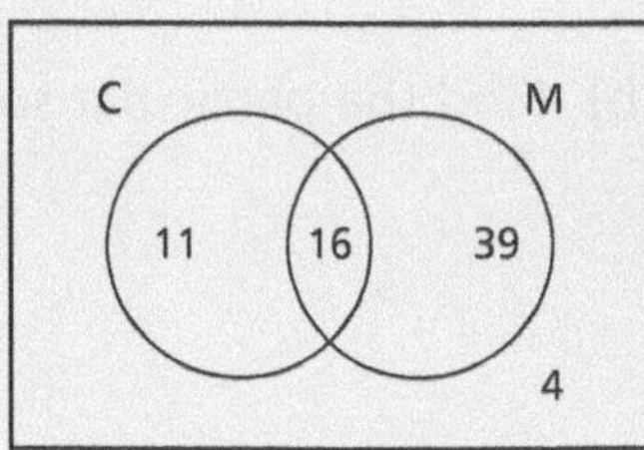

Probability Rules

$P(A \cup B) = P(A) + P(B) - P(A \cap B)$

$P(A') = 1 - P(A)$

$P(A \cap B) = P(B \mid A) \times P(A)$

Key Point

$P(A \cap B')$ = probability of **A and not** B.

1) P(A) = 0.4, P(B) = 0.5, P(A ∩ B) = 0.1. Find the probability of P(A ∪ B).

P(A ∪ B) = P(A) + P(B) – P(A ∩ B)

P(A ∪ B) = 0.4 + 0.5 – 0.1 = 0.8

2) A netball team has an important club match to play. Previous matches suggest that if the weather is fine then the chance of winning is $\frac{7}{8}$, but if the weather is not fine then the chance of winning is $\frac{1}{5}$.

The weather forecast predicts that the probability of fine weather for the match is $\frac{3}{4}$.

Calculate the probability that the team wins.

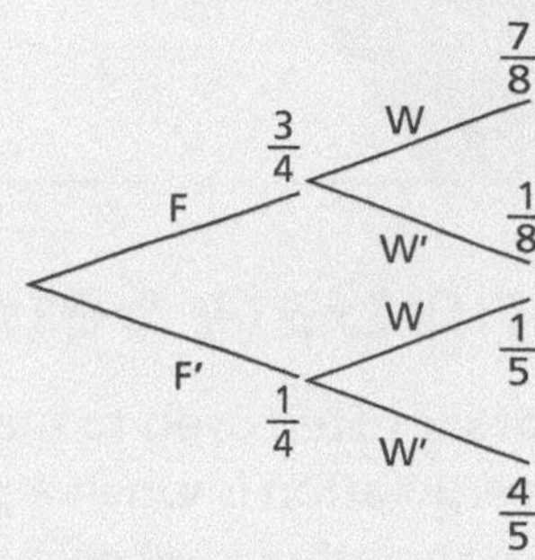

P(W) = P(F) × P(W | F) + P(F′) × P(W | F′)

$P(W) = \frac{3}{4} \times \frac{7}{8} + \frac{1}{4} \times \frac{1}{5} = \frac{113}{160}$

Find the correct paths along the probability tree and multiply along the branches.

Independent Events

- If A and B are **independent**, then P(A | B) = P(A).
- Therefore, if A and B are independent events, then P(A ∩ B) = P(A) × P(B).

The events A and B are independent.

P(A) = 0.4 and P(B) = 0.3

a) Find P(A ∪ B).

P(A ∪ B) = P(A) + P(B) – P(A ∩ B)

As the events are independent, P(A ∩ B) = P(A) × P(B)

= 0.4 × 0.3 = 0.12

P(A ∪ B) = 0.4 + 0.3 – 0.12 = 0.58

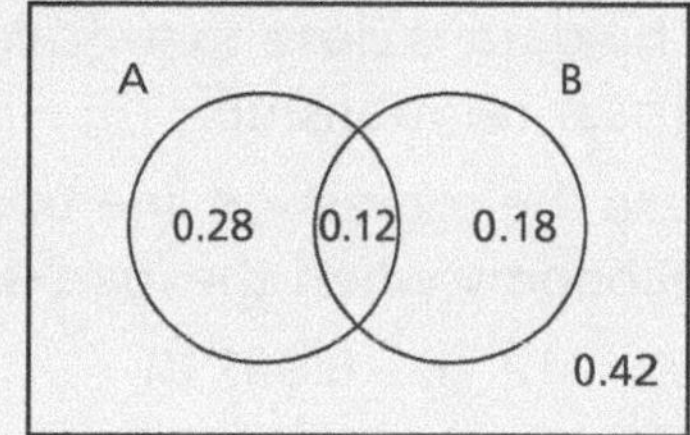

b) Find P(A ∩ B′).

P(A ∩ B′) = 0.28

Quick Test

1. The probability that a person is right-handed is 0.6. The probability that a person wears glasses is 0.2.
   Assuming that the two events are independent, find the probability that a person is both right-handed and wears glasses.
2. Event A is the probability of selecting a heart from a pack of 52 playing cards.
   Event B is the probability of selecting an ace from a pack of 52 playing cards.
   A card is selected at random from the pack. Find P(B | A).
3. P(A) = 0.7, P(B) = 0.4 and P(A ∪ B) = 0.82.
   Show that events A and B are independent.

Key Words

mutually exclusive
conditional probability
independent events

Modelling and Data Presentation

You must be able to:

- Use graphs to make predictions within the range of values of the explanatory variable and appreciate the dangers of extrapolation
- Use knowledge of logarithms to reduce a relationship of the form $y = ax^n$ into linear form to estimate a and n.
- Use modelling with probability, including critiquing assumptions made.

Using Data Sets to Make Predictions

- Data is often used to make predictions.
- **Interpolation** is when a prediction is made within the given data range.
- **Extrapolation** is when a prediction is made outside the given data range.
- Extrapolation is based on the assumption that the trend continues, which is not always the case, and therefore these predictions should be viewed with caution.

The scatter diagram shows information from a haulage company regarding the weight of the load, W, in kilograms and the fuel efficiency, Y, in miles per gallon.

W	100	101	102	103	104	105	106	107
Y	13	12.9	11.4	10.8	9.3	8.7	7.8	6.1

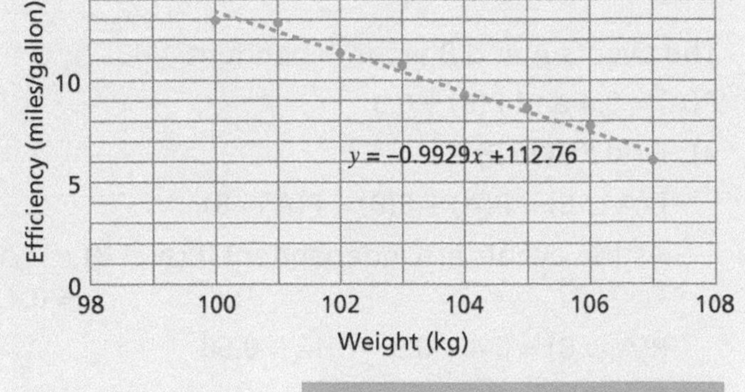

The equation of the regression line shown is
$Y = 112.76 - 0.9929W$

a) Explain why the use of a regression line is appropriate in this case.

The data is close to a straight line, suggesting a linear relationship.

b) Use the regression line to estimate the fuel economy when the load is 102.5 kg.

$Y = 112.76 - 0.9929W$

$Y = 112.76 - 0.9929 \times 102.5 = 10.99$ miles/gallon

Substitute in the value for W.

c) Use the regression line to estimate the fuel economy when the load is 110 kg.

$Y = 112.76 - 0.9929W$

$Y = 112.76 - 0.9929 \times 110 = 3.54$ miles/gallon

d) Which of the two estimates above is more reliable? Explain your answer.

The first estimate is more reliable as it is based on a value inside the given data range for the weight of the load. The second estimate is extrapolated, so is less reliable.

Key Point

Take note of the units used to form the regression line and the units in the question.

Changing the Variable to Form a Linear Relationship

- An exponential graph has y-values which become very large and are bunched together for smaller values of x; this makes interpolation difficult.
- Logs can be used to scale down the y-values:

  $y = ka^x$

  $\log y = \log(ka^x)$

  $\log y = \log(k) + \log(a^x)$

  $\log y = x\log(a) + \log(k)$
- This is now a linear relationship and so a line of regression can be used.

Key Point

Remember the log rules:

- $\log(ab) = \log a + \log b$
- $\log a^x = x\log a$

It is thought that the data below follows the model $y = ka^x$.

x	1	2	3	4	5
y	5.9	12	26	49	96

By transforming the graph into a linear function, find an estimate for k.

x	1	2	3	4	5
$\log y$	0.77	1.08	1.41	1.69	1.98

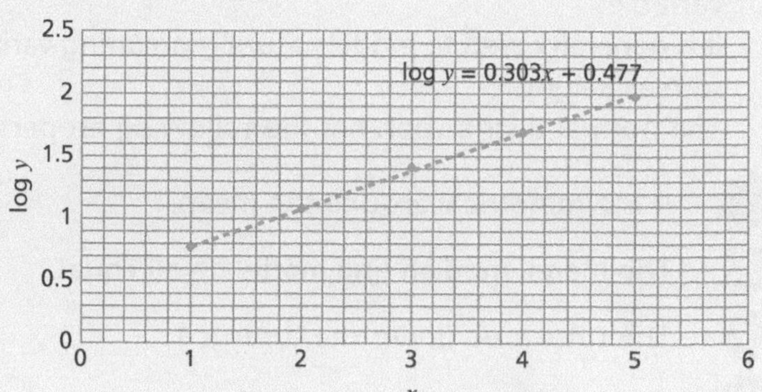

The intercept for the graph is 0.477.

The intercept is $\log(k)$ so $\log(k) = 0.477$, $k = e^{0.477}$, $k = 1.61$

Modelling with Probability

- Theoretical modelling is used to make predictions in certain situations.
- The theoretical probability might not match what is seen in real life. For example, if you rolled a fair, six-sided dice six times then you should expect to get one of each number. In reality this is very unlikely.
- The higher the number of trials carried out, the closer the real life results will be to the theory.
- The theory is based on an infinite number of trials.

A spinner has four faces numbered 1–4. The spinner is spun 20 times and the number of 1s is recorded.

a) Suggest a suitable model to this situation.

X represents the number of 1s recorded.

$X \sim B\left(20, \frac{1}{4}\right)$

b) State any assumptions you have made.

Assumed that the spinner is fair.

Assumed that all spins are independent.

Quick Test

1. The equation of a line of regression y on x is $y = 7x + 4.5$. The regression line was based on x-values in the range $6 \leq x \leq 20$.
   a) Use the line of regression to estimate a value of y when $x = 10$.
   b) Comment on the reliability of this estimate.

Key Words

interpolation
extrapolation

Statistical Distributions 1

You must be able to:

- Understand and use the normal distribution as a model and use the model to calculate probabilities
- Appreciate and use the notation $X \sim \mathrm{N}(\mu, \sigma^2)$ to represent a variable which can be approximated by a normal distribution
- Use the normal distribution to find means and standard deviations.

The Normal Distribution

- The normal distribution is used to model continuous variables.
- It is generally used to model naturally occurring variables, such as height.
- The normal distribution has the following properties:

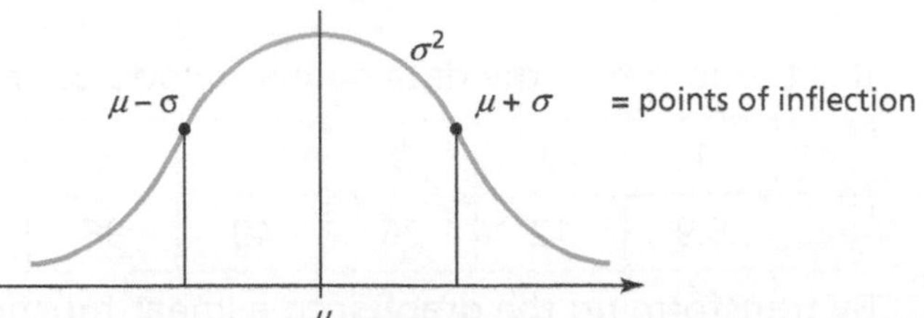

1. It is symmetrical around the mean.
2. The mean, median and mode are all equal.
3. The total area under the curve is 1.
4. The notation used is $\mathrm{N}(\mu, \sigma^2)$ where μ is the mean and σ^2 is the variance.
5. The **standard normal distribution** is represented by Z and $Z \sim \mathrm{N}(0, 1^2)$.
6. To code a normal distribution to the standard normal distribution, use $Z = \dfrac{X - \mu}{\sigma}$.

Key Point

When using the normal distribution tables on your calculator, remember the answer given is the area on the left of the Z value.

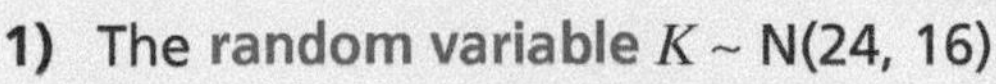

1) The **random variable** $K \sim \mathrm{N}(24, 16)$.

a) Find $\mathrm{P}(20 < K < 26)$.

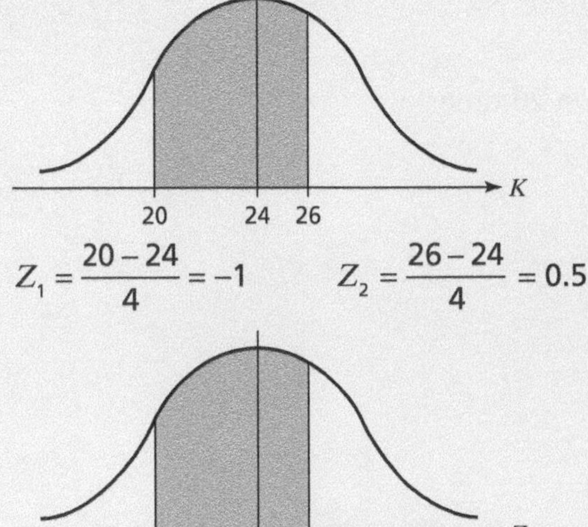

$Z_1 = \dfrac{20 - 24}{4} = -1 \qquad Z_2 = \dfrac{26 - 24}{4} = 0.5$

$\mathrm{P}(20 < K < 26) = \mathrm{P}(-1 < Z < 0.5)$

$= \mathrm{P}(Z < 0.5) - \mathrm{P}(Z < -1)$

$= 0.6915 - 0.1587$

$= 0.5328$

Key Point

When coding to the standard normal distribution, remember to divide by the standard deviation, not the variance.

Key Point

Always sketch a diagram.

Key Point

Use your calculator to find the probabilities.

b) Find the value b such that $P(K > b) = 0.005$.

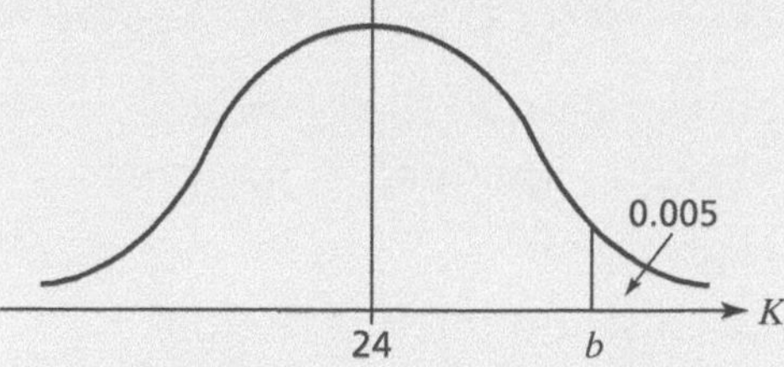

First find the value of Z which satisfies the same condition.

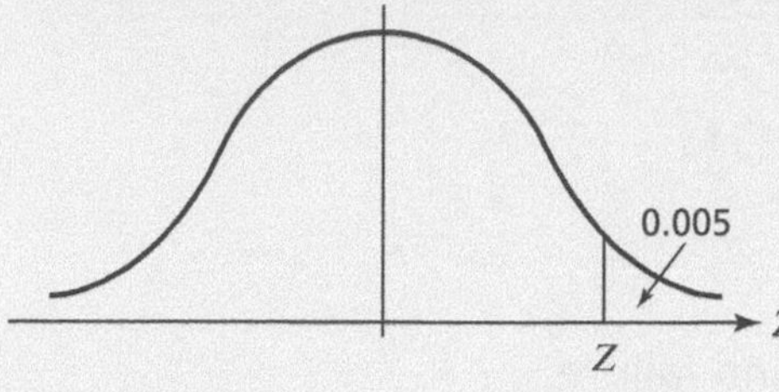

Using the percentage points table for the normal distribution
$Z = 2.5758$

$Z = \dfrac{X - \mu}{\sigma}$

$2.5758 = \dfrac{b - 24}{4}$, so $b = 34.3032$

2) A random variable $Y \sim N(\mu, \sigma^2)$.

Given that $P(Y < 24) = 0.10$ and $P(Y > 34) = 0.005$, find the value of μ and find the value of σ.

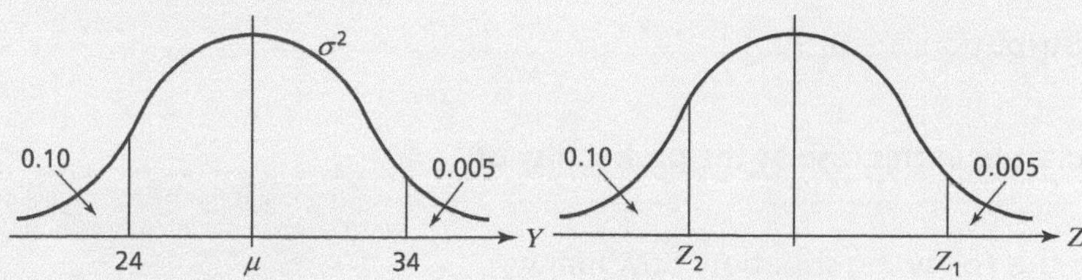

$Z_1 = 2.5758$ and $Z_2 = -1.2816$

$-1.2816 = \dfrac{24 - \mu}{\sigma}$ $\qquad 2.5758 = \dfrac{34 - \mu}{\sigma}$

$-1.2816\sigma = 24 - \mu$ and $2.5758\sigma = 34 - \mu$

$3.8574\sigma = 10$ ← Subtract the equations.

$\sigma = 2.5924$

$2.5758 \times 2.5924 = 34 - \mu$

$\mu = 27.32$

Key Point

Remember to use the percentage points for the normal distribution table when working backwards.

Key Point

The simultaneous equations can be solved as in core mathematics.

Quick Test

1. The random variable $X \sim N(40, 9)$.
   a) Find $P(X < 43)$.
   b) Find $P(X > 36)$.
2. The random variable $X \sim N(75, 6^2)$.
   Find the value of b such that $P(X > b) = 0.975$.
3. The random variable $Y \sim N(\mu, 25)$ and $P(Y < 18) = 0.9032$.
   Find the value of μ.

Key Words

standard normal distribution
random variable

Statistical Distributions 2

You must be able to:

- Use the normal distribution as an approximation to the binomial distribution and understand when the approximation is appropriate
- Understand and apply the continuity correction
- Know under what conditions a binomial distribution or a normal distribution might be a suitable model.

Selecting an Appropriate Distribution

- The binomial distribution is a discrete distribution.
- The conditions for a binomial distribution are:
  - A single trial has exactly two possible outcomes – success and failure.
  - This trial is repeated a fixed number, n, times.
  - The n trials are independent.
  - The probability of success, p, remains the same for each trial.
- The normal distribution is a continuous distribution.
- The normal distribution is used to represent naturally occurring variables.

1) A bag contains 30 red ball and 10 green balls. Ten balls are taken out of the bag without replacement. John suggests that this can be modelled by the binomial distribution $X \sim \mathrm{B}\left(10, \frac{3}{4}\right)$.
Explain why John is wrong.

The binomial model is not suitable in this case as the probability of success is not constant.

The ball is not being replaced and this changes the probability of success.

2) Suggest a suitable model for the following situation: how many times a person gets an even number when rolling a dice 12 times.

Let X represent the number of even numbers recorded when a dice is rolled 12 times.

$X \sim \mathrm{B}\left(12, \frac{1}{2}\right)$

Approximating the Binomial Distribution by a Normal

- If a random variable $X \sim \mathrm{B}(n, p)$ and n is large, completing calculations can be very difficult and most calculators will not be able to perform the necessary ones.
- If p is close to 0.5, the distribution in question is fairly symmetrical so a normal distribution can be used as an approximation.
- If n is large and p is close to 0.5, $X \sim \mathrm{N}(np, np(1-p))$.
- As the binomial distribution is used to model discrete random variables and the normal distribution is used to model continuous random variables, when approximating you need to apply the **continuity correction**.

1) If X is a discrete distribution and Y is a continuous distribution, apply the continuity correction to the following:

a) $P(X \leqslant 10)$

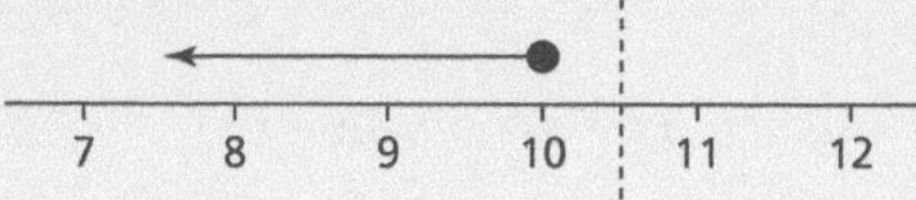

$P(X \leqslant 10) \sim P(Y < 10.5)$

You want to include 10 and values between 10 and 10.5, as they would round to 10 on a discrete scale.

b) $P(X \geqslant 5)$

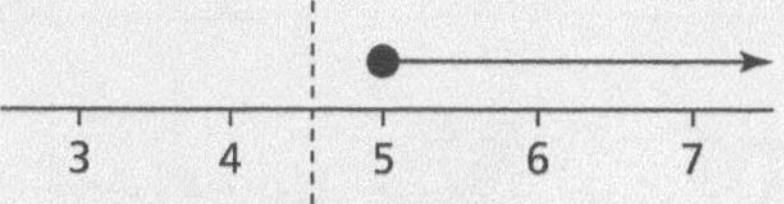

$P(X \geqslant 5) \sim P(Y > 4.5)$

c) $P(X < 12)$

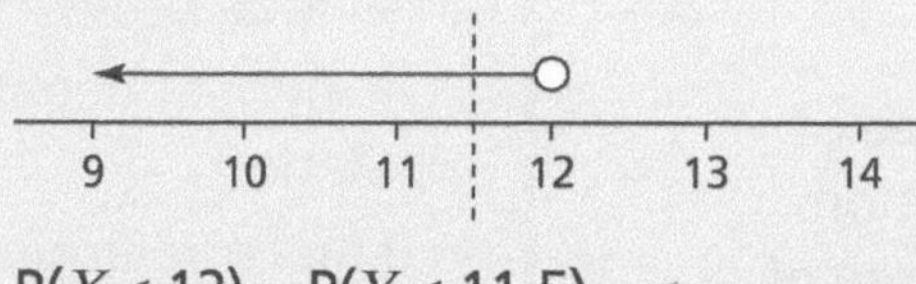

$P(X < 12) \sim P(Y < 11.5)$

You do not want to include 12 and values between 11.5 and 12, as they would round to 12 on a discrete scale.

2) A random variable $X \sim B(150, 0.4)$.
Use a normal approximation to estimate $P(50 \leqslant X < 70)$.

$np = 150 \times 0.4 = 60$

$np(1 - p) = 150 \times 0.4 \times 0.6 = 36$

$X \sim B(150, 0.4)$ so $Y \sim N(60, 36)$

$P(50 \leqslant X < 70) = P(49.5 < Y < 69.5)$

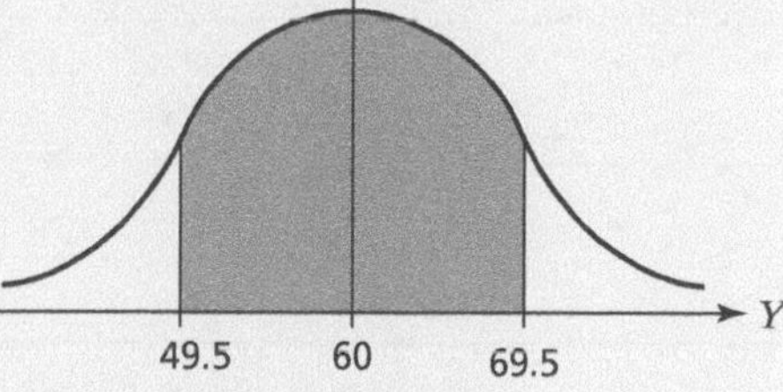

$$P(49.5 < Y < 69.5) = P\left(\frac{49.5 - 60}{6} < Z < \frac{69.5 - 60}{6}\right)$$

$$= P(-1.75 < Z < 1.58)$$

$P(Z < 1.58) - P(Z < -1.75)$

$0.9429 - 0.0401 = 0.9028$

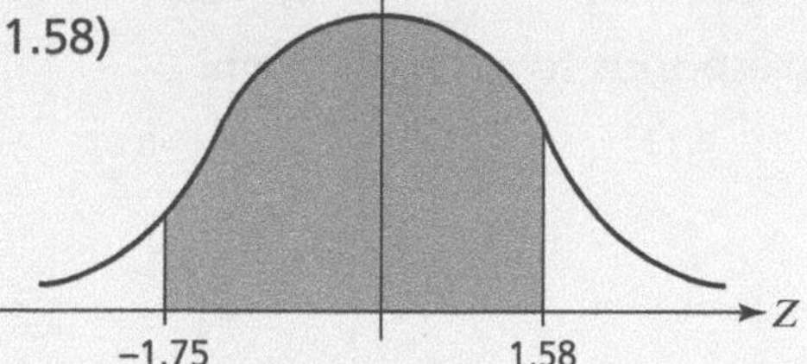

Key Point

If $X \sim B(n, p)$, then:

$\mu = np$

$\sigma^2 = np(1 - p)$

Key Point

There is no definitive answer to how large n needs to be or how close to 0.5 p should be. The larger n and the closer p is to 0.5, the better the approximation.

Quick Test

1. If X is a discrete distribution and Y is a continuous distribution, apply the continuity correction to $P(X \leqslant 14)$.
2. Suggest a suitable model for the following situation: how many heads a person gets when flipping a fair coin 40 times.
3. A random variable $X \sim B(200, 0.2)$. Use a normal approximation to estimate $P(25 \leqslant X < 35)$.

Key Words

continuity correction

Hypothesis Testing 1

You must be able to:
- Calculate a value of r using your calculator
- Interpret a given correlation coefficient using a given p-value or critical value
- Apply a hypothesis test to a correlation coefficient.

Product Moment Correlation Coefficient

- The **product moment correlation coefficient** (PMCC), r, is a measure of the strength of correlation between two variables.
- $|r| \leqslant 1$
- Values of r between 0 and 1 indicate a greater or lesser degree of positive correlation. The closer to zero, the worse the correlation.
- Values of r between 0 and –1 indicate a greater or lesser degree of negative correlation. The closer to zero, the worse the correlation.

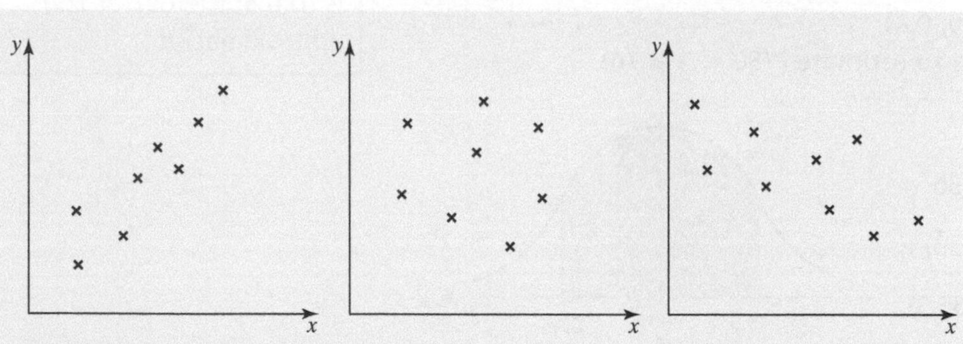

A student calculates the PMCC for each set of data above.

The values were: **a)** 0.91 **b)** 0.12 **c)** –0.87

Write down the value which corresponds to each diagram.

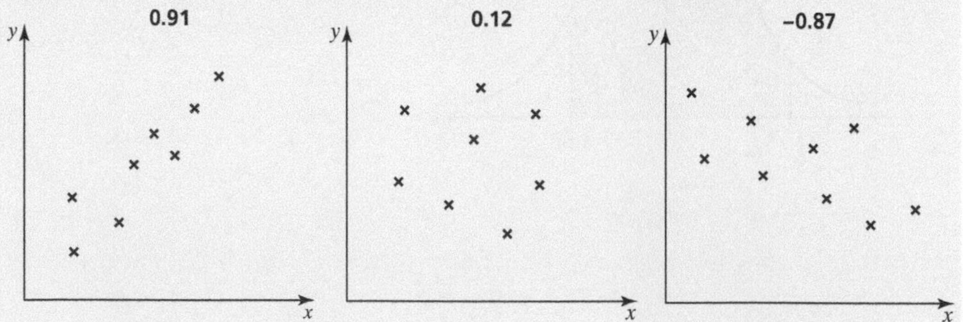

Hypothesis Testing Whether or Not the Correlation Coefficient is Zero

- When you take samples from a population and calculate the PMCC, you could get a different result for each sample.
- A **hypothesis test** helps to determine if the result from a particular sample is significant enough to suggest the PMCC is greater or less than zero.

Key Point

$r = 1$ perfect positive correlation

$r = -1$ perfect negative correlation

$r = 0$ no correlation

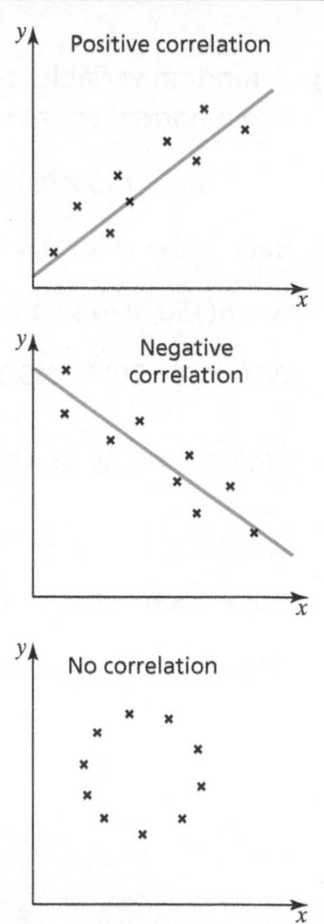

Key Point

If the correlation between two variables is close to zero, you can be satisfied there is no linear correlation. A hypothesis test helps to decide how close to zero a value needs to be before you can say there is no linear correlation.

- The steps involved are:

1. State your **null hypothesis.**
2. Interpret the question and determine the **alternate hypothesis.**
3. Calculate r as required.
4. Find the **critical value** from the table. Note the critical value represents the minimum value of r which would be considered a significant result. Anything greater than the critical value is considered significant.
5. Test to see if your value is significant.
6. Draw a conclusion.

Key Point

p represents the PMCC of the population. r represents the PMCC of the sample.

1) The PMCC between 25 pairs of observations is $r = 0.54$. Using a 5% significance level, test whether or not p differs from zero.

1. State your null hypothesis: $H_0: p = 0$
2. Determine the alternate hypothesis: $H_1: p \neq 0$
3. Calculate r: r is already given to be 0.54
4. Find the critical values from the table.
   The sample size is 25 and the significance level is 0.05, so the critical value is 0.3961.
5. Test to see if your value is significant:
   $0.54 > 0.3961$, therefore this value is significant.
6. Draw a conclusion:
   0.54 is a significant result and therefore there is enough evidence to reject the null hypothesis. There is enough evidence to suggest the correlation is not zero.

The wording of the question states 'differs from 0', therefore this is a **two-tailed test**.

This is two-tailed so the significance level is halved.

Product Moment Coefficient Level					Sample size, n
0.10	0.05	0.025	0.01	0.005	
0.2841	0.3598	0.4227	0.4921	0.5368	22
0.2774	0.3515	0.4133	0.4815	0.5256	23
0.2711	0.3438	0.4044	0.4716	0.5151	24
0.2653	0.3365	0.3961	0.4622	0.5052	25
0.2598	0.3297	0.3882	0.4534	0.4958	26
0.2546	0.3233	0.3809	0.4451	0.4869	27

2) For seven pairs of observations, the PMCC is –0.61. Test, at a 5% significance level, whether or not there is evidence of negative correlation.

$H_0: p = 0 \quad H_1: p < 0$

$r = -0.61$

Critical value = –0.6694

$-0.61 > -0.6694$

There is insufficient evidence to reject the null hypothesis. There is no evidence to suggest the correlation is less than zero.

Key Point

The null hypothesis is always $H_0: p = 0$.

$H_1: p \neq 0$, two-tailed test

$H_1: p > 0$, one-tailed test

$H_1: p < 0$, one-tailed test

Quick Test

1. Describe what the PMCC measures.
2. For 15 pairs of observations, the PMCC is 0.75. Test, at a 1% significance level, whether or not there is evidence of positive correlation.
3. For 20 pairs of observations, the PMCC is –0.42. Test, at a 5% significance level, whether or not there is evidence of negative correlation.

Key Words

product moment correlation coefficient
hypothesis test
null hypothesis
alternate hypothesis
critical value

Hypothesis Testing 2

You must be able to:

- Conduct a statistical hypothesis test for the mean of a normal distribution with known, given or assumed variance and interpret the results in context.

Testing Hypotheses About the Mean of a Normal Distribution

- The mean is one of the **population parameters** of a normal distribution.
- A hypothesis test can be used to determine whether or not the mean of a sample differs significantly from the mean of the population.
- If several samples are taken from the same population, different answers will be achieved for the value of $\bar{X}$. The values of $\bar{X}$ will form a distribution where $\bar{X} \sim \text{N}\left(\mu, \frac{\sigma^2}{n}\right)$. Therefore if $X \sim \text{N}(\mu, \sigma^2)$, then $\bar{X} \sim \text{N}\left(\mu, \frac{\sigma^2}{n}\right)$.
- The **test statistic** for this hypothesis test is $Z = \dfrac{\bar{X} - \mu}{\frac{\sigma}{\sqrt{n}}}$.
- The steps are as follows:

1. Identify or calculate the sample mean, $\bar{X}$.
2. Identify the population mean, μ.
3. Write down the null and the alternate hypothesis.
4. Calculate the value of the test statistic.
5. Either find the critical region or calculate the appropriate probability based on the significance level.
6. Conclude if your result is significant.
7. Make a conclusion in context to the question.

Key Point

$\bar{X}$ represents the sample mean.

μ represents the population mean.

1) A company sells packets of sweets. The weight of sweets in a packet follows a normal distribution with a standard deviation of 1.5 g. The company claims that the weight of a packet of sweets is 22 g.

A random sample of 16 packets is taken as part of the company's quality control process and the mean of the sample is found to be 21.1 g. The quality control manager thinks there is enough evidence to suggest that the mean weight of the packets has dropped below 22 g and production should be stopped.

Test, at a 5% significance level, whether or not there is enough evidence to support his claim.

1. The sample mean is 21.1
2. The population mean is 22.
3. $H_0: \mu = 22 \qquad H_1: \mu < 22$
4. $Z = \dfrac{\bar{X} - \mu}{\frac{\sigma}{\sqrt{n}}} = \dfrac{21.1 - 22}{\frac{1.5}{4}} = -2.4$
5. $P(\bar{X} < 21.1) = P(Z < -2.4) = 0.0082$
6. $0.0082 < 0.05$, so this is a significant result.
7. There is enough evidence to support the quality control manager's claim and production should be stopped until the problem is resolved.

Identify or calculate the sample mean, $\bar{X}$.

Identify the population mean, μ.

Write down the null and the alternate hypothesis.

Calculate the value of the test statistic.

The population standard deviation σ is 1.5 and n is 16.

Either find the critical region or calculate the appropriate probability.

Conclude if your result is significant.

Make a conclusion in context to the question.

2) The IQ of a population is found to be normally distributed with a mean of 100 and a standard deviation of 25. A psychologist wishes to test the theory that taking a cod liver oil tablet improves your performance in the test.

Thirty-six students were given a cod liver oil tablet and then took the test. The sample mean for the group was 105. By finding the critical region, test at the 5% significance level whether or not there is enough evidence to support the psychologist's claim.

The critical region is $Z > 1.6449$ (this is a one-tailed test).

The sample mean is 105.

The population mean is 100.

$H_0: \mu = 100 \qquad H_1: \mu > 100$

$$Z = \frac{\bar{X} - \mu}{\frac{\sigma}{\sqrt{n}}} = \frac{105 - 100}{\frac{25}{6}} = 1.2$$

$1.2 < 1.6449$, so this result is not significant. There is insufficient evidence to reject the null hypothesis. The psychologist's claim is not supported by this evidence.

This can be found by using the percentage points table to find which value of Z has an area of 0.05 to the right of the value.

Key Point

Read the question carefully to determine if the test is one-tailed or two-tailed.

Quick Test

1. A random sample is taken from a population with mean 21 and variance 2.25. A sample of 20 is taken from the population and the sample mean is calculated to be 21.2. Test, at a 5% significance level, $H_0: \mu = 21 \qquad H_1: \mu \neq 21$
2. A random sample is taken from a population with mean 120 and variance 4. A sample of 30 is taken from the population and the sample mean is calculated to be 118. Test, at a 5% significance level, $H_0: \mu = 120 \qquad H_1: \mu > 120$
3. A random sample is taken from a population with mean 85 and standard deviation 4. A sample of 25 is taken from the population. Using a 5% significance level, find the critical region for the test statistic $\bar{X}$ where $H_0: \mu = 85$ and $H_1: \mu \neq 85$.

Key Words

population parameter
test statistic

Practice Questions

Set Notation and Conditional Probability

1 Given P(A | B) = 0.6, P(A | B′) = 0.2 and P(B) = 0.7, find:

a) P(A) [2]

b) P(A ∪ B) [2]

c) P(A ∪ B′) [2]

d) P(B | A) [2]

e) P(A′ | B′) [2]

2 The Venn diagram shows information about employees at a factory. The event F is being female and the event T is being part-time.

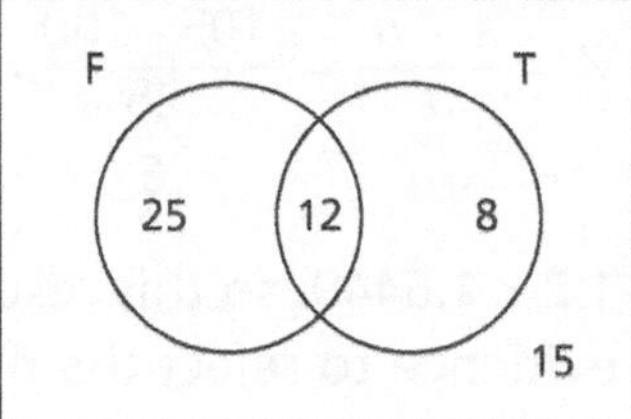

One of the factory employees is chosen at random. Write down:

a) the probability that the person selected is female [1]

b) the probability that the person selected is part-time [1]

c) the probability that the person is female given that they are part-time [2]

d) the probability that the person is female and works part-time. [2]

Total Marks .... / 16

Modelling and Data Presentation

1 The graph shows data collected in an experiment to test the deflection when the mass is increased.

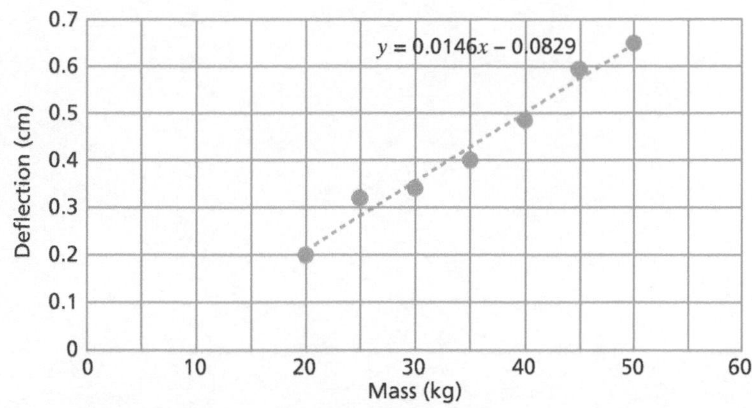

a) Estimate the deflection when the mass is 47 kg. [1]

b) Estimate the deflection when the mass is 60 kg. [1]

c) Comment on which of the above estimates is more reliable and explain your answer. [2]

2 A six-sided dice is rolled 300 times and the number of 6s is recorded.

a) Suggest a suitable model to represent this situation. [3]

b) State any assumptions you have made. [2]

3 It is thought that the data below follows the model $y = ka^x$.

x	1	2	3	4	5
y	18	54	162	486	1458

By transforming y into a linear function, find an estimate for k. [4]

Total Marks .... / 13

Practice Questions

Statistical Distributions 1 & 2

1 Let X be a continuous random variable. Find:

a) $P(X < 62)$ if $X \sim N(60, 4^2)$ [3]

b) $P(X > 54)$ if $X \sim N(48, 9)$ [3]

c) $P(X < 23)$ if $X \sim N(30, 5^2)$ [3]

2 The heights of a group of women are normally distributed with a mean of 165 cm and a standard deviation of 5 cm. A woman is selected at random from this group.

a) Find the probability that she is taller than 170 cm. [4]

b) Find the height that fewer than 5% of women in this group are taller than. [4]

3 The heights of bonsai trees are normally distributed. Given that 10% of trees are taller than 60 cm and 5% are shorter than 30 cm, find the mean and the standard deviation of the heights. [6]

4 The waiting time in a dental surgery is normally distributed with a standard deviation of 4.5 minutes. Given that the probability of waiting more than 20 minutes is 0.0005, find:

a) the mean waiting time [3]

b) the probability of waiting fewer than 10 minutes. [2]

Total Marks .... / 28

Hypothesis Testing 1 & 2

1. A random sample is taken from a population believed to have mean 26 and variance 9. A sample of 10 is taken from the population and the sample mean is calculated to be 28.

   Test, at a 5% significance level, H_0: $\mu = 26$ H_1: $\mu \neq 26$ [5]

2. The results of a class test are approximated by a normal distribution $X \sim N(65, 100)$. After the introduction of a new revision technique, the results of eight students have a mean of 70. Is there evidence that the results have improved? [5]

3. String produced in a factory is designed so that 10-metre lengths have breaking strengths that are normally distributed with mean 7.5 kg and standard deviation 1.2 kg. The manufacturer claims that constructing the string in a different way (using a coating) will increase its mean breaking strength but leave the standard deviation unaltered. To test this claim, the coating was applied to thirty 10-metre lengths and the breaking strength was found to be 7.6 kg.

   Using a 5% level of significance and stating your hypotheses clearly, test whether the manufacturer's claim is justified. [5]

4. The product moment correlation coefficient was calculated as –0.56, based on 25 pairs of data.

   a) Test whether this is significantly less than zero, using a 5% level of significance. [3]

   b) Test whether this is significantly different to zero, using a 5% level of significance. [3]

5. Dan read a gardening article which suggested that talking to your tomato plants can improve yield. He took a sample of 16 plants and calculated the product moment correlation coefficient (PMCC) between the time (in minutes) spent talking to the plant and the number of tomatoes yielded. The PMCC was calculated to be 0.69.

   Test, at a 5% significance level, if there is enough evidence to support the claim in the article. [6]

Total Marks .... / 27

Review Questions

Iterative Methods and Finding Roots

1. Show that the equation $x^2 - \sin 2x = 0$ has a root in the interval $0.9 < \alpha < 1.0$. From your results, suggest whether the root is likely to be closer to 0.9 or 1.0. [6]

2. Show that the equation $f(x) = \sin x - x^3 = 0$ has a root α such that $0.5 < \alpha < 1$.

   By using the iteration $x_{n+1} = \sqrt[3]{\sin x}$ with $x_0 = 1$, write down x_1, x_2, x_3 and x_4, and hence find α to 3 decimal places. [8]

3. By rearranging the equation $f(x) = x^2 + x - 1 = 0$ into a suitable form, use an iteration formula to find an approximation for a root α to 2 decimal places. Start with $x_0 = 0.5$. [5]

Total Marks .... / 19

Numerical Integration and Solving Problems in Context

1. Use the trapezium rule with four strips to estimate the value for $\int_0^{\frac{\pi}{4}} \sqrt{1+\sec x}\, dx$. [6]

2. Consider the function $f(x) = e^{2x} - x^2 - 5$. Apply the Newton-Raphson process twice, with a starting estimate of $x_0 = 1$, to find a root α to 2 decimal places such that $f(\alpha) = 0$. [6]

3 Show that a root α of the equation $x^3 + 4x - 12 = 0$ exists in the interval $1.5 < x < 2$. Find α to 3 significant figures by applying the Newton-Raphson iteration once, taking $x_0 = 1.75$ as your starting value. [9]

Total Marks .... / 21

Proof by Contradiction and Further Trigonometric Identities

1 Prove (using the addition formulae) the trigonometric identity $\sin 3x \equiv 3\sin x - 4\sin^3 x$. [5]

2 If $a^2 + b^2 = c^2$, where a, b and c are integers, prove by contradiction that at least one of a or b must be even. [8]

3 Prove the identity $\sec 2x + \tan 2x \equiv \frac{\cot x + 1}{\cot x - 1}$. Hence show $\cot 15° = 2 + \sqrt{3}$. [15]

Total Marks .... / 28

Using Vectors in Kinematics

You must be able to:

- Differentiate and integrate vectors with respect to time
- Use calculus in kinematics for motion in two dimensions
- Derive *suvat* formulae for constant acceleration in two dimensions
- Use vectors to solve problems.

Calculus in Kinematics

- The **position vector r** of a moving particle is $\mathbf{r} = f(t)\mathbf{i} + g(t)\mathbf{j}$.
- To find the **velocity vector v**, you differentiate the **i** and **j** components in turn. So $\mathbf{v} = \frac{d\mathbf{r}}{dt} = f'(t)\mathbf{i} + g'(t)\mathbf{j}$.
- To find the **acceleration vector a**, you again differentiate each component in turn. So $\mathbf{a} = \frac{d\mathbf{v}}{dt} = f''(t)\mathbf{i} + g''(t)\mathbf{j}$.

Key Point

You should recall that **i** is a unit vector in the direction of the x-axis, and **j** is a unit vector in the direction of the y-axis.

A particle moves in a plane such that at time t its position vector **r** is given by $\mathbf{r} = (2t^3 - t)\mathbf{i} + (t^2 + 5t)\mathbf{j}$.

a) Find the particle's velocity after three seconds.

$\mathbf{v} = \frac{d\mathbf{r}}{dt} = (6t^2 - 1)\mathbf{i} + (2t + 5)\mathbf{j}$

At $t = 3$, $\mathbf{v} = (6 \times 3^2 - 1)\mathbf{i} + (2 \times 3 + 5)\mathbf{j} = (53\mathbf{i} + 11\mathbf{j})\,\text{ms}^{-1}$

b) Find the particle's speed after three seconds.

Speed $= |\mathbf{v}| = \sqrt{53^2 + 11^2} = 54.1\,\text{ms}^{-1}$

c) Find the particle's **direction of motion** after three seconds.

So $\theta = \arctan\left(\frac{11}{53}\right) = 11.7°$

The direction of motion is therefore at an angle of 11.7° to the horizontal.

The particle's direction of motion is the direction of the velocity vector.

d) Find the particle's acceleration after three seconds.

$\mathbf{a} = \frac{d\mathbf{v}}{dt} = 12t\mathbf{i} + 2\mathbf{j}$ At $t = 3$, $\mathbf{a} = (12 \times 3)\mathbf{i} + 2\mathbf{j} = (36\mathbf{i} + 2\mathbf{j})\,\text{ms}^{-2}$

- Since integration is the reverse of differentiation: $\mathbf{r} = \int \mathbf{v}\,dt$ and $\mathbf{v} = \int \mathbf{a}\,dt$

Key Point

Remember that when you integrate a vector, you must always add a constant of integration in vector form.

A particle P moves in a plane such that at time t its acceleration vector **a** is given by $\mathbf{a} = t^2\mathbf{i} + (3 - t)\mathbf{j}$. The velocity of P after one second is $(2\mathbf{i} + 3\mathbf{j})\,\text{ms}^{-1}$ and it starts from a fixed origin O.

Find the velocity vector of P.

$\mathbf{v} = \int \mathbf{a}\,dt = \int (t^2\mathbf{i} + (3-t)\mathbf{j})\,dt = \frac{t^3}{3}\mathbf{i} + \left(3t - \frac{t^2}{2}\right)\mathbf{j} + \mathbf{c}$

$2\mathbf{i} + 3\mathbf{j} = \frac{1}{3}\mathbf{i} + \frac{5}{2}\mathbf{j} + \mathbf{c}$, so $\mathbf{c} = \frac{5}{3}\mathbf{i} + \frac{1}{2}\mathbf{j}$

Substituting $t = 1$, $\mathbf{v} = 2\mathbf{i} + 3\mathbf{j}$.

$\therefore \mathbf{v} = \frac{t^3}{3}\mathbf{i} + \left(3t - \frac{t^2}{2}\right)\mathbf{j} + \frac{5}{3}\mathbf{i} + \frac{1}{2}\mathbf{j}$ $\therefore \mathbf{v} = \left(\frac{t^3}{3} + \frac{5}{3}\right)\mathbf{i} + \left(3t - \frac{t^2}{2} + \frac{1}{2}\right)\mathbf{j}$

Proof of Two *suvat* Formulae

- You need to be able to prove the following two *suvat* formulae. These may be required when **a** is a **constant vector**.

- Suppose a particle P has constant acceleration **a** at time t and initial velocity **u**. Then $\mathbf{v} = \int \mathbf{a}\, dt = \mathbf{a}t + \mathbf{c}$.
  - Substituting $t = 0$, $\mathbf{v} = \mathbf{u}$ gives $\mathbf{u} = \mathbf{a} \times 0 + \mathbf{c}$.
  - So $\mathbf{c} = \mathbf{u}$ and therefore $\mathbf{v} = \mathbf{u} + \mathbf{a}t$.
  - Now $\mathbf{r} = \int \mathbf{v}\, dt = \int (\mathbf{u} + \mathbf{a}t)\, dt \quad = \mathbf{u}t + \frac{1}{2}\mathbf{a}t^2 + \mathbf{k}$
  - If you assume that at time $t = 0$, the displacement of P is 0, then $0 = \mathbf{k}$ and so $\mathbf{r} = \mathbf{u}t + \frac{1}{2}\mathbf{a}t^2$.
- You should be able to apply these formulae to motion in two dimensions. For example, if a particle's initial velocity is $(\mathbf{i} - 3\mathbf{j})\,\text{ms}^{-1}$ and it has acceleration $(-\mathbf{i} + 4\mathbf{j})\,\text{ms}^{-2}$, then after two seconds:
  - the velocity of the particle is: $\mathbf{v} = \mathbf{u} + \mathbf{a}t = (\mathbf{i} - 3\mathbf{j}) + (-\mathbf{i} + 4\mathbf{j}) \times 2$
    $= (\mathbf{i} - 3\mathbf{j}) + (-2\mathbf{i} + 8\mathbf{j}) = (-\mathbf{i} + 5\mathbf{j})\,\text{ms}^{-1}$
  - the displacement of the particle is:
    $\mathbf{r} = \mathbf{u}t + \frac{1}{2}\mathbf{a}t^2 = (\mathbf{i} - 3\mathbf{j}) \times 2 + \frac{1}{2}(-\mathbf{i} + 4\mathbf{j}) \times 2^2 = (2\mathbf{i} - 6\mathbf{j}) + (-2\mathbf{i} + 8\mathbf{j}) = 2\mathbf{j}$ m

Notice that the displacement here is a vector **2j** m, but if the distance had been asked for, the answer would be 2 m.

Key Point

The types of questions you may encounter in 2D kinematics are those relating to the movement of particles.

Problem Solving

At time t seconds, particles A and B have position vectors

$$\mathbf{r}_A = \begin{pmatrix} t^2 - 3t - 12 \\ \dfrac{3t^2}{2} - 6t + 1 \end{pmatrix} \text{ and } \mathbf{r}_B = \begin{pmatrix} \dfrac{t^2}{2} + t + 12 \\ t^2 + 1 \end{pmatrix} \text{ respectively.}$$

You can also work using column vectors.

a) Find the time(s) when particles A and B meet, and their position vector at these time(s).

When two particles collide, both **i** components must be equal, and both **j** components must be equal.

$$t^2 - 3t - 12 = \frac{t^2}{2} + t + 12$$

Equating **i** components.

So $\dfrac{t^2}{2} - 4t - 24 = 0$ Or $t^2 - 8t - 48 = 0$ $\therefore (t + 4)(t - 12) = 0$

$t + 4 = 0$ gives a negative solution for t, so only valid answer is $t = 12$.

$\dfrac{3t^2}{2} - 6t + 1 = 216 - 72 + 1 = 145$ and $t^2 + 1 = 145$

Substituting $t = 12$ in both the **j** components as a check.

Therefore the particles do meet when $t = 12$.

When $t = 12$, $t^2 - 3t - 12 = 144 - 36 - 12 = 96$

The particles therefore meet at $\mathbf{r} = \begin{pmatrix} 96 \\ 145 \end{pmatrix}$

b) Find the velocity vector of A and find the time when A is travelling in a southerly direction.

When a particle is travelling in a southerly (or northerly) direction, the **i** component must be equal to zero.

$$\mathbf{v}_A = \frac{d\mathbf{r}_A}{dt} = \begin{pmatrix} 2t - 3 \\ 3t - 6 \end{pmatrix}$$

Setting $2t - 3 = 0$ and solving, $t = \frac{3}{2}$ s. When $t = \frac{3}{2}$, $\mathbf{v}_A = \begin{pmatrix} 0 \\ -\frac{3}{2} \end{pmatrix}$

The minus sign shows the particle is travelling south.

Quick Test

1. A particle's position vector at time t is given by $\mathbf{r} = \left(\frac{1}{2}t^2\right)\mathbf{i} + (3t)\mathbf{j}$.
   Find its speed after two seconds and show its acceleration is constant.
2. A particle's acceleration vector is $\mathbf{a} = 2t\mathbf{j}\,\text{ms}^{-2}$. Given its initial velocity is $(\mathbf{i} - \mathbf{j})\,\text{ms}^{-1}$, find its speed when $t = 2$.
3. A particle's position vector at time t is given by $\mathbf{r} = (3 + 2t)\mathbf{i} + (t^2)\mathbf{j}$. Find the angle its direction of motion makes with vector **i** after two seconds.

Key Words

position vector
velocity vector
acceleration vector
direction of motion
constant vector

Projectiles

You must be able to:
- Solve problems involving projectiles
- Derive formulae for time of flight, range and greatest height
- Derive the formula for the equation of the path of flight of a projectile.

Projectiles and Vectors

- Consider a particle projected with speed u at an angle α to the horizontal. Since the x- and y-axes indicate the directions of **positive displacement**, the particle experiences an acceleration of $-g$ in the direction of the y-axis. Notice there is no force on the particle in the x-axis direction, so it is not accelerating in this direction.
- There are four main equations you can use, obtained by considering the **horizontal** and **vertical motion** separately (see table, right).
- Without t, you can also apply $v^2 = u^2 + 2as$ in a vertical direction.

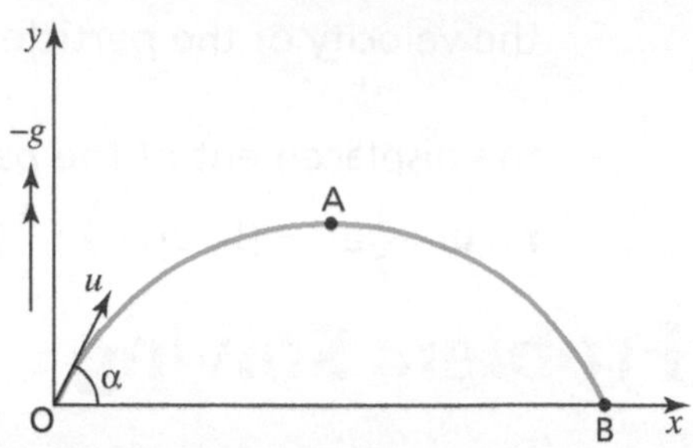

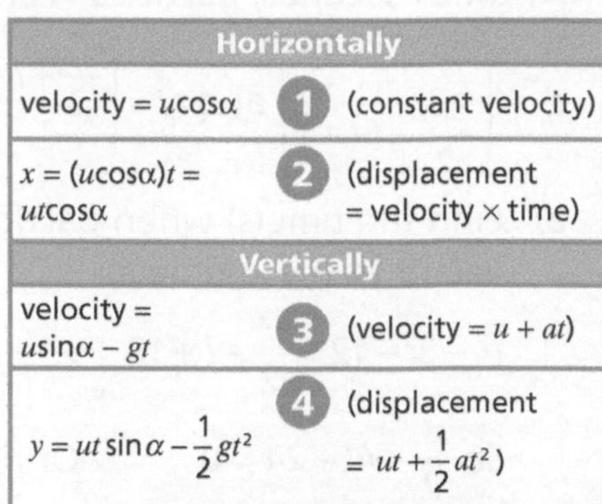

Horizontally		
velocity = $u\cos\alpha$	1	(constant velocity)
$x = (u\cos\alpha)t = ut\cos\alpha$	2	(displacement = velocity × time)
Vertically		
velocity = $u\sin\alpha - gt$	3	(velocity = $u + at$)
$y = ut\sin\alpha - \frac{1}{2}gt^2$	4	(displacement = $ut + \frac{1}{2}at^2$)

Time and Range of Flight

- In the case above, the vertical displacement is zero at B.

Therefore $ut\sin\alpha - \frac{1}{2}gt^2 = 0$

$$t\left(u\sin\alpha - \frac{1}{2}gt\right) = 0 \Rightarrow t = 0 \text{ (at O) or } u\sin\alpha - \frac{1}{2}gt = 0 \text{ (at B)}$$

$$\Rightarrow \quad u\sin\alpha = \frac{1}{2}gt \quad \Rightarrow t = \frac{2u\sin\alpha}{g}$$

- So the formula for the **time of flight** of the particle is $t = \frac{2u\sin\alpha}{g}$
- To find the **horizontal range**, substitute this value of t into equation 2:

$$x = u\cos\alpha\left(\frac{2u\sin\alpha}{g}\right) = \frac{2u^2\sin\alpha\cos\alpha}{g} \Rightarrow x = \frac{u^2\sin 2\alpha}{g}$$

Since $\sin 2\alpha \equiv 2\sin\alpha\cos\alpha$

- So the formula for the particle's range of flight is $x = \frac{u^2\sin 2\alpha}{g}$.

Notice that to make the range of flight as far as possible, you need to maximise $\frac{u^2\sin 2\alpha}{g}$.
The maximum value of $\sin 2\alpha$ is 1, when $2\alpha = 90°$ or $\alpha = 45°$. Therefore, to maximise the range, the projectile should be fired at an angle of 45°.

Greatest Height

- At the point of **greatest height**, at A in the case above, the vertical velocity is instantaneously zero.
- Therefore $u\sin\alpha - gt = 0 \quad \Rightarrow \quad t = \frac{u\sin\alpha}{g}$
- So the formula for the time of flight to the highest point is $t = \frac{u\sin\alpha}{g}$

Notice that this is exactly half the value of the time to B.

- To find the greatest height, substitute this value of t into equation 4:

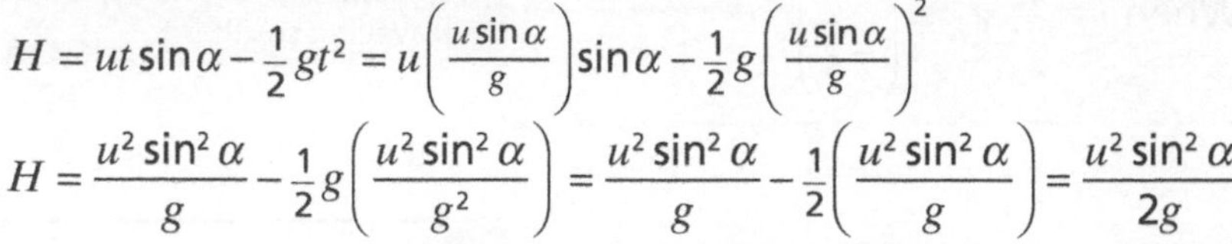

$$H = ut\sin\alpha - \frac{1}{2}gt^2 = u\left(\frac{u\sin\alpha}{g}\right)\sin\alpha - \frac{1}{2}g\left(\frac{u\sin\alpha}{g}\right)^2$$

$$H = \frac{u^2\sin^2\alpha}{g} - \frac{1}{2}g\left(\frac{u^2\sin^2\alpha}{g^2}\right) = \frac{u^2\sin^2\alpha}{g} - \frac{1}{2}\left(\frac{u^2\sin^2\alpha}{g}\right) = \frac{u^2\sin^2\alpha}{2g}$$

- So the formula for the particle's greatest height is $H = \frac{u^2\sin^2\alpha}{2g}$.

A ball is projected with speed 60 ms⁻¹ from the top of a cliff 100 m high towards the sea, at an angle of elevation of 50°, as shown right.

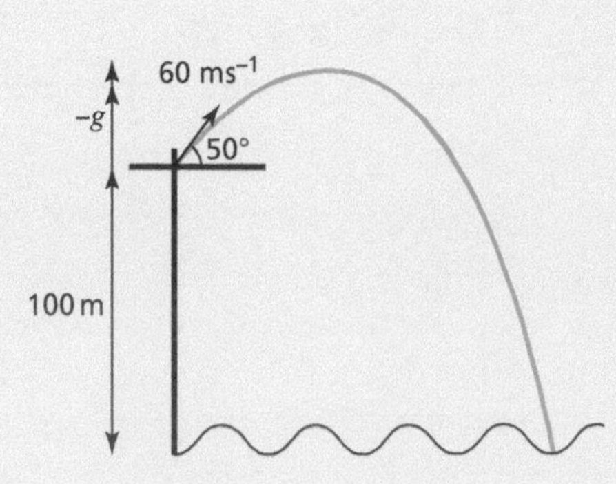

a) Find the greatest height above the cliff reached by the ball.

$$H = \frac{u^2\sin^2\alpha}{2g} = \frac{60^2 \times \sin^2 50°}{2 \times 9.8} = 108 \text{ m}$$

b) Find the speed and the angle at which the ball hits the sea.

When the ball hits the sea, its vertical displacement is –100.

So applying $v^2 = u^2 + 2as$ vertically: $v^2 = 60^2 + 2 \times (-9.8) \times (-100)$

$v^2 = 5560 \Rightarrow v = -74.57\,\text{ms}^{-1}$.

Ball's horizontal velocity is $u\cos\alpha = 60\cos 50° = 38.57\,\text{ms}^{-1}$

So the ball hits the sea with a speed of $\sqrt{74.57^2 + 38.57^2} = 83.9\,\text{ms}^{-1}$

The angle with which it hits the sea is

$\arctan\left(\frac{74.57}{38.57}\right) = 62.7°$ below the horizontal.

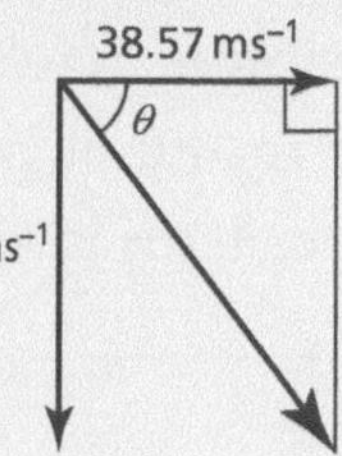

When applying $v^2 = u^2 + 2as$ here, remember upwards is the positive direction. Therefore $u = +60$ and $a = -9.8$. You take v to be negative as the final vertical speed is in a downwards direction.

Equation of Flight

- To find the equation of the path of the flight, find y in terms of x. Make t the subject from equation 2, and substitute into equation 4:

  $x = ut\cos\alpha \Rightarrow t = \frac{x}{u\cos\alpha}$

  – Substituting this into $y = ut\sin\alpha - \frac{1}{2}gt^2$ gives $y = u\left(\frac{x}{u\cos\alpha}\right)\sin\alpha - \frac{1}{2}g\left(\frac{x}{u\cos\alpha}\right)^2$

  – Therefore $y = x\tan\alpha - \frac{gx^2}{2u^2\cos^2\alpha}$

  – Or $y = x\tan\alpha - \frac{gx^2\sec^2\alpha}{2u^2}$ or $y = x\tan\alpha - \frac{gx^2}{2u^2}(1+\tan^2\alpha)$

Key Point

Although you are not expected to memorise these formulae, you may be required to derive them, as shown left. The final three formulae are all equivalent, though the latter one is likely to be the most useful, as only one trigonometric function (tanα) is included.

A particle is projected from a point O on a horizontal plane, with speed 50 ms⁻¹, at an angle of elevation α. Given that the particle passes through the point with coordinates (40, 12), find the two possible values for α.

$12 = 40\tan\alpha - \frac{9.8 \times 40^2}{2 \times 50^2}(1+\tan^2\alpha)$

$12 = 40\tan\alpha - 3.136(1 + \tan^2\alpha)$

$12 = 40\tan\alpha - 3.136 - 3.136\tan^2\alpha$

$3.136\tan^2\alpha - 40\tan\alpha + 15.136 = 0$

$\tan\alpha = \frac{-(-40) \pm \sqrt{(-40)^2 - 4 \times 3.136 \times 15.136}}{2 \times 3.136}$

$\Rightarrow \tan\alpha = 0.3903$ or $\tan\alpha = 12.36$, so $\alpha = 21.3°$ or $\alpha = 85.4°$

You can use the equation $y = x\tan\alpha - \frac{gx^2}{2u^2}(1+\tan^2\alpha)$ and substitute in the values for x, y and u.

Using the quadratic formula.

Projectiles and Vectors

- You also need to be able to work with equations 1–4 in vector form.
- Suppose a particle is projected from a point O with speed $u\mathbf{i} + v\mathbf{j}$ ms⁻¹. Then equations 1–4 become as listed right.
- Velocity $v = u\mathbf{i} + (v - gt)\mathbf{j} = \begin{pmatrix} u \\ v - gt \end{pmatrix}$
- Displacement $s = ut\mathbf{i} + (vt - \frac{1}{2}gt^2)\mathbf{j} = \begin{pmatrix} ut \\ vt - \frac{1}{2}gt^2 \end{pmatrix}$

1. horizontal velocity $= u$
2. horizontal displacement $= ut$
3. vertical velocity $= v - gt$
4. vertical displacement $= vt - \frac{1}{2}gt^2$

Quick Test

1. A particle is projected from the ground with speed 40 ms⁻¹ at 35° to the horizontal. Find the time taken for it to reach its maximum height.
2. A particle is projected with speed $5\mathbf{i} + 15\mathbf{j}$ ms⁻¹. Find the distance the particle is from its starting point after two seconds.
3. Find the angle at which a particle needs to be projected if it has an initial speed of 35 ms⁻¹ and reaches a maximum height of 50 m.

Key Words

positive displacement
horizontal motion
vertical motion
time of flight
horizontal range
greatest height
vector form

Resolving Forces and Forces in Equilibrium

You must be able to:

- Resolve forces in two dimensions
- Find the resultant force of two or more forces
- Solve problems involving particles in equilibrium.

Resolving Forces in Two Dimensions

- It is often helpful in mechanics to **resolve** a set of forces into **components**, usually horizontal and vertical. When considering objects on planes, you will also need to be able to resolve forces into components along the plane, and perpendicular to the plane.

A lawnmower is being pulled along the ground with a force of 25 N, at an angle of 30° as shown (right). Resolve the 25 N force into a horizontal component X and a vertical component Y.

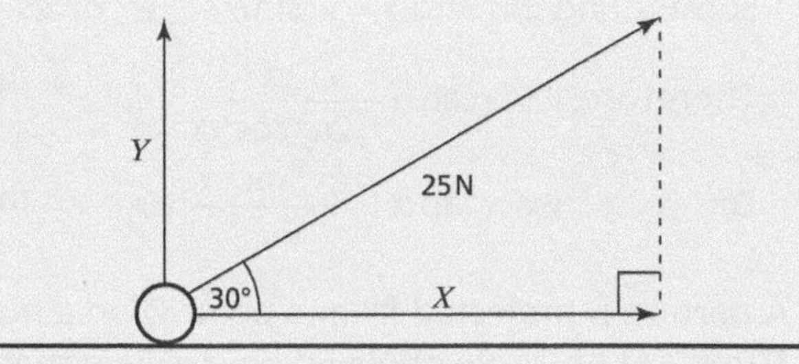

By using trigonometry, $\cos 30° = \frac{X}{25}$ and $\sin 30° = \frac{Y}{25}$.

Therefore $X = 25\cos 30° = \frac{25\sqrt{3}}{2}$ N

and $Y = 25\sin 30° = 12.5$ N

This resolved force can also be written in vector form as $F = \frac{25\sqrt{3}}{2}\mathbf{i} + 12.5\mathbf{j} = \begin{pmatrix} \frac{25\sqrt{3}}{2} \\ 12.5 \end{pmatrix}$

- In general, if a force F acts at an angle θ to the horizontal, the component **adjacent** to the angle is $F\cos\theta$ and the component **perpendicular** to this is $F\sin\theta$.

A 5 kg particle rests on a plane inclined at an angle of 20° to the horizontal. Find the component of the weight of the particle perpendicular to the plane (P) and the component of the weight parallel to the plane (Q).

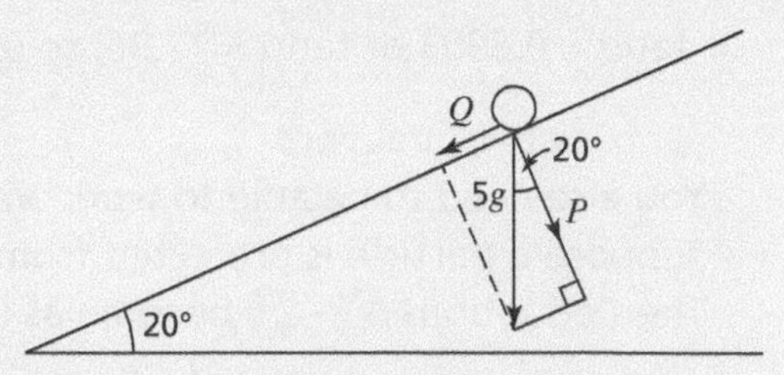

By using trigonometry, $\cos 20° = \frac{P}{5g}$ and $\sin 20° = \frac{Q}{5g}$.

Therefore $P = 5g\cos 20° = 46.0$ N and $Q = 5g\sin 20° = 16.8$ N

- In general, if a mass m rests on a plane inclined at an angle θ to the horizontal, the component of weight into the plane is $mg\cos\theta$ and the component down the plane is $mg\sin\theta$.

The Resultant of Two or More Forces

- To find the **resultant force** of two or more forces, split the forces into two perpendicular components, combine these components and calculate the resultant direction and magnitude.

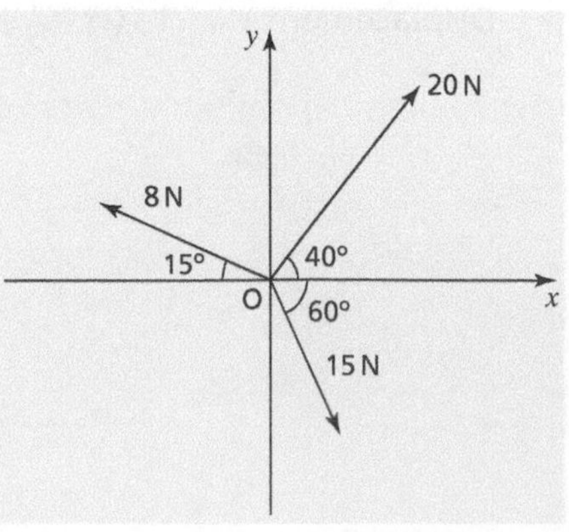

Three forces of 20 N, 8 N and 15 N act on a particle at O as shown in the diagram (right). Find the magnitude and direction of the resultant force.

Components along Ox: $20\cos40° + 15\cos60° - 8\cos15° = 15.093$

Components along Oy: $20\sin40° - 15\sin60° + 8\sin15° = 1.9359$

Or using vectors: $\begin{pmatrix} x_r \\ y_r \end{pmatrix} = \begin{pmatrix} 20\cos40 \\ 20\sin40 \end{pmatrix} + \begin{pmatrix} 15\cos60 \\ -15\sin60 \end{pmatrix} + \begin{pmatrix} -8\cos15 \\ 8\sin15 \end{pmatrix} = \begin{pmatrix} 15.093 \\ 1.9359 \end{pmatrix}$

By Pythagoras, the resultant force $F = \sqrt{15.093^2 + 1.9359^2} = 15.2\text{ N}$

and $\theta = \arctan\left(\frac{1.9359}{15.093}\right) = 7.3°$

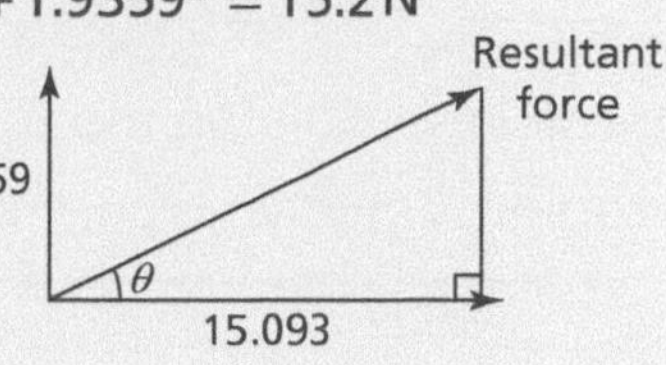

The resultant force is therefore of magnitude 15.2 N, acting at an angle of 7.3° to Ox.

Particles in Equilibrium

- If a particle is in **equilibrium**, then the resultant force on the particle is zero. In practice, you generally have to resolve forces in any two perpendicular directions to solve the given problem.

Three forces of 10 N, 8 N and F N act on a particle at O, as shown in the diagram. Given that the particle is in equilibrium, find the value of F and the value of θ.

$F\cos\theta + 10\cos40 - 8\cos15 = 0 \Rightarrow F\cos\theta = 8\cos15 - 10\cos40$ ← Resolve horizontally.

$F\sin\theta - 10\sin40 + 8\sin15 = 0 \Rightarrow F\sin\theta = 10\sin40 - 8\sin15$ ← Resolve vertically.

$\tan\theta = \frac{10\sin40 - 8\sin15}{8\cos15 - 10\cos40} = 65.07$ and $\theta = 89.1°$

$F = \frac{8\cos15 - 10\cos40}{\cos89.1} = 4.36\text{ N}$

- For a particle in equilibrium on a plane, you should resolve forces perpendicular to the plane and parallel to the plane.

A particle rests in equilibrium on a plane inclined at 40° under the action of the forces as shown in the diagram. Find the value of P and the value of θ.

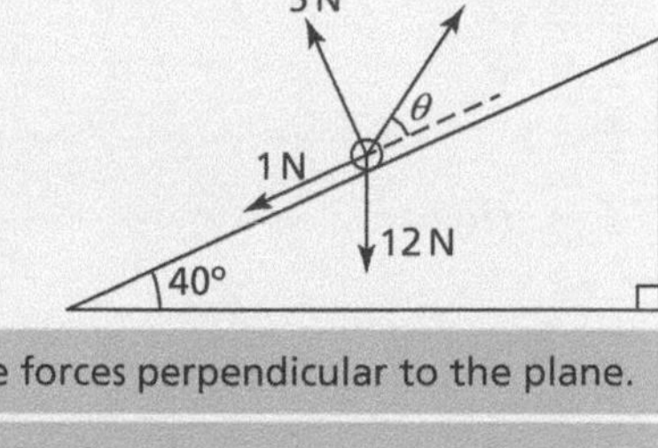

$5 + P\sin\theta = 12\cos40 \Rightarrow P\sin\theta = 12\cos40 - 5$ **(1)** ← Resolve forces perpendicular to the plane.

$P\cos\theta = 1 + 12\sin40$ **(2)** ← Resolve forces parallel to the plane.

Equation **(1)** divided by equation **(2)** gives:

$\frac{P\sin\theta}{P\cos\theta} = \frac{12\cos40 - 5}{1 + 12\sin40}$

$\tan\theta = \frac{12\cos40 - 5}{1 + 12\sin40} = 0.481 \Rightarrow \theta = \arctan 0.481 = 25.7°$

$P = \frac{12\cos40 - 5}{\sin\theta} = \frac{12\cos40 - 5}{\sin25.7} = 9.67\text{ N}$ ← Substituting into equation **(1)**.

Quick Test

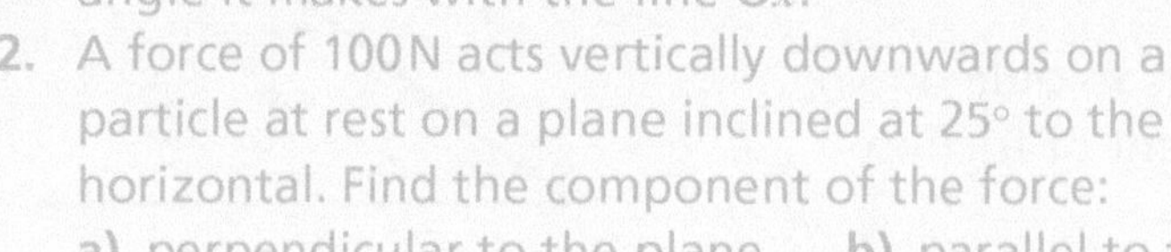

1. Considering the three forces of 4 N, 15 N and 6 N, shown right, find the resultant force F and the angle it makes with the line Ox.
2. A force of 100 N acts vertically downwards on a particle at rest on a plane inclined at 25° to the horizontal. Find the component of the force:
   a) perpendicular to the plane b) parallel to the plane.

Key Words

resolve
components
adjacent
perpendicular
resultant force
equilibrium

Friction and Motion on an Inclined Plane

You must be able to:

- Understand and use $F \leq \mu R$ for friction
- Understand the idea of limiting friction
- Solve problems involving motion on an inclined plane
- Solve problems involving connected particles and rough planes.

Friction

- **Friction** opposes motion and depends on the two materials in contact.
- Consider a block resting on a plane. When the plane is horizontal, there is no frictional force. As the angle of inclination θ increases from θ_1 to θ_2, the frictional force between the block and the plane will increase until it reaches a maximum value. At this point the friction is 'limiting'. The block will then be on the point of sliding down the plane.
- In general, $F \leq \mu R$ where R is the reaction force between the plane and the block. μ (> 0) is a constant, called the coefficient of friction.
- In general, the greater the value of μ, the greater the 'roughness' between the two surfaces. μ is usually less than 1, though can take higher values for particularly **rough** contact surfaces.
- In the case of **limiting friction**, you have $F = \mu R$.

Friction = 0

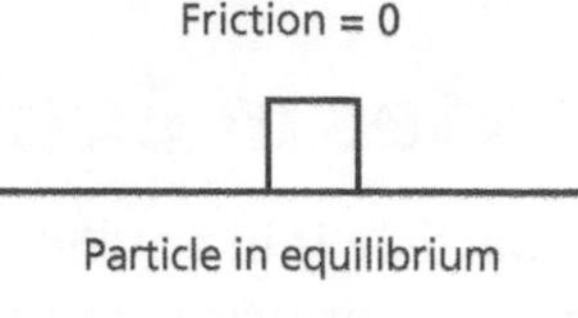

Particle in equilibrium

0 < Friction < F_max

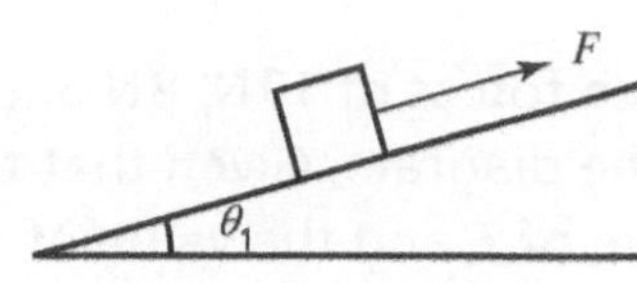

Particle in equilibrium

Friction = F_max

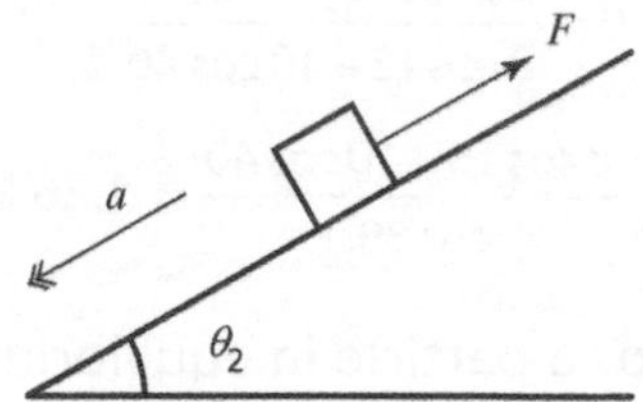

Particle in equilibrium, but on point of sliding down plane

A 2 kg mass rests on a rough horizontal floor. The coefficient of friction between the mass and the floor is $\frac{1}{5}$.

Find the horizontal force P required so that the mass is on the point of moving.

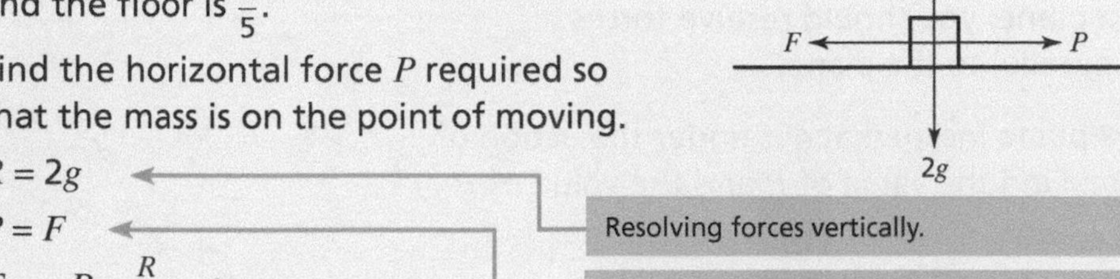

$R = 2g$ — Resolving forces vertically.

$P = F$ — Resolving forces horizontally.

$F = \mu R = \frac{R}{5}$ — Friction is limiting.

So $P = F = \frac{R}{5} = \frac{2g}{5} = 3.92\,\text{N}$

Key Point

The force of friction will always act in a direction so as to oppose the motion of the object, or to oppose the anticipated direction of motion if the surface was '**smooth**'.

Motion on an Inclined Plane

- When a particle is moving on an inclined plane, forces perpendicular to the plane are in equilibrium, since there is no motion in this direction.
- If the particle is moving, or on the point of moving, then friction is limiting and you can apply the equation $F = \mu R$.
- If the particle is moving up (or down) the plane, then you can apply Newton's Second Law (the equation of motion), $F = ma$, in the direction of displacement. Note that 'F' in this equation stands for 'resultant force in the direction of motion', and not 'friction'.

A particle of mass m kg is released from rest and starts to slide down a plane inclined at 60° to the horizontal. The coefficient of friction between the particle and the plane is $\frac{1}{3}$.

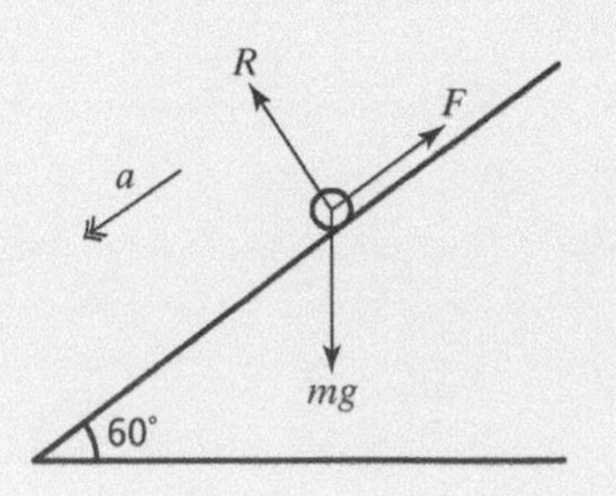

a) Find the subsequent acceleration of the particle.

Resolving perpendicular to the plane: $R = mg\cos 60° = \frac{mg}{2}$ **(1)**

Friction limiting: $F = \mu R = \frac{R}{3}$ **(2)**

Equation of motion down the plane: $mg\sin 60° - F = ma$ **(3)**

Taking equation **(3)**: $mg\sin 60° - F = ma$

$\frac{mg\sqrt{3}}{2} - \frac{R}{3} = ma \Rightarrow \frac{mg\sqrt{3}}{2} - \frac{mg}{6} = ma \Rightarrow \frac{g\sqrt{3}}{2} - \frac{g}{6} = a \Rightarrow a = 6.85\text{ ms}^{-2}$

b) Find the acceleration of the particle were the plane smooth.

Equation **(3)** would be $mg\sin 60° = ma$, giving $a = g\sin 60° = \frac{g\sqrt{3}}{2} = 8.49\text{ ms}^{-2}$

Were the plane to have been smooth, equation **(1)** would remain the same (though this would not be required here) and friction $(F) = 0$.

Key Point

Show the direction of acceleration on your force diagram. This will indicate the direction of positive displacement when applying the equation of motion.

Connected Particles

- For connected particles on smooth planes, write down the **equation of motion** for each particle. Solving these equations simultaneously will allow you to find values for the acceleration and the tension in the string.
- On an inclined plane you will also need to resolve the weight of the particle on the plane in order to find the resultant force on that particle (in the direction of its acceleration).

A particle P of mass 8 kg is attached to one end of a light **inextensible string**. P rests on a smooth fixed plane inclined at an angle θ to the horizontal, where $\tan\theta = \frac{3}{4}$. A particle Q of mass 5 kg is attached to the other end of the string, which passes over a smooth, fixed pulley.

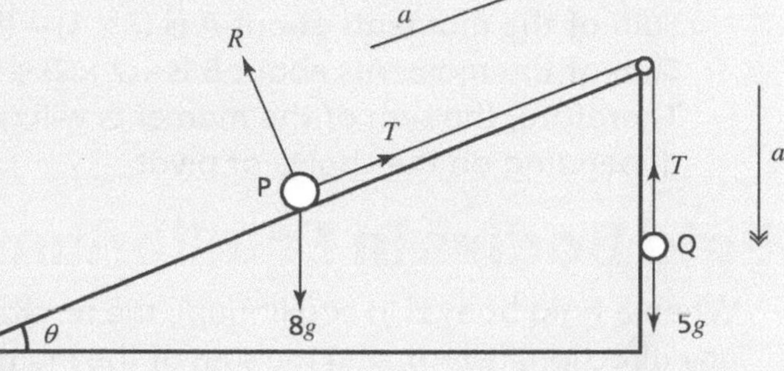

a) Find the acceleration of the system.

$\tan\theta = \frac{3}{4} \Rightarrow \sin\theta = \frac{3}{5}$ and $\cos\theta = \frac{4}{5}$, as shown by a right-angled triangle:

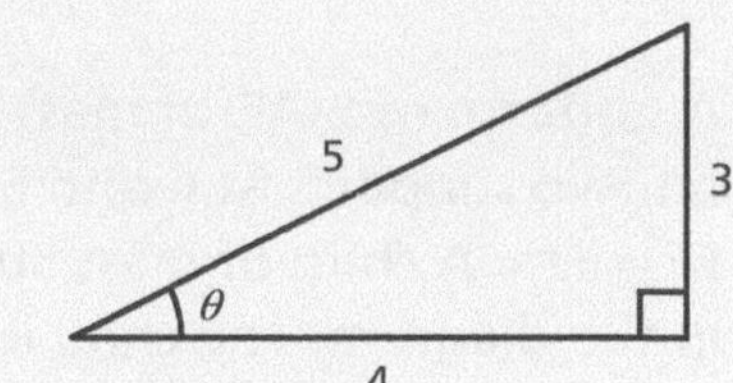

Equation of motion of Q: $5g - T = 5a$ **(1)**

Equation of motion of P: $T - 8g\sin\theta = 8a$

$T - \frac{24g}{5} = 8a$ **(2)**

Adding equations **(1)** and **(2)** gives $5g - \frac{24g}{5} = 13a$

So $\frac{g}{5} = 13a$ and $a = \frac{g}{65} = 0.151\text{ ms}^{-2}$

b) Find the tension in the string.

Substituting back into equation **(2)** gives $T - \frac{24g}{5} = \frac{8g}{65}$, so $T = 48.2\text{ N}$

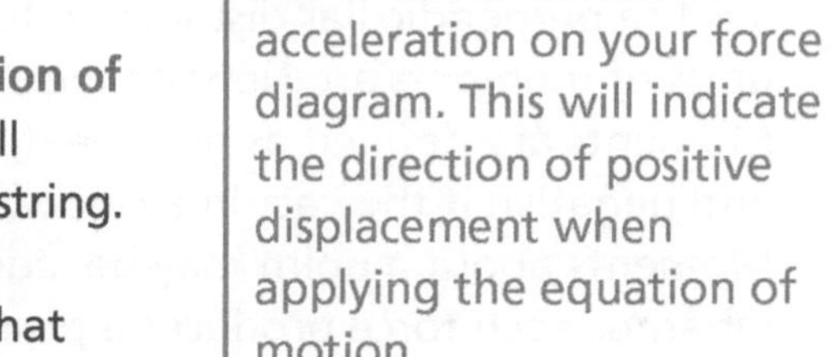

Quick Test

1. A particle of mass 0.5 kg slides down a rough plane inclined at 30° to the horizontal. Given that its acceleration is 3 ms^{-2}, find the coefficient of friction.
2. A particle is projected up a rough plane inclined at 20° to the horizontal. Given that the coefficient of friction between the particle and the plane is 0.4, find the deceleration of the particle.

Key Words

friction
rough (surface)
limiting friction
smooth (surface)
equation of motion
inextensible (string)

Moments

You must be able to:

- Understand and use moments in static contexts
- Apply moments to problems involving the equilibrium of rigid bodies
- Apply moments to 'ladder problems'.

Moments and Sums of Moments

If you imagine a piece of string connecting point P and a point on the line of force, the force would turn the string anticlockwise about P. So, the moment of the force is positive.

- The **moment** of a force can be thought of as the force's turning effect.
- The moment of a force F about a fixed point P is defined as Fd, where d is the perpendicular distance to the line of action of the force. The units of moments are Newton metres, or Nm.
- Moments are defined as positive if they act in an anticlockwise sense, and negative if they act in a clockwise sense.
- Moments about a point may be 'added', but take careful note of whether each force produces a positive or a negative moment.
- In the diagram (right), consider point A as the **pivot**. Then the 3 N force is contributing a clockwise (negative) moment about A:
  - Sum of the moments about A is $(2 \times 1) - (3 \times 2) + (4 \times 6) = 20$ Nm
  - Sum of the moments about B is $-(2 \times 2) + (3 \times 1) + (4 \times 3) = 11$ Nm
  - Therefore, the sum of the moments will generally change, depending on the choice of pivot.

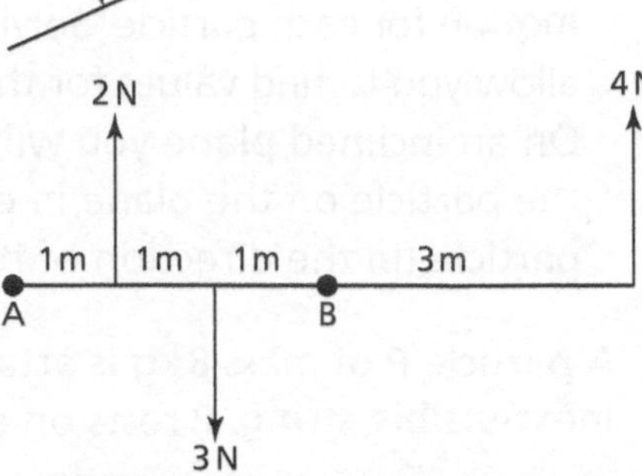

Key Point

If a rod is uniform, you can take its weight to be acting through its centre of mass, i.e. its midpoint.

Rigid Bodies in Equilibrium

- When a **rigid body** is in equilibrium, the resultant force on the body (in any direction) is zero, **and** the sum of the moments about any point is zero. Both these points can be applied when solving rigid body problems.

A **uniform rod**, AF, of mass 30 kg and length 6 m rests on two supports at C and E, where AC = 1.5 m and EF = 1 m. A child of mass 50 kg stands at B, as shown in the diagram. The child is modelled as a particle, and the child and rod are in equilibrium.

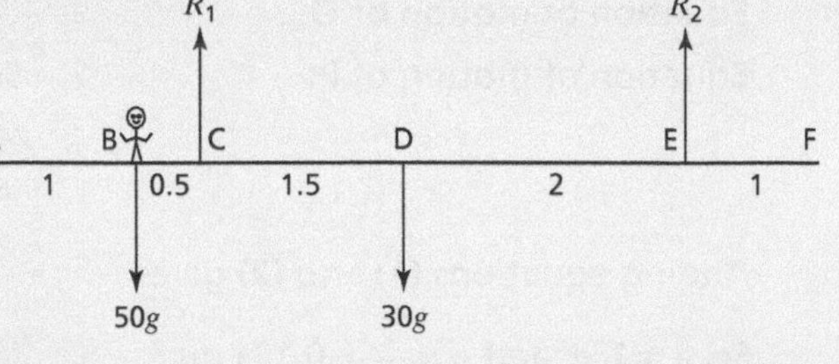

a) Find the magnitude of the reaction force exerted on the rod at E.

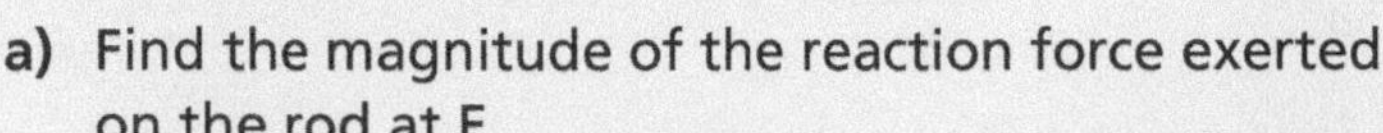

Resolving forces vertically: $R_1 + R_2 = 50g + 30g = 80g$

Taking moments about C: $0.5 \times 50g - 1.5 \times 30g + R_2 \times 3.5 = 0 \Rightarrow 3.5R_2 = 20g$, so $R_2 = \frac{40g}{7}$

b) Find the magnitude of the reaction force exerted on the rod at C.

Substituting into the first equation: $R_1 + \frac{40g}{7} = 80g$, so $R_1 = \frac{520g}{7}$

c) The child now moves to a position G on the rod so that it is on the point of turning about C. Find the distance AG.

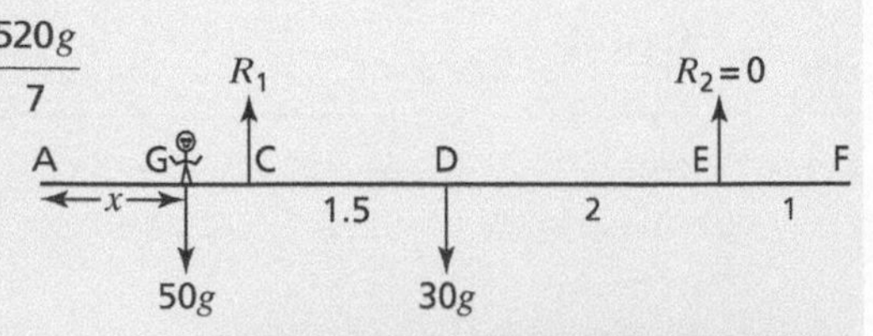

Let the distance AG = x

The rod is on the point of turning about C, so $R_2 = 0$.

Taking moments about C: $50g(1.5 - x) = 1.5 \times 30g \Rightarrow 1.5 - x = 0.9$, so $x = 0.6$ m

Here you should take moments about C, as R_1 will then not be included in your moments equation.

Ladder Problems

- Ladder problems usually involve a ladder resting against a (smooth or rough) wall with the foot of the ladder resting on rough ground.
- Where the wall is smooth, there will be only a reaction force at that point on the ladder, but if it is rough, there will be both a reaction force and a frictional force. The frictional force will be directed vertically upwards to oppose that point of the ladder slipping vertically down.

A uniform ladder of mass 20 kg and length $2l$ m rests against a rough wall with the foot of the ladder resting on rough ground. The ladder rests at an angle of 60° to the horizontal, and μ is the coefficient of friction between ladder and ground, and ladder and wall. Given that the ladder is on the point of slipping, show that $\mu = 2 - \sqrt{3}$.

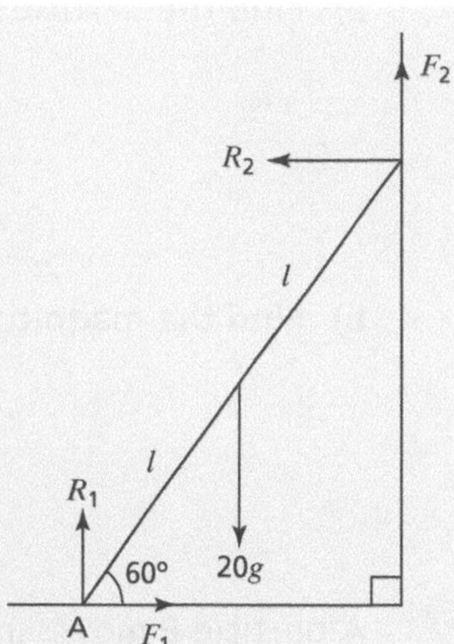

Friction is limiting: $F_1 = \mu R_1$ **(1)**

Friction is limiting: $F_2 = \mu R_2$ **(2)**

Resolving horizontally: $F_1 = R_2$ **(3)**

Resolving vertically: $F_2 + R_1 = 20g$ **(4)**

Moments about A: $20g \times l\cos 60° = F_2 \times 2l\cos 60° + R_2 \times 2l\sin 60°$ **(5)**

$10gl = F_2 l + R_2 l\sqrt{3} \Rightarrow 10g = F_2 + R_2\sqrt{3}$

$10g = \mu R_2 + R_2\sqrt{3}$ **(6)**

From equation **(3)**: $R_2 = F_1 = \mu R_1 = \mu(20g - F_2) = \mu(20g - \mu R_2)$

$R_2 = 20g\mu - \mu^2 R_2 \Rightarrow R_2(1 + \mu^2) = 20g\mu \Rightarrow R_2 = \dfrac{20g\mu}{1+\mu^2}$

Substituting into equation **(6)**: $10g = \dfrac{20g\mu^2}{1+\mu^2} + \dfrac{20g\mu\sqrt{3}}{1+\mu^2}$

$1 = \dfrac{2\mu^2}{1+\mu^2} + \dfrac{2\mu\sqrt{3}}{1+\mu^2} \Rightarrow 1 + \mu^2 = 2\mu^2 + 2\mu\sqrt{3}$

$\mu^2 + 2\sqrt{3}\mu - 1 = 0$

Using the quadratic formula: $\mu = \dfrac{-2\sqrt{3} \pm \sqrt{(2\sqrt{3})^2 - 4\times 1\times(-1)}}{2} \Rightarrow \mu = \dfrac{-2\sqrt{3} \pm 4}{2}$

Since $\mu > 0$: $\mu = \dfrac{4 - 2\sqrt{3}}{2} = 2 - \sqrt{3}$

Quick Test

1. Find the sum of the moments about A in the diagram.
2. Consider the rod example (page 112). At what distance from A should the child stand to ensure the reactions at C and E have the same value?
3. Consider the ladder example (above), but with it resting against a smooth, rather than a rough wall. What will the new value of μ be?

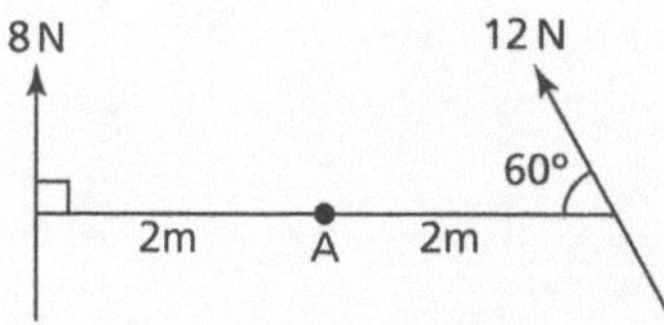

Key Words

moment
pivot
rigid body
uniform rod

Practice Questions

Using Vectors in Kinematics

1. A particle's position vector at time t is given by $\mathbf{r} = (3 - t^2)\mathbf{i} + (t^3 - t)\mathbf{j}$.

   a) Find the particle's speed after three seconds. [4]

   b) Find the magnitude of its acceleration after one second. [4]

2. A particle P moves in a plane such that at time t its velocity vector $\mathbf{v}$ is given by $\mathbf{v} = (3t - 1)\mathbf{i} + (t + 2)\mathbf{j}$.

   a) Find the particle's direction of motion after two seconds. [2]

   b) Find the magnitude of its acceleration. [3]

3. A particle P moves in a plane such that at time t its velocity vector $\mathbf{v}$ is given by $\mathbf{v} = (4t - 1)\mathbf{i} + (t^2 + 2t + c)\mathbf{j}$.

   a) Given that after one second the speed of P is $\sqrt{34}$ ms^{-1}, find the possible values for c. [5]

   b) Given that $c > 0$ and at time $t = 0$ particle P has position vector $\mathbf{i} - \mathbf{j}$, find the general position vector of P at time t. [4]

Total Marks .... / 22

Projectiles

1 A particle P is projected from a point on the ground with speed 65 ms^{-1} at an angle of 30° to the horizontal. Find the horizontal range of the particle. [5]

2 Dorothy hits a tennis ball with speed 20 ms^{-1}, at a height of 3 m. She intends for the ball to land at a horizontal distance of 18 m from her. At which angle(s) should she project the ball? [9]

3 A ball is projected from the top of a cliff at point O, with speed 10 ms^{-1}, at an angle of 30° to the horizontal. T seconds later it lands in the sea at a point P.

Given that the angle of depression of P from O is 60°, find the time of the ball's flight. [7]

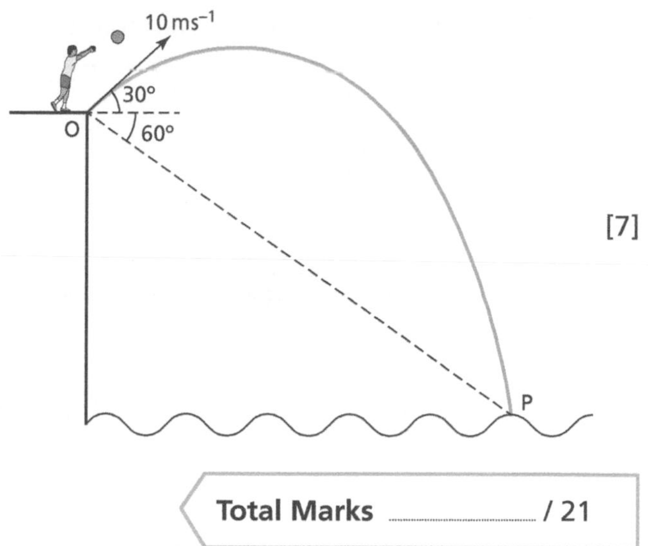

Total Marks .... / 21

Resolving Forces and Forces in Equilibrium

1 In this diagram, four forces are acting on a particle at O. Given that the particle is in equilibrium, find the magnitude of the force P and the angle θ. [7]

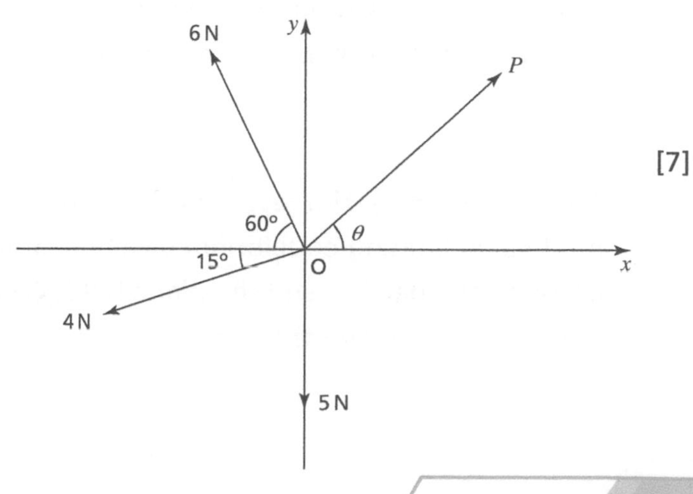

2 A particle of mass 5 kg rests in equilibrium on a smooth plane inclined at 30° to the horizontal. It is held in place by an inextensible string, making an angle of 20° to the plane.

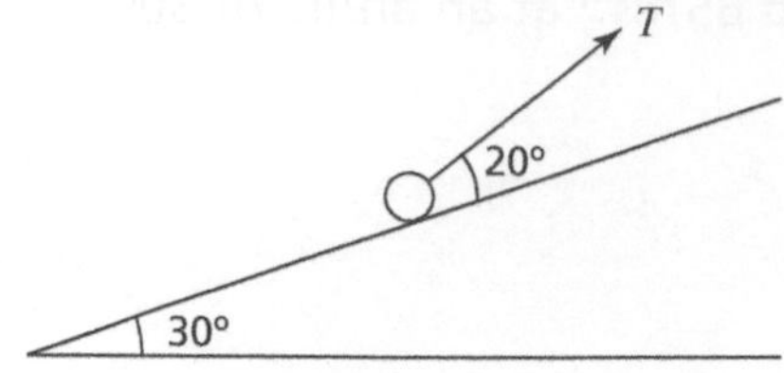

a) Find the tension in the string. [3]

b) Find the normal reaction force exerted by the plane on the particle. [4]

3 A mass of 2 kg is suspended by two light strings from two fixed points on the ceiling. A horizontal force of 5 N is applied to the mass. Given that the mass is in equilibrium, find the tensions in each string.

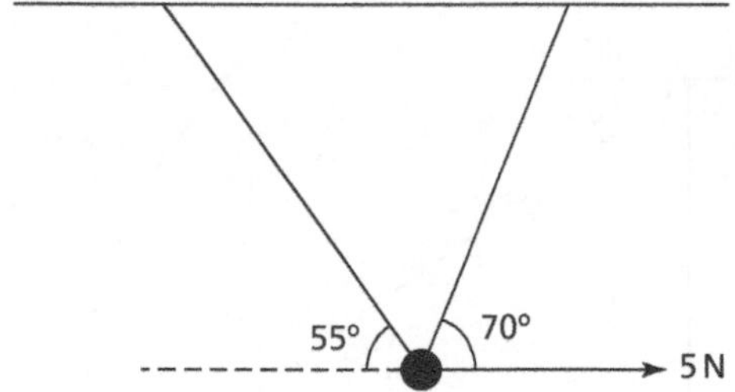

[9]

Total Marks .... / 23

Friction and Motion on an Inclined Plane

1 A particle of mass 1.5 kg slides down a rough plane inclined at 40° to the horizontal. Given that the coefficient of friction between the particle and the plane is 0.45, calculate its acceleration. [6]

2 A particle of mass 3 kg rests on a rough plane inclined at an angle of 25° to the horizontal. It is held in limiting equilibrium by a horizontal force of P N. The coefficient of friction between the particle and the plane is 0.3. Given that the particle is on the point of moving **down** the plane, find the value of P. [7]

3 A particle of mass m lies in equilibrium on a rough plane inclined at an angle of $\theta°$ to the horizontal. Show that $\mu \geqslant \tan\theta$. [5]

Total Marks .... / 18

Moments

1 In the diagram, AC is a light rod. Find the total moment of the system of forces:

a) about A. [2]

b) about B. [2]

c) about C. [2]

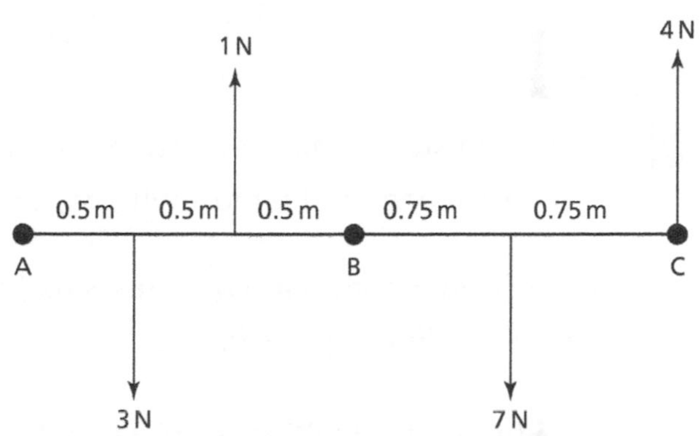

2 A uniform rod AB of length $2l$ is smoothly hinged at a point O on a vertical wall. Given that the rod is resting in equilibrium, find the value of the force F. [5]

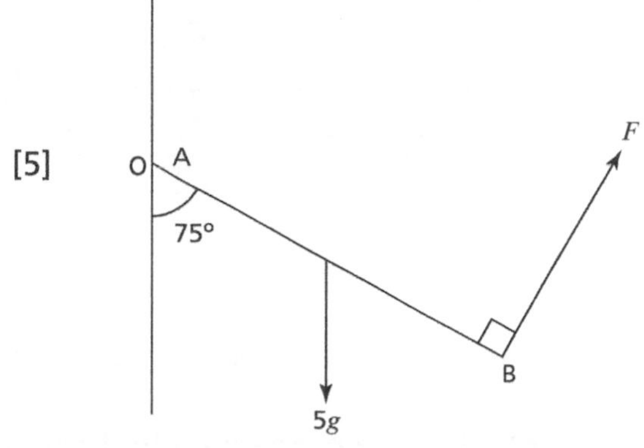

3 A uniform ladder of mass 15 kg and length $2l$ m rests against a smooth wall, with the foot of the ladder resting on rough ground. The ladder is inclined at an angle of 70° to the horizontal. A man of mass 80 kg begins to climb the ladder, which begins to slip when he reaches three-quarters of the way up it.

Find the value of μ, the coefficient of friction between the ladder and the ground. [8]

Total Marks .... / 19

Review Questions

Set Notation and Conditional Probability

1 Two events, A and B, have the following probabilities associated with them.

$P(A) = \frac{1}{3}$ $\qquad$ $P(B) = \frac{1}{2}$ $\qquad$ $P(A' \cap B) = \frac{1}{2}$

a) Find $P(A \mid B)$. [3]

b) State the relationship between A and B. [1]

c) Calculate $P(A' \cap B')$ and explain what is meant by this probability. [3]

2 A market stall completes a stocktake of cat food. The seller records packets in three different flavours – chicken, beef and lamb – with 24, 38 and 40 packets of each respectively.

Each flavour comes in two sizes, small and medium. 10 of the chicken packets are medium; the corresponding figures for beef and lamb are 12 and 17 respectively.

a) Complete the two-way table below. [3]

	Chicken	Beef	Lamb	Total
Small				
Medium				
Total				

The seller selects a packet at random from the store cupboard.

b) Find the probability that the packet is chicken flavoured. [2]

c) Find the probability that the packet is lamb flavoured and a small size. [2]

d) Find the probability that the packet is beef flavoured given that it is medium size. [2]

Total Marks .... / 16

Modelling and Data Presentation

1 On a separate sheet of paper, sketch scatter diagrams with at least eight points to illustrate the following:

a) Data with a product moment correlation coefficient of –1. [1]

b) Data with a product moment correlation coefficient of 0.8. [1]

c) Data with a product moment correlation coefficient of 0.1. [1]

2 A coin is tossed 100 times and the number of heads is recorded.

a) Suggest a suitable model to represent this situation. [2]

b) State any assumptions you have made. [2]

3 It is thought that the data below follows the model $y = ka^x$.

x	1	2	3	4	5
y	8	32	128	512	2048

By transforming y into a linear function, find an estimate for k. [4]

4 Henry's Bakery claims that it sells iced fingers to one-third of customers. 120 customers visit Henry's Bakery on any given day.

a) Suggest a distribution which could be used to model the number of iced fingers sold and state the value of any parameters. [2]

b) State any assumptions you have made for the model to be valid. [1]

Total Marks .... / 14

Review Questions

Statistical Distributions 1 & 2

1 Let X be a continuous random variable. Find:

a) $P(X < 52)$ if $X \sim N(50, 4^2)$ [3]

b) $P(X > 20)$ if $X \sim N(24, 9)$ [3]

c) $P(14 < X < 23)$ if $X \sim N(20, 6^2)$ [3]

2 A random variable $X \sim B(200, 0.4)$. Using a suitable approximation, find:

a) $P(X < 65)$ [3]

b) $P(X > 90)$ [2]

c) $P(75 < X < 85)$ [2]

3 In a multiple-choice test, there are four possible answers per question. Given that there are 80 questions on the test, use a suitable approximation to find the probability of getting more than 30 correct if the answer to each question is chosen at random. [6]

4 The random variable $X \sim N(30, \sigma^2)$.

Given that $P(X > 34) = 0.05$, find the value of σ. [5]

Total Marks .... / 27

Hypothesis Testing 1 & 2

1. A random sample is taken from a population believed to have mean 19 and variance 4. A sample of 100 is taken from the population and the sample mean is calculated to be 20.

   Test, at a 1% significance level, H_0: $\mu = 19$ H_1: $\mu \neq 19$ [5]

2. St Martin's School had GCSE results in 2016 with a mean pass rate of 8.7 and a standard deviation of 1.2. Some new initiatives are introduced to improve the pass rate and in 2017 the pass rate increased to 9.1. One hundred students sat GCSEs at the school in 2017.

   The Head claims the new initiatives have improved results. Test, at the 5% significance level, whether or not there is enough evidence to support the Head's claim. [7]

3. Jam is produced at a factory. The manufacturer claims that the amount of jam in a jar has a mean of 250 g and a standard deviation of 6 g. A customer complains that the manufacturer is overstating the amount of jam in a jar and reports it to Trading Standards.

   A quality control officer visits the factory and takes a sample of 20 jars, which has a mean of 248 g. Using a 5% significance level, test if there is enough evidence to support the customer's complaint. [7]

4. The product moment correlation coefficient was calculated as 0.61, based on 16 pairs of data.

   a) Test whether the correlation is significantly greater than zero, using a 5% level of significance. [3]

   b) Test whether the correlation is significantly different to zero, using a 5% level of significance. [3]

Total Marks .... / 25

Review Questions

Using Vectors in Kinematics

1 The position vector at time t, of a particle P, is given by $\mathbf{r} = (t^2 + t + 1)\mathbf{i} + \left(4t - \frac{3t^2}{2}\right)\mathbf{j}$.

a) Find the time t_1 when P is travelling in a north-easterly direction. [5]

b) Find the distance of P from its starting point at the time $t = t_1$. [2]

2 A particle P has constant acceleration $(\mathbf{i} - 2\mathbf{j})\,\text{ms}^{-2}$. Find the displacement of P from its original position after three seconds, given that its final velocity is $(12\mathbf{i} + 20\mathbf{j})\,\text{ms}^{-1}$. [3]

3 Two particles, P and Q, move in a plane so that after t seconds their respective position vectors are given as $\mathbf{r}_P = \begin{pmatrix} \frac{t^2}{2} - 2t \\ 2t - t^2 \end{pmatrix}$ and $\mathbf{r}_Q = \begin{pmatrix} 5t - \frac{t^2}{2} \\ t^2 - 11t \end{pmatrix}$.

a) Find the velocity vectors of both P and Q after t seconds. [4]

b) Find the value of t when particles P and Q are travelling parallel to each other. [5]

Total Marks .... / 19

Projectiles

1 A particle P is projected from a point on the ground with speed $12\mathbf{i} + 20\mathbf{j}\ \text{ms}^{-1}$.

a) Find the time at which the particle reaches its maximum height. [3]

b) Find the distance the particle is from its starting point after four seconds. [3]

2 A particle P is initially projected from a point on horizontal ground, with speed $u\ \text{ms}^{-1}$, at an angle of $\alpha°$ to the horizontal. Given that P reaches a maximum height of 86 m and has a range of 150 m, find its angle of projection $\alpha°$ and hence its initial speed u. [9]

3 A particle P is projected from an initial point on horizontal ground, with speed u ms^{-1}, at an angle of $\alpha°$ to the horizontal.

a) Show that the particle hits the ground again after time $T = \dfrac{2u\sin\alpha}{g}$. [3]

b) Given that after time $\dfrac{2T}{3}$, the particle is travelling in a direction perpendicular to its initial direction of motion, show that $\alpha = 60°$. [8]

Total Marks .... / 26

Resolving Forces and Forces in Equilibrium

1 A mass m kg is suspended by light string from a fixed point on the ceiling. A horizontal force, P, is applied to the mass, resulting in a tension of 12 N in the string.

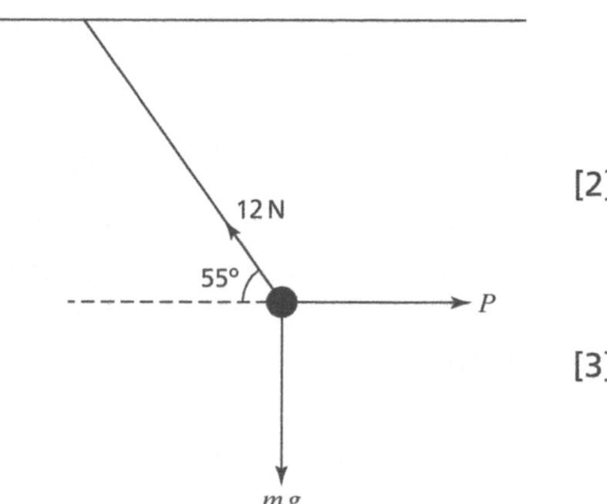

Given that the mass is in equilibrium, find:

a) the force, P [2]

b) the mass, m. [3]

2 A mass of 12 kg is held in place on a smooth inclined plane by a horizontal force, X. The plane makes an angle of 25° with the horizontal. Given that the mass is in equilibrium, find:

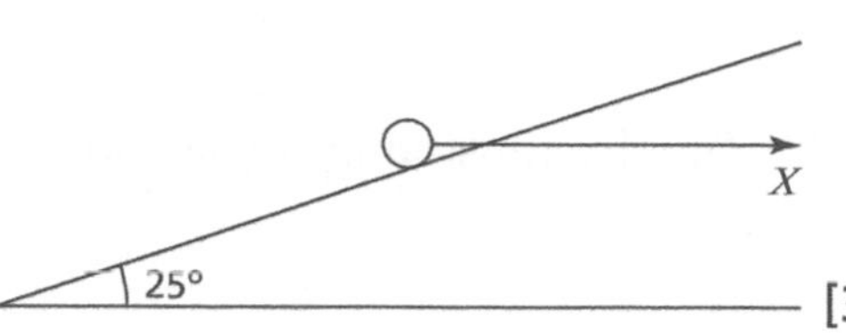

a) the value of X [3]

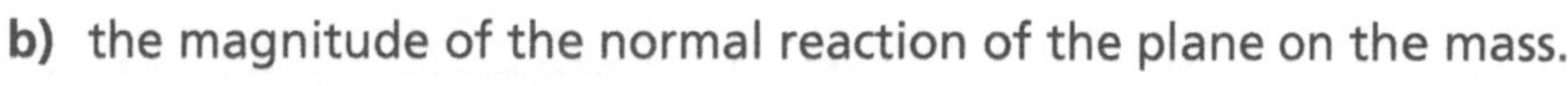

b) the magnitude of the normal reaction of the plane on the mass. [4]

Review Questions

3 A mass of m kg is suspended by two light strings from two fixed points on the ceiling. The tensions in the two strings are 12 N and T N respectively. Given that the mass is in equilibrium, find the value of T and the value of m. [7]

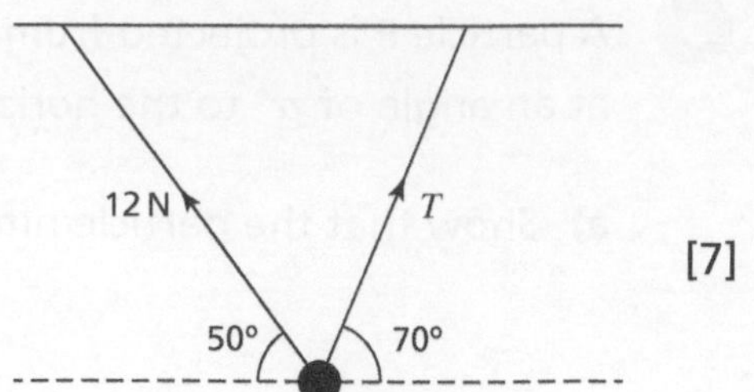

Total Marks .... / 19

Friction and Motion on an Inclined Plane

1 A constant horizontal force of 8 N is applied to a particle of mass 2.5 kg lying on a rough horizontal floor. Given that the particle is on the point of moving, find the coefficient of friction between the particle and the floor. [5]

2 A particle is projected up a rough plane inclined at 30° to the horizontal. Given that the coefficient of friction between particle and plane is 0.2, find the deceleration of the particle and show that after the particle comes to instantaneous rest, it will start to slide back down the plane. [10]

3 Two particles, P and Q, of masses 4 kg and 1.5 kg respectively, are connected by a light **inextensible** string. P rests on a rough horizontal table and Q hangs vertically, with the string passing over a smooth pulley. The coefficient of friction between P and the table is 0.2. The system is released from rest.

Find:

a) the acceleration of the system. [10]

b) the tension in the string. [1]

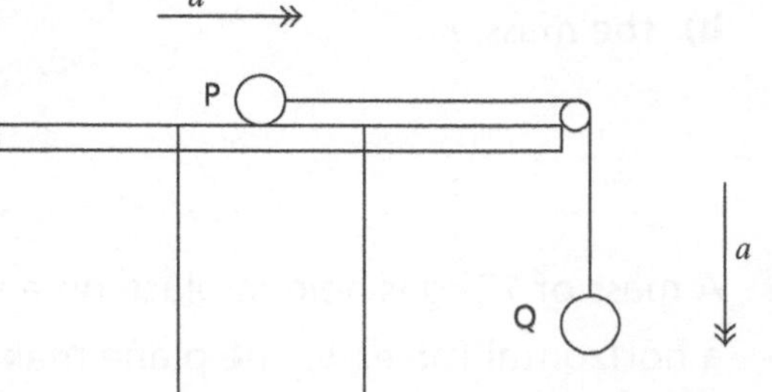

Total Marks .... / 26

Moments

1 A non-uniform rod AE of mass 70 kg and length 3 m rests on two supports B and D, where AB = 0.5 m and DE = 1.6 m. A child of mass 20 kg stands on the rod at C, where BC = 0.7 m. Given that the reaction force at D is twice the reaction force at B, find the distance of the centre of mass of the rod from A. [8]

2 The diagram shows a light rod AD in equilibrium. Find the value of the force P and find the distance CD. [6]

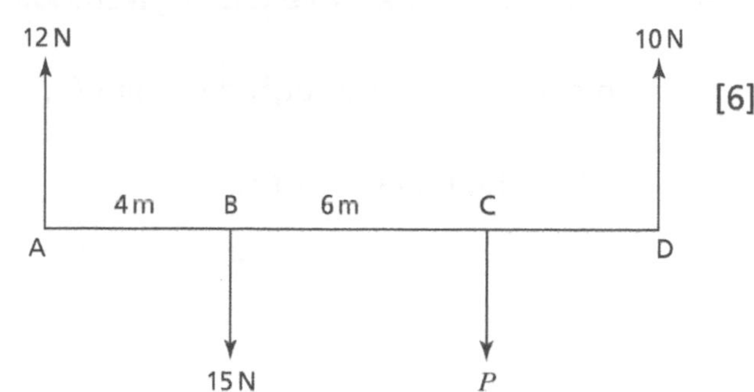

3 A non-uniform rod AB of mass 5 kg and length $3l$ has its centre of mass at a point X on the rod, where AX = l. The rod rests horizontally, supported by two strings as shown in the diagram. Find the tensions in the strings T_1 and T_2, and the value of θ. [14]

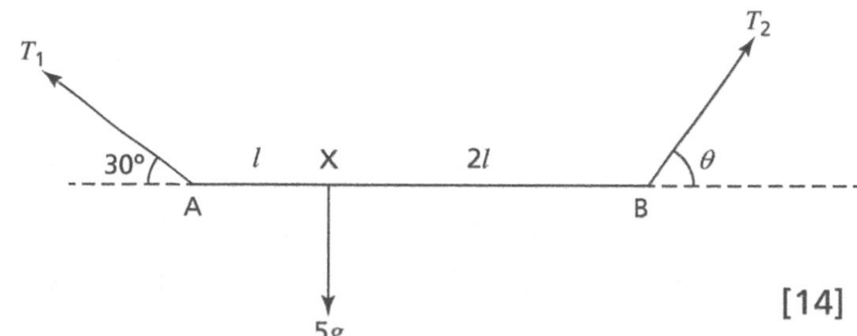

Total Marks .... / 28

Mixed Questions

Pure Mathematics

1 Differentiate:

a) $\sin^2(3x)$ [3]

b) $\tan^2(2x)$ [3]

2 The gradient of a curve C at a given point (x, y) is given by the equation $\frac{dy}{dx} = x^2 + 2x$.

The curve passes through the point (1, 2).

a) Find the equation of C. [3]

b) Find the area of the region bounded by the curve, the x-axis and the lines $x = -2$ and $x = 0$. [4]

3 In the diagram, AB is a tangent to the radius OA, and BC is a tangent to the radius OC. The two shaded sections have equal area. Find the value (in radians) of the angle AOC. [6]

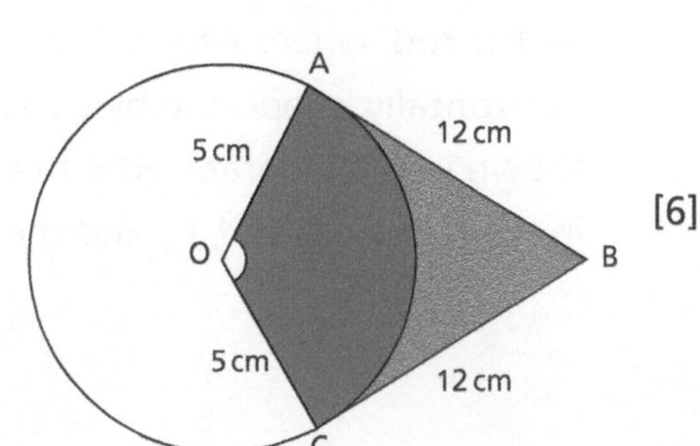

4 $f(x) = x - \frac{3}{x+2} + 1$

a) Show that $f(x) = \frac{x^3 + 2x^2 - 4x + 1}{x^2 + x - 2}$. [4]

b) Show that $f(x) = \frac{x^2 + 3x - 1}{x + 2}$. [4]

5 Given the functions $f: x \mapsto 3x + 2$ and $g: x \mapsto \frac{3}{x+4}$

a) Find the function $f^{-1}(x)$, stating the domain and the range. [3]

b) Find the value of gf(–3). [2]

c) Solve the equation $f(x) + g(x) = gf(-3)$. [4]

6 A curve has equation $y^2 + y = 10 - x^2$. Find the equation of the tangent to the curve at the point (–2, 2). [3]

Mixed Questions

7 Sketch a graph and use it to solve the inequality $|-2x+3| \leqslant -x^2 + 2x + 3$. [7]

8 Find the first four terms of the expansion of $\frac{3x-11}{(x-1)(x+3)}$ and state the range of values of x for which the expansion is valid. [8]

9 Use integration by parts to show that $\int_2^4 x\ln x \, dx = 7\ln 4 - 3$. [4]

10 Prove the trigonometric identity $\cot\left(x - \frac{\pi}{4}\right) = \frac{\sin x + \cos x}{\sin x - \cos x}$. [5]

11 A curve has parametric equation $x = t^2$, $y = t^3$.

Find the equation of the normal when $t = 1$. [6]

12 Find the coordinate of the point A on the y-axis that is equidistant from the points B(1, 2, 0) and C(2, −1, 2). [6]

13 A sequence is given by $u_n = \frac{1}{2}n + 3$.

a) Describe the sequence as increasing, decreasing or periodic. [1]

b) Write the first five terms of the sequence. [1]

c) Write a recursive formula to describe the sequence. [1]

d) Find $\sum_{1}^{5} u_n$ [1]

14 Fredrick and Amber spend n weeks training to run a 10 km race for charity. Fredrick starts his training by running 1 km the first week, 1.5 km the second week, 2 km the third week, and so on. Amber starts by running 2.8 km the first week and increases her run by d km each week.

a) Given they both run the full distance of the race on the final week of training, find the value of d. [3]

b) Fredrick and Amber both received sponsorship pledges for each kilometre run during their training. Fredrick raised £m per km run and Amber raised £5.50 per km run. Given that they raised the same amount of money, find the value of m. [3]

15 **a)** Find $\int x\cos x \, dx$ [3]

b) Find $\int \cos^2 y \, dy$ [3]

c) Hence find the general solution of the differential equation $\frac{dy}{dx} = x\cos x \sec^2 y$, $0 < y < \frac{\pi}{2}$. [2]

16 A curve has parametric equations $x = 2\sec(t)$, $y = 3\cos(t)$, $t \in \mathbb{R}: 0 \leq t < \frac{\pi}{2}$.

a) Express the parametric equations as a Cartesian equation $y = f(x)$, stating the domain and the range of $f(x)$. [4]

b) Sketch the curve $y = f(x)$. [2]

17 A sequence is given by $u_{n+1} = (u_n)^2 - 3$.

a) Find the first four terms when $u_1 = 3$. [1]

b) Find the first five terms when $u_1 = \sqrt{3}$. [1]

c) Comment on the relation between the answers to part **a)** and **b)**. [1]

18 Solve the equation $\dfrac{1+\sin^2 x}{\cos^2 x} = 3$ in the range $0 \leqslant \theta \leqslant 2\pi$. [7]

Mixed Questions

19 Show that the equation $f(x) = 3^{-x} + x^3 - 1 = 0$ has a root α such that $0.8 < \alpha < 0.9$.

By using a suitable iteration formula of the form $x_{n+1} = g(x_n)$ with $x_0 = 0.9$, write down x_1, x_2 and x_3, and hence determine α to 3 decimal places. [8]

20 Amare and Uma are buying their first house. They take out a 30-year mortgage for £200 000 at a fixed interest rate of 4%, which the bank charges on the amount owed at the beginning of each year. Amare and Uma pay £1000 per month.

a) Find the amount they owe after paying for three years. [3]

b) Write down a single calculation to show the amount they owe after paying for three years. [4]

c) Hence, write a formula to find the amount owed to the bank, D, after n years and simplify it. [2]

21 Find the exact values of the following, without using a calculator:

a) $\arctan\left(\sin\frac{3\pi}{2}\right)$ [2]

b) $\sec\left(\arctan\sqrt{3}\right)$ [2]

c) $\sec\left(\arcsin\left(-\frac{1}{2}\right)\right)$ [2]

22 Use the trapezium rule with three strips to estimate the value for $\int_0^{\frac{\pi}{3}} \cos\left(\sqrt{\sin x}\right)\,dx$. [6]

23 The path of a circus performer being ejected from a cannon can be modelled by the parametric equations $x = t(30\cos(\theta))$, $y = -4.9t^2 + t(30\sin(\theta))$. The performer will be caught in a net that is 15 m long starting at 75 m from the cannon.

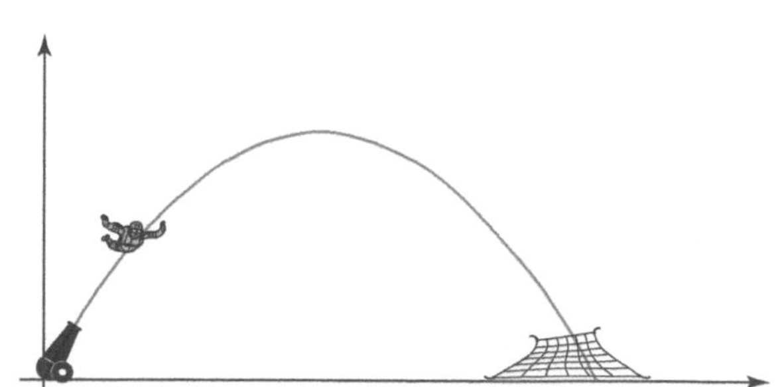

Find the minimum and maximum values of θ so that the performer lands safely in the net. [7]

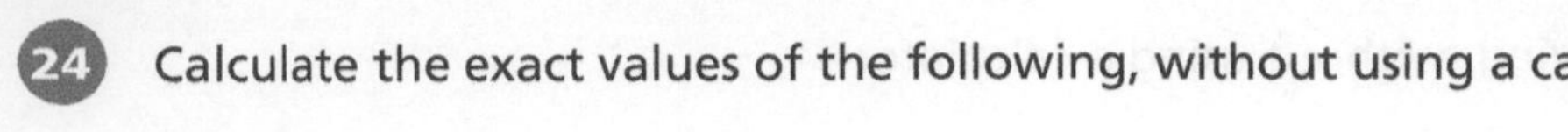

Mixed Questions

24 Calculate the exact values of the following, without using a calculator:

a) $\sin 70° \cos 10° - \sin 10° \cos 70°$ [2]

b) $\dfrac{\sqrt{3} + \tan 75°}{1 - \sqrt{3} \tan 75°}$ [3]

c) $\sin 75° \cos 75°$ [4]

25 Express $3\cos 2x - \sin 2x$ in the form $R\cos(2x + \alpha)$, where $R > 0$ and $0 \leqslant \alpha \leqslant \dfrac{\pi}{4}$.

Hence solve the equation $3\cos 2x - \sin 2x = 1$ in the range $0 \leqslant x \leqslant \pi$. [11]

Total Marks .... / 165

Statistics and Mechanics

1 A uniform ladder of mass 12 kg and length $2l$ m rests against a smooth wall with the foot of the ladder resting on rough ground. The ladder is inclined at an angle of 60° to the horizontal. A child of mass 50 kg begins to climb the ladder.

Given that μ, the coefficient of friction between the ladder and the ground, is equal to 0.3, find in terms of l the distance the child can climb up the ladder before it begins to slip. [8]

2 The velocity vector at time t of a particle P is given by $\mathbf{v} = (3t - 2)\mathbf{i} + (6 - 2t)\mathbf{j}$.

a) Find the time t_1 when P is travelling in a northerly direction. [2]

b) Find the time t_2 when P is travelling in an easterly direction. [2]

c) Find the displacement of P between the times when $t = t_1$ and $t = t_2$. [3]

Mixed Questions

3 The weights of a group of males are normally distributed with mean 85 kg and standard deviation 2.6 kg. A random sample of 10 of these males is selected.

a) Write down the distribution of $\bar{M}$, the mean weight in kilograms, of this sample. [1]

b) Find $P(\bar{M} < 83.5)$. [3]

The mean weight of the sample was 87 kg.

c) Stating your hypotheses clearly, and using a 5% level of significance, test whether or not the mean weight of these men is greater than 85 kg. [4]

4 Gemma does $\frac{3}{5}$ of the work that comes into an office and John does the rest. If 55% of Gemma's work is perfect and 35% of John's work is perfect, find the probability that:

a) a job will be done perfectly [2]

b) a job will have been done by Gemma if it was done perfectly. [3]

5 A particle P is projected with speed $20\,\text{ms}^{-1}$ at an angle of depression of 10° from the top of a cliff, 200 m high. It is directed towards the sea.

a) Find the time taken for P to reach the sea. [4]

b) Find the angle the particle makes with the surface when it hits the sea. [5]

6 Here is a scatter diagram where the line of best fit crosses the $\ln(y)$-axis at 3 and the x-axis at $4\frac{1}{3}$.

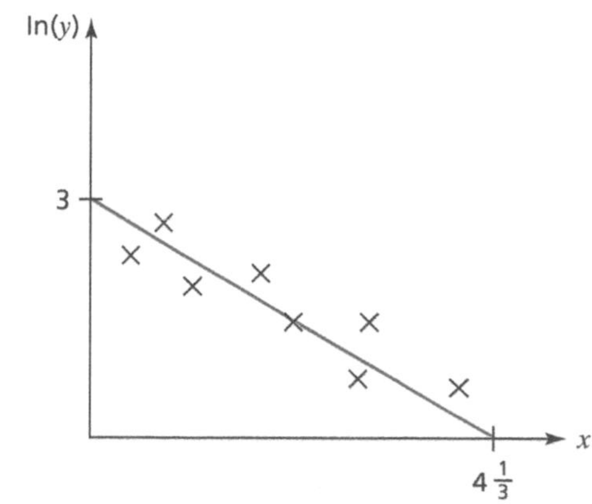

a) Find the equation of the line of best fit. [2]

b) Express y as a function of x, where $y = ab^x$ and a and b are constants to be found. Give a and b correct to 1 significant figure. [3]

Mixed Questions

7 The continuous random variable X is normally distributed with mean 200 and variance 196.

a) Find $P(X < 180)$. [3]

b) Find k such that $P(200 - k \leqslant Y \leqslant 200 + k) = 0.516$. [5]

8 A particle of mass 1.5 kg rests on a rough plane inclined at an angle of 30° to the horizontal. It is held in limiting equilibrium by a horizontal force of P N. The coefficient of friction between the particle and the plane is 0.45. Given that the particle is on the point of moving up the plane, find the value of P. [7]

9 The four forces in the diagram below are in equilibrium.

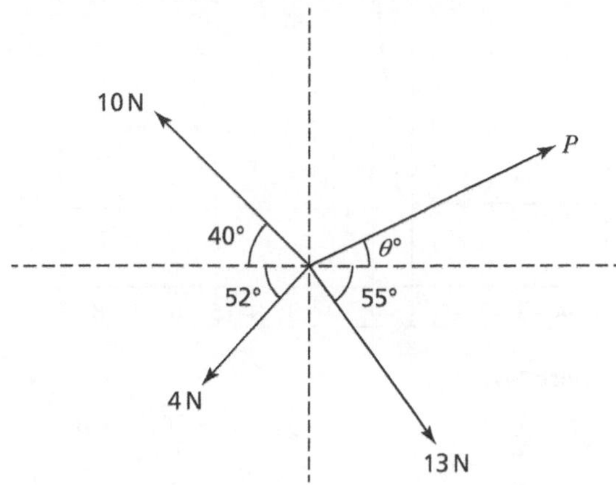

Find the values of P and θ. [8]

Total Marks .... / 65

Answers

Pages 6–23 Revise Questions

Page 7 Quick Test

1. f(x): domain $\{x \in \mathbb{R}: x \neq 2\}$, range $\{y \in \mathbb{R}: y \neq 0\}$
   g(x): domain $\mathbb{R}$, range $\{y \in \mathbb{R}: y \geqslant 3\}$
2. $\text{fg}(x) = \frac{5}{x^2+1}, \text{gf}(x) = \frac{25}{(x-2)^2} + 3$
3. $\text{g}^{-1}(x) = \sqrt{x-3}$, domain $\{x \in \mathbb{R}: x \geqslant 3\}$, range $\{y \in \mathbb{R}: y > 0\}$

Page 9 Quick Test

1. a)

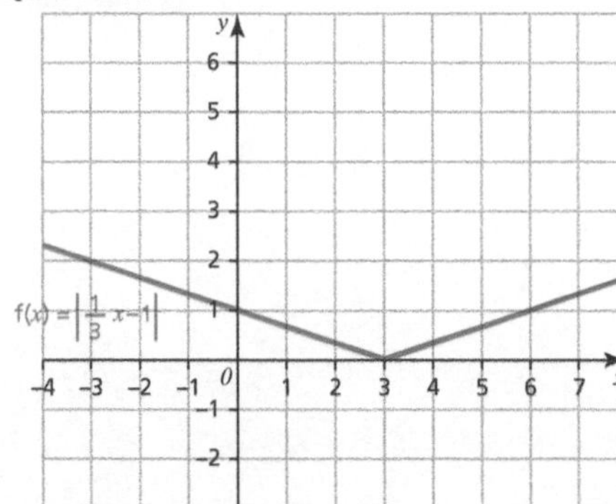

b)

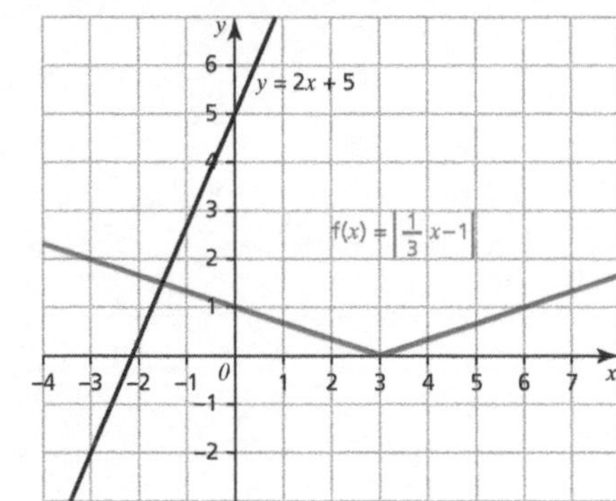

The graph shows the point of intersection is at the part of f(x) that has been reflected over the x-axis, so solve for $-\text{f}(x) = 2x + 5$.

$$-\left(\frac{1}{3}x - 1\right) = 2x + 5$$

$$-\frac{7}{3}x = 4$$

$$x = -\frac{12}{7}$$

2. a) and b)

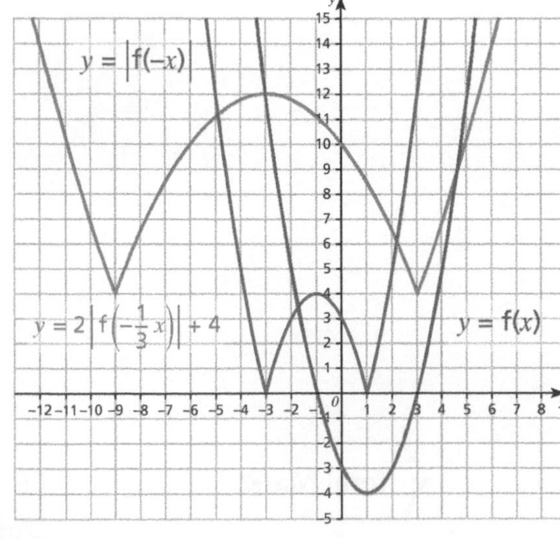

3.

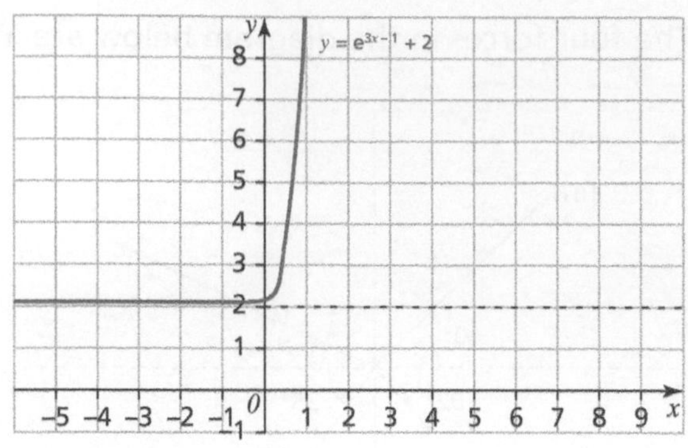

Page 11 Quick Test

1. $\frac{3x^2 + 7x + 6}{(x-1)(x+3)}$
2. $\frac{(2x+1)^2(3x-1)}{3x+2}$
3. $\frac{3}{x+1} - \frac{2}{(x-2)^2}$

Page 13 Quick Test

1. a) $y = \frac{15}{x}$

b)

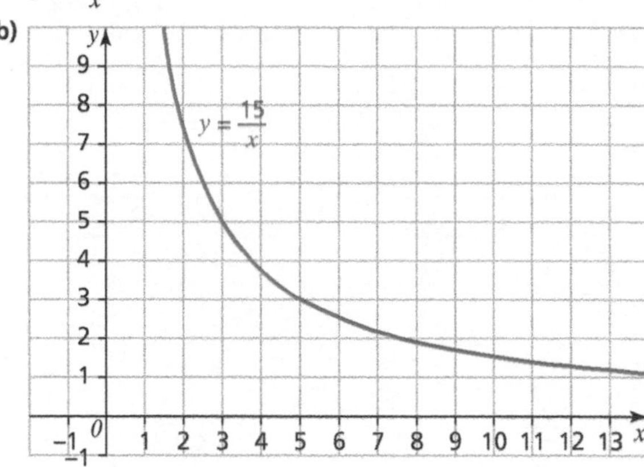

2.

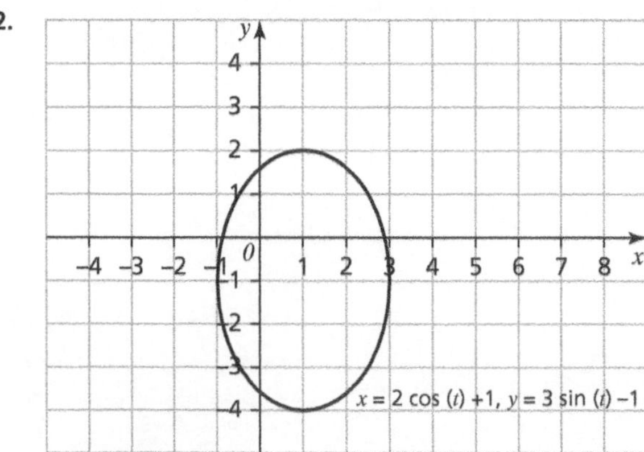

Page 15 Quick Test

1. $\left(\frac{4}{25}, 0\right)$
2. a) 4 metres b) 2.19 metres and 19.74 metres

Page 17 Quick Test

1. 2, 6, 18, 54, 162
2. a) The sequence is periodic with a period of 2.
   b) The sequence is decreasing.
   c) The sequence is increasing.
3. $\sum_{1}^{7} 2n + 3 = 77$

Page 19 Quick Test

1. a) $-\frac{7}{4}, -\frac{3}{2}, -\frac{5}{4}, -1, -\frac{3}{4}$ b) 9th term c) 1062.50
2. 1245

Page 21 Quick Test

1. a) 400, −100, 25, $-\frac{25}{4}$, $\frac{25}{16}$ b) $\frac{5242875}{16384}$
   c) Convergent, 320

Page 23 Quick Test

1. $\frac{1}{\sqrt{1-2x}} = 1 + x + \frac{3x^2}{2} + \frac{5x^3}{2}$, valid for $|x| < \frac{1}{2}$
2. $(2x-3)^{-2} = \frac{1}{9} + \frac{4x}{27} + \frac{4x^2}{27} + \frac{32x^3}{243}$, valid for $|x| < \frac{3}{2}$
3. $\frac{5x-7}{(x-1)(x-2)} = -\frac{7}{2} - \frac{11x}{4} - \frac{19x^2}{8} - \frac{35x^3}{16}$, valid for $|x| < 1$

Pages 24–29 Practice Questions

Page 24: Composite and Inverse Functions

1. a) $f^2(x) = ff(x) = (x+3)+3 = x+6$ **[1]**, domain $\mathbb{R}$, range $\mathbb{R}$ **[1 for both]**
   $f(x^2) = x^2 + 3$ **[1]**, domain $\mathbb{R}$, range $\mathbb{R}$ **[1 for both]**
   b) $f^2(x) = f(x^2)$
   $x + 6 = x^2 + 3$ **[1]**
   $0 = x^2 - x - 3$
   $x = \frac{1 \pm \sqrt{(-1)^2 - (4 \times 1 \times -3)}}{2 \times 1} = \frac{1+\sqrt{13}}{2}$ **[1]**
   $\frac{1-\sqrt{13}}{2}$ **[1]**
   c) Finding the inverse function is equivalent to changing the subject of the formula.
   Let $y = x + 3$, $x = y - 3$, then $f^{-1}(x) = x - 3$ **[1]**
2. $fg(x) = 2\ln(e^{3x})$ **[1]** $= 2 \times 3x = 6x$ **[1]**
   $gf(x) = e^{3(2\ln(x))}$ **[1]** $= e^{6\ln(x)} = x^6$ **[1]**

Recall that e^x is the inverse operation of $\ln(x)$, so $e^{\ln(x)} = x$ and $\ln(e^x) = x$.

Page 24: Modulus and Exponential Functions

1.

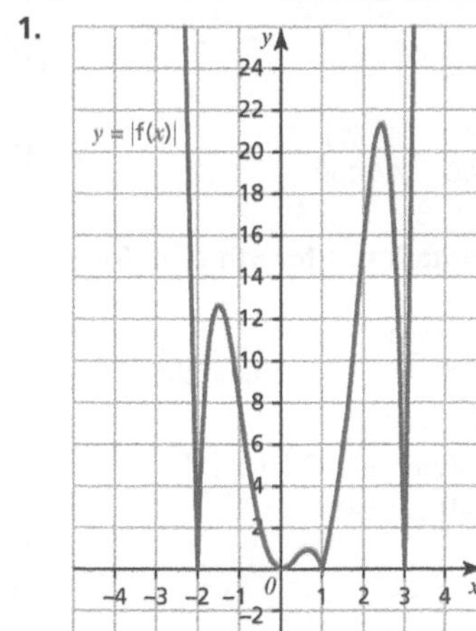

[1 for general shape; 1 for reflection above the x-axis]

2.

The graph of $y = 2\sin(3x) - 1$ is a transformation of $y = \sin(x)$. It is a vertical stretch by a factor of 2, a horizontal stretch by a factor of $\frac{1}{3}$ and a translation by $\begin{pmatrix} 0 \\ -1 \end{pmatrix}$.

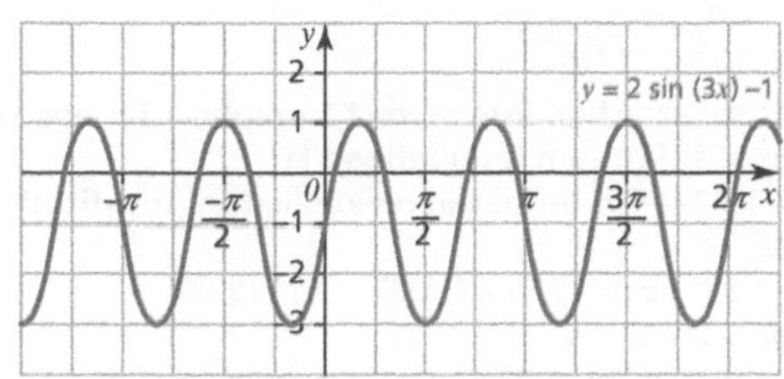

[1 for general shape of sin curve; 1 for vertical stretch; 1 for horizontal stretch; 1 for translation]

3.

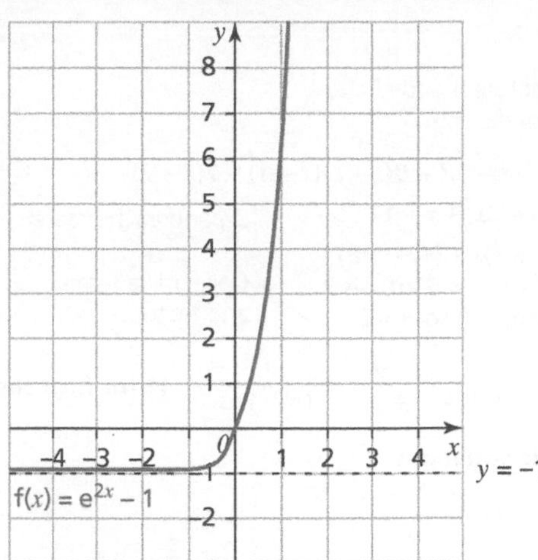

[1 for y-intercept; 1 for asymptote; 1 for general shape]
y-intercept: $y = e^{2 \times 0} - 1 = 1 - 1 = 0$
When $x \to \infty$, $y \to \infty$. When $x \to -\infty$, $y \to -1$, so there is an asymptote at $y = -1$.

Page 25: Algebraic Fractions

1. $\frac{(2x+1)(x^2-2x-3)}{(3x-4)(x+2)} \times \frac{x^2-4}{2x^2-5x-3}$
   $= \frac{(2x+1)(x+1)(x-3)}{(3x-4)(x+2)} \times \frac{(x+2)(x-2)}{(2x+1)(x-3)}$

Factorise the numerators and denominators of both fractions.

[1 for each factorisation up to a maximum of 3]

Cancel where possible.

$\frac{\cancel{(2x+1)}(x+1)\cancel{(x-3)}}{(3x-4)\cancel{(x+2)}} \times \frac{\cancel{(x+2)}(x-2)}{\cancel{(2x+1)}\cancel{(x-3)}}$

$= \frac{(x+1)(x-2)}{3x-4}\left(= \frac{x^2-x-2}{3x-4}\right)$

[1 for correct numerator; 1 for correct denominator. Answer does not need to be expanded.]

2. $(3x^4 + 2x^2 - x + 2) \div (x - 1)$

$$\begin{array}{r} 3x^3 + 3x^2 + 5x + 4 \\ x-1 \overline{) \; 3x^4 + 0x^3 + 2x^2 - x + 2} \\ -(3x^4 - 3x^3) \\ 3x^3 + 2x^2 \\ -(3x^3 - 3x^2) \\ 5x^2 - x \\ -(5x^2 - 5x) \\ 4x + 2 \\ -(4x - 4) \\ 6 \end{array}$$

[1 for method of algebraic division; 1 for correct quotient; 1 for correct remainder]

$\frac{(3x^4 + 2x^2 - x + 2)}{x-1} = 3x^3 + 3x^2 + 5x + 4 + \frac{6}{x-1}$

[1 for final answer in the form of a sum]

3. $\frac{5x^2+3x+13}{(x-1)^2(x+2)} \equiv \frac{A}{x+2} + \frac{B}{x-1} + \frac{C}{(x-1)^2}$
   $5x^2 + 3x + 13 \equiv A(x-1)^2 + B(x+2)(x-1) + C(x+2)$ **[1]**

Multiply each term by $(x-1)^2(x+2)$.

To find A, substitute $x = -2$.

$(5 \times (-2)^2) + (3 \times -2) + 13$
$= A(-2-1)^2 + B(-2+2)(1-1) + C(-2+2)$
$27 = 9A \Rightarrow A = 3$ **[1]**

To find C, substitute $x = 1$.

$(5 \times 1^2) + (3 \times 1) + 13 = A(1-1)^2 + B(1+2)(1-1) + C(1+2)$

$21 = 3C$, so $C = 7$ **[1]**

To find B, substitute A and C, then equate coefficients.

$5x^2 + 3x + 13 = 3(x - 1)^2 + B(x + 2)(x - 1) + 7(x + 2)$

$= Bx^2 + Bx - 2B + 3x^2 + x + 17$ Expand each bracket. **[1]**

$= (B + 3)x^2 + (B + 1)x + (17 - 2B)$ **[1]**

The coefficient of x^2 is 5, of x is 3 and the constant is 3, so $B = 2$. Collect like terms and factorise. **[1]**

$$\frac{5x^2 + 3x + 13}{(x-1)^2(x+2)} \equiv \frac{3}{x+2} + \frac{2}{x-1} + \frac{7}{(x-1)^2}$$ **[1 for final answer]**

Page 26: Parametric Equations

1. a) $x = \sqrt{t}$
   $t = x^2$ **[1]**
   $y = 2t + 3$
   $y = 2x^2 + 3$ **[1]** $\{x \in \mathbb{R}: 0 \leq x \leq 3\}$ **[1]** $\{y \in \mathbb{R}: 3 \leq y \leq 21\}$ **[1]**

   b)

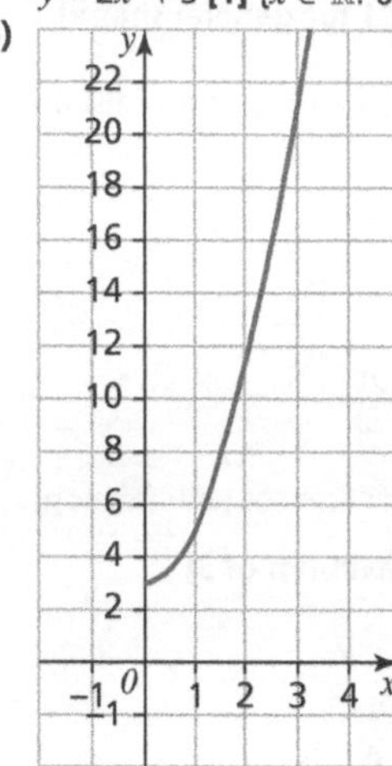

[1 for general shape with correct y-intercept at (0, 3); 1 for domain and range]

2. Recall that parametric equations $x = a + r\cos(t)$, $y = b + r\sin(t)$ describe a circle with centre (a, b) and radius r.

The parametric equations given describe a circle with centre (−2, 1) and radius 3. **[1]**

The domain of $\left\{t \in \mathbb{R}: 0 \leq t \leq \frac{3\pi}{2}\right\}$ will give $\frac{3}{4}$ of a circle.

The range of $x = -2 + 3\cos(t)$, $\left\{t \in \mathbb{R}: 0 \leq t \leq \frac{3\pi}{2}\right\}$ and hence the domain of the circle is $\{x \in \mathbb{R}: -5 \leq x \leq 1\}$. **[1]**

The range of $y = 1 + 3\sin(t)$, $\left\{t \in \mathbb{R}: 0 \leq t \leq \frac{3\pi}{2}\right\}$ and hence the domain of the circle is $\{y \in \mathbb{R}: -2 \leq x \leq 4\}$. **[1]**

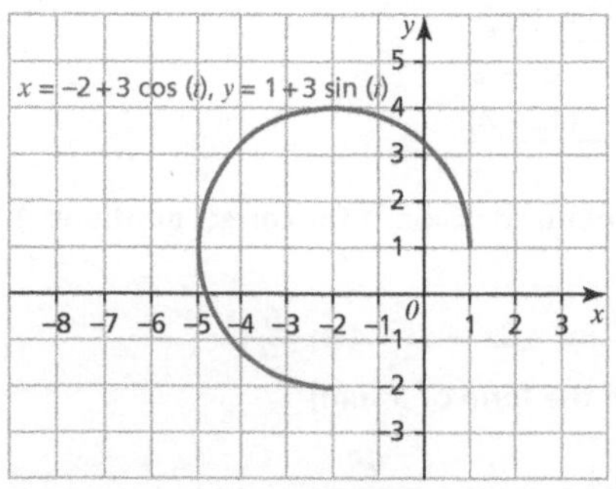

[1 for correct general shape with centre and radius correct; 1 for correct domain and range]

Alternatively, either convert the parametric equations to a Cartesian equation using the trigonometric identity $\sin^2(\theta) + \cos^2(\theta) \equiv 1$ or substitute values of t and plot the resulting (x, y) points.

Page 26: Problems Involving Parametric Equations

1. a) The figure skater starts at $t = 0$, $x = 4\cos(0) + 2 = 6$ **[1]**; $y = 3\sin(0) = 0$ **[1]**

   b) Find the points of intersection of the line $y = 1$ with the curve.

   $3\sin(2t) = 1$ **[1]**

   $\sin^{-1}\left(\frac{1}{3}\right) = 2t$

   $t = 0.17$ **[1]**

   Sketching the line $y = 1$ onto the parametric graph shows that the line $y = 1$ will intersect the curve four times. Using the properties of sine, the value of $\sin(2x)$ is the same as the value of $\sin\left(\frac{\pi}{2} - x\right)$, $\sin(\pi + x)$ and $\sin\left(\frac{3\pi}{2} - x\right)$.

   So the times are $t = 0.17$ s, $t = \frac{\pi}{2} - 0.17 = 1.40$ s, $t = 0.17 + \pi = 3.31$ s, $t = \frac{3\pi}{2} - 0.17 = 4.54$ s **[1 for all four times]**

   The x-coordinates are then: $x = 4\cos(0.17) + 2 = 5.94$, $x = 4\cos(1.40) + 2 = 2.68$, $x = 4\cos(3.31) + 2 = -1.94$ and $x = 4\cos(4.54) + 2 = 1.31$ **[2 for all correct; 1 for at least two correct]**

2. Find the values of t where the curve intersects the x-axis.

$x = 2e^{2t}$

$\frac{x}{2} = e^{2t}$ It's easier not to solve for x in this case and to instead substitute $\frac{x}{2}$ for e^{2t}. **[1]**

$y = e^{4t} - 3e^{2t} + 2$

$y = \left(\frac{x}{2}\right)^2 - 3\left(\frac{x}{2}\right) + 2$ **[1]**

$\frac{x^2}{4} - \frac{3x}{2} + 2 = 0$

$x^2 - 6x + 8 = 0$

$(x - 4)(x - 2) = 0$ **[1]**

When $x = 4$, $2e^{2t} = 4$

$e^{2t} = 2$ **[1]**

$2t = \ln(2)$

$t = \frac{1}{2}\ln(2)$ **[1]**

When $x = 2$, $2e^{2t} = 2$

$e^{2t} = 1$

$2t = \ln(1)$

$2t = 0$, so $t = 0$ **[1]**

Page 27: Types of Sequences

1. a) Increasing **[1]**
   b) 1, 3, 5, 7, 9 **[1]**
   c) $u_{n+1} = u_n + 2$, first term 1 **[1]**
   d) $\sum_{1}^{5} u_n = 1 + 3 + 5 + 7 + 9 = 25$ **[1]**
2. $\sum_{1}^{5}\left(\frac{1}{2n}\right)^2 = \frac{1}{4} + \frac{1}{16} + \frac{1}{36} + \frac{1}{64} + \frac{1}{100} = \frac{5269}{14400}$ **[1]**
3. $\sum_{1}^{5} \frac{2}{n+1}$ **[1 for sigma notation; 1 for nth term formula]**

Page 28: Arithmetic Sequences and Series

1. $u_{30} = 2500 + (49 \times 500) = £27000.00$ Find the 30th term. **[1]**
   $S_{30} = \frac{30}{2}(2500 + 27000) = £442500.00$ **[1]**
2. $u_{10} \Rightarrow 152 = a + 9d$ (1) **[1]**, $u_{50} \Rightarrow -208 = a + 49d$ (2) **[1]**
   Subtracting (1) from (2): $-360 = 40d$, $d = -9$ **[1]**
   $152 = a + (9 \times -9)$, $a = 233$ **[1]**
   $u_n = 233 - 9(n - 1)$ **[1]**
3. $u_n = 35000 + 400(n - 1)$ **[1]**
   $u_{10} = 35000 + (400 \times 9) = £38600$ **[1]**

Page 28: Geometric Sequences and Series

1. a) Substitute $n = 1$, $n = 2$, $n = 3$, $n = 4$ and $n = 5$ to get $\frac{1}{32}, \frac{1}{8}, \frac{1}{2}, 2, 8$ **[1]**
   b) $S_{10} = \frac{\frac{1}{32}(4^{10} - 1)}{(4 - 1)} = \frac{349\,525}{32}$
   [1 for correct substitution into correct formula; 1 for answer]
   c) Divergent ($|r| > 1$) **[1]**; sum is undefined **[1]**
2. a) $a = 1250$, $r = 1.03$ **[1 for both, can be implied by substitution into formula]**
   $u_n = 1250 \times (1.03)^n$; $u_5 = 1250 \times (1.03)^5 = £1449.09$ **[1]**
   b) $10\,000 = 1250 \times (1.03)^n$ **[1]**
   $8 = (1.03)^n$
   $\log_{10}(8) = n\log_{10}(1.03)$ **[1]**

$n = \frac{\log_{10}(8)}{\log_{10}(1.03)} = 70.35$; at the end of 71 years **[1]**

3. a) $a = 30000$, $r = 1.06$ **[1]**
$u_5 = 30000(1.06)^4 = 37874.31$ **[1]**
b) $75000 < 30000(1.06)^{n-1}$
$2.5 < 1.06^{n-1}$
$\log_{10}(2.5) < (n-1)\log_{10}(1.06)$
$n > \frac{\log_{10}(2.5)}{\log_{10}(1.06)} + 1 = 16.725$ **[1]**
After 17 years, he will make more than £75000. **[1]**

Page 29: Binomial Sequences

1. The value of $a = -1$, the value of $b = 3$.

Take care when the expression is written in the form $(bx + a)^n$.

$$(a+bx)^n = a^n\left(1 + \frac{n \times \frac{b}{a}x}{1!} + \frac{n(n-1)\left(\frac{b}{a}x\right)^2}{2!} + \frac{n(n-1)(n-2)\left(\frac{b}{a}x\right)^3}{3!} + \cdots\right)$$

$$(3x-1)^{-3} = (-1)^{-3}\left(1 + \frac{-3 \times -3x}{1!} + \frac{-3(-3-1)(-3x)^2}{2!} + \frac{-3(-3-1)(-3-2)(-3x)^3}{3!} + \cdots\right)$$ **[1]**

$= -1(1 + 9x + 54x^2 + 270x^3 + \ldots)$ **[1]** $= -1 - 9x - 54x^2 - 270x^3 + \ldots$ **[1]**

Valid for $|x| < \frac{1}{3}$ **[1]**

2. $\frac{7x+1}{(x+4)(2x-1)} \equiv \frac{A}{x+4} + \frac{B}{2x-1}$
$7x + 1 = A(2x-1) + B(x+4)$ **[1]**
Substitute $x = \frac{1}{2}: 7 \times \frac{1}{2} + 1 = B\left(\frac{1}{2} + 4\right) \Rightarrow \frac{9}{2} = \frac{9}{2}B \Rightarrow B = 1$
Substitute $x = -4: 7 \times -4 + 1 = A(2 \times -4 - 1) \Rightarrow -27 = -9A \Rightarrow A = 3$
[1 for A and B]
Write in index form
$\frac{7x+1}{(x+4)(2x-1)} \equiv \frac{3}{x+4} + \frac{1}{2x-1} = 3(x+4)^{-1} + (2x-1)^{-1}$

Expand $3(x+4)^{-1} = 3\left(4^{-1}\left(1 + \frac{-1 \times \frac{1}{4}x}{1!} + \frac{-1(-1-1)\left(\frac{1}{4}x\right)^2}{2!} + \frac{-1(-1-1)(-1-2)\left(\frac{1}{4}x\right)^3}{3!} + \cdots\right)\right)$ **[1]**

$= \frac{3}{4}\left(1 - \frac{x}{4} + \frac{x^2}{16} - \frac{x^3}{64}\right) = \frac{3}{4} - \frac{3x}{16} + \frac{3x^2}{64} - \frac{3x^3}{256}$ **[1]**

Valid for $|x| < 4$

Expand $(2x-1)^{-1} = -1^{-1}\left(1 + \frac{-1 \times -2x}{1!} + \frac{-1(-1-1)(-2x)^2}{2!} + \frac{-1(-1-1)(-1-2)(-2x)^3}{3!} + \cdots\right)$ **[1]**

$= -1(1 + 2x + 4x^2 + 8x^3) = -1 - 2x - 4x^2 - 8x^3$ **[1]**

Valid for $|x| < \frac{1}{2}$

Simplify $3(x+4)^{-1} + (2x-1)^{-1}$

$= \frac{3}{4} - \frac{3x}{16} + \frac{3x^2}{64} - \frac{3x^3}{256} - 1 - 2x - 4x^2 - 8x^3$

$= -\frac{1}{4} - \frac{35x}{16} - \frac{253x^2}{64} - \frac{2051x^3}{256} + \cdots$ **[1]**

Valid for $|x| < \frac{1}{2}$ **[1]**

3. a) **[1]**
$$\frac{2}{x+3} = 2(x+3)^{-1} = 2\left(3^{-1}\left(1 + \frac{-1 \times \frac{1}{3}x}{1!} + \frac{-1(-1-1)\left(\frac{1}{3}x\right)^2}{2!} + \frac{-1(-1-1)(-1-2)\left(\frac{1}{3}x\right)^3}{3!} + \cdots\right)\right)$$

$= \frac{2}{3}\left(1 - \frac{x}{3} + \frac{x^2}{9} - \frac{x^3}{27} + \cdots\right)$ **[1]** $= \frac{2}{3} - \frac{2x}{9} + \frac{2x^2}{27} - \frac{2x^3}{81} + \cdots$ **[1]**

Valid for $|x| < 3$ **[1]**

b) Let $x = 0.02$, then $\frac{2}{x+3} = \frac{2}{3.02} = \frac{200}{302}$ **[1]**

$\frac{200}{302} = \frac{2}{3} - \frac{2 \times 0.02}{9} + \frac{2 \times (0.02^2)}{27} - \frac{2 \times (0.02^3)}{81} + \ldots$

$= \frac{2}{3} - \frac{1}{225} + \frac{1}{33750} - \frac{1}{5062500}$ **[1]**

$= \frac{3352649}{5062500} \approx 0.66225$ (to 5 d.p.) **[1]**

Pages 30–41 Revise Questions

Page 31 Quick Test

1. 480°
2. 10.7 cm²
3. 25.2 cm

Page 33 Quick Test

1. $\sqrt{82} = 9.06$
2. $\sqrt{131} = 11.4$
3. $t = \pm 2\sqrt{2}$

Page 35 Quick Test

1. $\frac{1 - \sec\theta}{\tan^2\theta} \approx \frac{1 - \left(1 + \frac{\theta^2}{2}\right)}{\theta^2} = -\frac{\frac{\theta^2}{2}}{\theta^2} = -\frac{1}{2}$
2. $\frac{2\sqrt{3}}{3}$
3. $-\frac{\pi}{4}$

Page 37 Quick Test

1. $\frac{\pi}{12}, \frac{7\pi}{12}, \frac{13\pi}{12}, \frac{19\pi}{12}$
2. LHS
$= \operatorname{cosec}^4\theta - \cot^4\theta$
$= (\operatorname{cosec}^2\theta + \cot^2\theta)(\operatorname{cosec}^2\theta - \cot^2\theta)$
$= \operatorname{cosec}^2\theta + \cot^2\theta$
$= 1 + 2\cot^2\theta$
= RHS
3. $\frac{2\pi}{3}, \frac{4\pi}{3}, \frac{\pi}{2}, \frac{3\pi}{2}$

Page 39 Quick Test

1. $\frac{\sin 2\theta \cos 2\theta}{2}$
2. $\theta = \frac{\pi}{12}$ or $\frac{5\pi}{12}$
3. $\frac{4\sqrt{2}}{9}$

Page 41 Quick Test

1. a) $\sqrt{89}\sin(x + 0.559)$ b) $-\sqrt{89} < f(x) < \sqrt{89}$
2. $\sqrt{2}\cos\left(x - \frac{\pi}{4}\right)$
3. 0.869

Pages 42–45 Practice Questions

Page 42: Radians, Arc Lengths and Areas of Sectors

1. Use $P = 2r + r\theta$ **[1]**
$P = 20, r = 4 \Rightarrow \theta = 3$ **[1]**
$A = \frac{1}{2}r^2\theta = \frac{1}{2} \times 16 \times 3 = 24\text{ cm}^2$ **[2]**
2. a) Using the cosine rule: **[1]**
$\cos AOB = \frac{12^2 + 12^2 - 7^2}{2 \times 12 \times 12} = 0.8300$ **[1]**
$\angle AOB = 0.592^c$ **[1]**
b) Area of segment $= \frac{1}{2}r^2(\theta - \sin\theta)$ **[1]**
$= \frac{1}{2} \times 12^2 \times (0.592 - \sin 0.592) = 2.45\text{ cm}^2$ **[1]**
3. Use $A = \frac{1}{2}r^2(\theta - \sin\theta)$ **[1]**

$A = 50, \theta = \frac{3\pi}{4} \Rightarrow r = 7.787$ cm **[1]**

$S = r\theta = 7.787 \times \frac{3\pi}{4} = 18.3$cm **[2]**

Page 42: Vectors in 3D

1. $\sqrt{(2-(-1))^2 + (1-(-2))^2 + (3-4)^2} = \sqrt{3^2 + 3^2 + (-1)^2}$ **[2]**
$= \sqrt{19} = 4.36$ **[1]**

2. $(-8)^2 + p^2 + 10^2 = 200$ **[2]**
$\Rightarrow p^2 = 36$ **[1]**
$\Rightarrow p = \pm 6$ **[1]**

3.
a) $\overrightarrow{AB} = \begin{pmatrix} 4t \\ 3 \\ 1-t \end{pmatrix} - \begin{pmatrix} t-1 \\ 3t \\ 2 \end{pmatrix} = \begin{pmatrix} 3t+1 \\ 3-3t \\ -1-t \end{pmatrix}$ **[2]**

b) $|\overrightarrow{AB}|^2 = (3t+1)^2 + (3-3t)^2 + (-1-t)^2$ **[2]**
$= 9t^2 + 6t + 1 + 9t^2 - 18t + 9 + t^2 + 2t + 1$
$= 19t^2 - 10t + 11$ **[1]**
$|\overrightarrow{AB}|$ is minimum when $\frac{d|\overrightarrow{AB}|^2}{dt} = 0$ **[1]**
Differentiate $|\overrightarrow{AB}|^2$ to give $38t - 10$ **[1]**
$\Rightarrow t = \frac{5}{19}$ **[1]**

Page 43: Trigonometric Ratios

1. a) $\sin\frac{11\pi}{6} = \sin\left(2\pi - \frac{\pi}{6}\right) = -\sin\frac{\pi}{6} = -\frac{1}{2}$ **[2]**
b) $\tan\frac{13\pi}{4} = \tan\left(2\pi + \pi + \frac{\pi}{4}\right) = \tan\left(\pi + \frac{\pi}{4}\right) = \tan\frac{\pi}{4} = 1$ **[2]**
c) $\cos 4\pi = \cos 0 = 1$ **[1]**

2. a) $\arcsin\left(\sin\frac{9\pi}{4}\right) = \arcsin\left(\frac{\sqrt{2}}{2}\right) = \frac{\pi}{4}$ **[2]**
b) $\arccos\left(\sin\left(-\frac{5\pi}{6}\right)\right) = \arccos\left(-\frac{1}{2}\right) = \frac{2\pi}{3}$ **[2]**
c) $\arctan\left(\tan\frac{17\pi}{6}\right) = \arctan\left(-\frac{\sqrt{3}}{3}\right) = \frac{5\pi}{6}$ **[2]**

3. $\frac{x\sin 2x}{1-\cos x} \approx \frac{x(2x)}{1-\left(1-\frac{x^2}{2}\right)}$ **[1]**
$= \frac{2x^2}{\frac{x^2}{2}} = 4$ **[1]**

Page 44: Further Trigonometric Equations and Identities

1. $\cos^4 x - \sin^4 x = (\cos^2 x + \sin^2 x)(\cos^2 x - \sin^2 x)$ **[1]**
$= 1 \times (\cos^2 x - \sin^2 x)$ **[1]**
$= \cos^2 x - (1 - \cos^2 x)$ **[1]**
$= 2\cos^2 x - 1$

2. $(1 + \tan^2 x) - 6\tan x + 4 = 0$ **[1]**
$\tan^2 x - 6\tan x + 5 = 0$
$(\tan x - 5)(\tan x - 1) = 0$ **[2]**
$\tan x = 5 \Rightarrow x = 1.37$ or $x = \pi + 1.37 = 4.51$ **[2]**
$\tan x = 1 \Rightarrow x = \frac{\pi}{4}, \frac{5\pi}{4}$ **[2]**

3. $\tan x + \cot x$
$= \frac{\sin x}{\cos x} + \frac{\cos x}{\sin x}$ **[1]**
$= \frac{\sin^2 x + \cos^2 x}{\sin x \cos x}$ **[1]**
$= \frac{1}{\sin x \cos x}$ **[1]**
$= \sec x \operatorname{cosec} x$

4. $\cot\left(x - \frac{\pi}{4}\right) = 1 \Rightarrow \tan\left(x - \frac{\pi}{4}\right) = 1$ **[1]**
$x - \frac{\pi}{4} = \frac{\pi}{4} \Rightarrow x = \frac{\pi}{2}$ **[2]**
Or $x - \frac{\pi}{4} = \pi + \frac{\pi}{4} \Rightarrow x = \frac{3\pi}{2}$ **[2]**

Page 45: Using Addition Formulae and Double Angle Formulae

1. $\sin A = \frac{3}{5} \Rightarrow \cos A = \sqrt{1 - \sin^2 A} = \frac{4}{5}$ **[2]**
$\sin B = -\frac{2}{5} \Rightarrow \cos B = -\sqrt{1 - \sin^2 B} = -\frac{\sqrt{21}}{5}$ **[2]**
$\sin(A - B) = \sin A \cos B - \cos A \sin B$ **[1]**
$= \left(\frac{3}{5}\right)\left(-\frac{\sqrt{21}}{5}\right) - \left(\frac{4}{5}\right)\left(-\frac{2}{5}\right)$
$= \frac{8 - 3\sqrt{21}}{25}$ **[1]**

2. $\tan x$ is negative and x is obtuse, so x lies in the second quadrant. **[1]**
$\tan x = -\frac{7}{24} \Rightarrow \sin x = \frac{7}{25}$ and $\cos x = -\frac{24}{25}$ **[2]**
$\sin 2x = 2\sin x \cos x$
$= 2 \times \frac{7}{25} \times \left(-\frac{24}{25}\right) = -\frac{336}{625}$ **[2]**

3. $2\sin x \cos x = 3\cos^2 x$
$2\sin x \cos x - 3\cos^2 x = 0$
$\cos x(2\sin x - 3\cos x) = 0$ **[1]**
$\cos x = 0 \Rightarrow x = \frac{\pi}{2}$ or $x = \frac{3\pi}{2}$ **[2]**
$2\sin x - 3\cos x = 0 \Rightarrow 2\sin x = 3\cos x \Rightarrow \frac{\sin x}{\cos x} = \frac{3}{2} \Rightarrow \tan x = \frac{3}{2}$ **[1]**
$\Rightarrow x = 0.983$ or $x = 4.12$ **[2]**

Other methods can be used.

Page 45: Problem Solving Using Trigonometry

1. $5\cos x + 12\sin x = R\cos(x - \alpha)$
$= R\cos x \cos\alpha + R\sin x \sin\alpha$
$5 = R\cos\alpha$ **[1]**
$12 = R\sin\alpha$ **[1]**
$R^2 = 5^2 + 12^2 = 169 \Rightarrow R = 13$ **[2]**
$\frac{R\sin\alpha}{R\cos\alpha} = \tan\alpha = \frac{12}{5} \Rightarrow \alpha = 1.18^c$ **[2]**
$5\cos x + 12\sin x = 13\cos(x - 1.18^c)$

2. $7\cos x + 24\sin x = R\cos(x - \alpha)$
$= R\cos x \cos\alpha + R\sin x \sin\alpha$
$7 = R\cos\alpha$ **[1]**
$24 = R\sin\alpha$ **[1]**
$R^2 = 7^2 + 24^2 = 625 \Rightarrow R = 25$ **[2]**
$\frac{R\sin\alpha}{R\cos\alpha} = \tan\alpha = \frac{24}{7} \Rightarrow \alpha = 1.287^c$ **[2]**
$7\cos x + 24\sin x = 25\cos(x - 1.287^c)$
$7\cos x + 24\sin x = 5$
$25\cos(x - 1.287^c) = 5$
$\cos(x - 1.287^c) = 0.2$ **[1]**
$x - 1.287 = 1.369 \Rightarrow x = 2.66$ **[2]**
$x - 1.287 = 2\pi - 1.369 \Rightarrow x = 6.20$ **[2]**

3. From question 2, $f(x) = 8 + 7\cos x + 24\sin x$
$= 8 + 25\cos(x - 1.287^c)$ **[1]**
Minimum value is $8 - 25 = -17$ **[1]**
Maximum value is $8 + 25 = 33$ **[1]**
Maximum occurs when $\cos(x - 1.287^c) = 1$, i.e. $x = 1.287^c$ **[2]**

Pages 46–51 Review Questions

Page 46: Composite and Inverse Functions

1. a) Finding the inverse function is equivalent to changing the subject of the formula. Let $y = x^2 - 4$, $\sqrt{y + 4} = x$, then $f^{-1}(x) = \sqrt{x + 4}$ **[1]**
The domain is $\{x \in \mathbb{R}: x \geqslant -4\}$. The range is $\{y \in \mathbb{R}: y \geqslant 0\}$. **[1 for both]**

b) First sketch the graph of $f(x)$, remembering the domain of $\{x \in \mathbb{R}: x \geqslant 0\}$, then reflect the graph over the line $y = x$.

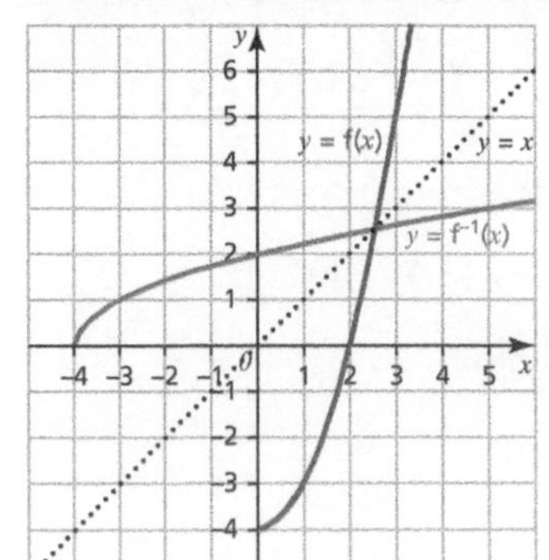

[1 for $y = f(x)$; 1 for $y = f^{-1}(x)$; 1 for correct domains on both curves]

c) Since $y = f^{-1}(x)$ is a reflection of $y = f(x)$ over the line $y = x$, both curves intersect with the line $y = x$ at the same place, so the solutions to $f(x) = f^{-1}(x)$ are the same as the solutions to $f(x) = x$, which is easier to solve in this case.

$f(x) = x$

$x^2 - 4 = x$ **[1]**

$x^2 - x - 4 = 0$

$x = \frac{1 \pm \sqrt{(-1)^2 - (4 \times 1 \times -4)}}{2 \times 1}$ **[1]** $= \frac{1 \pm \sqrt{17}}{2}$

$x = \frac{1 + \sqrt{17}}{2}$ **[1]**

Remember that the domain of $y = f(x)$ is $\{x \in \mathbb{R}: x \geqslant 0\}$, therefore $x = \frac{1 - \sqrt{17}}{2}$ is not a possible answer.

2. a) $fg(x) = \frac{2(2x - 4) + 5}{3} = \frac{4x - 3}{3}$ **[1]**

$fg(11) = \frac{(4 \times 11) - 3}{3} = \frac{41}{3}$ **[1]**

b) $gf(x) = 2\left(\frac{2x + 5}{3}\right) - 4 = \frac{4x + 10 - 12}{3} = \frac{4x - 2}{3}$ **[1]**

$fg(x) = gf(x)$

$\frac{4x - 3}{3} = \frac{4x - 2}{3}$ **[1]**

$-3 = -2$, no solutions **[1]**

Page 46: Modulus and Exponential Functions

1.

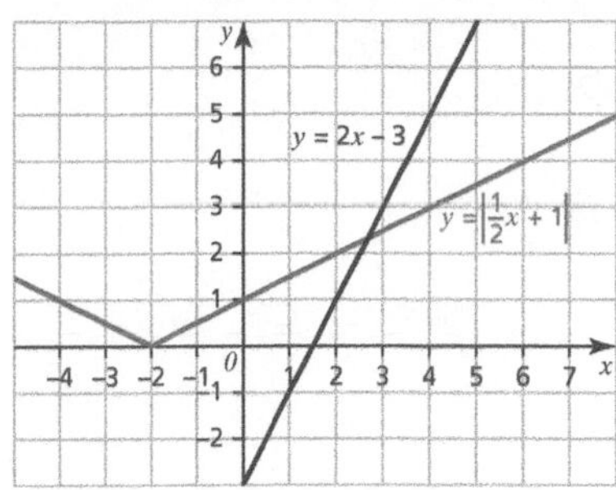

[1 for absolute value; 1 for line]

The lines intersect at one point.

$\frac{1}{2}x + 1 = 2x - 3$ **[1]**

$\frac{3}{2}x = 4$, so $x = \frac{8}{3}$ **[1]**

2. y-intercept: $y = e^{(4 \times 0) + 1} - 2 = e - 2 \approx 0.72$
When $x \to \infty$, $y \to \infty$. When $x \to -\infty$, $y \to -2$, so there is an asymptote at $y = -2$.

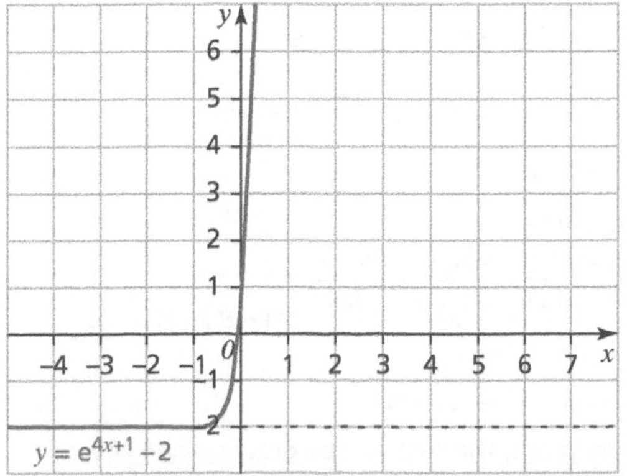

[1 for y-intercept; 1 for asymptote; 1 for general shape]

3. Sketch the graph of each transformation step by step.

$y = 2\left|f\left(\frac{1}{2}x + 1\right)\right| + 2$ is a horizontal stretch by a factor of 2 **[1]**; a translation in the x-direction of -1 **[1]**; then the absolute value **[1]**; then a vertical stretch by a scale factor of 2 **[1]** and a translation of +2 in the y-direction **[1]**. **[1 for general shape]**

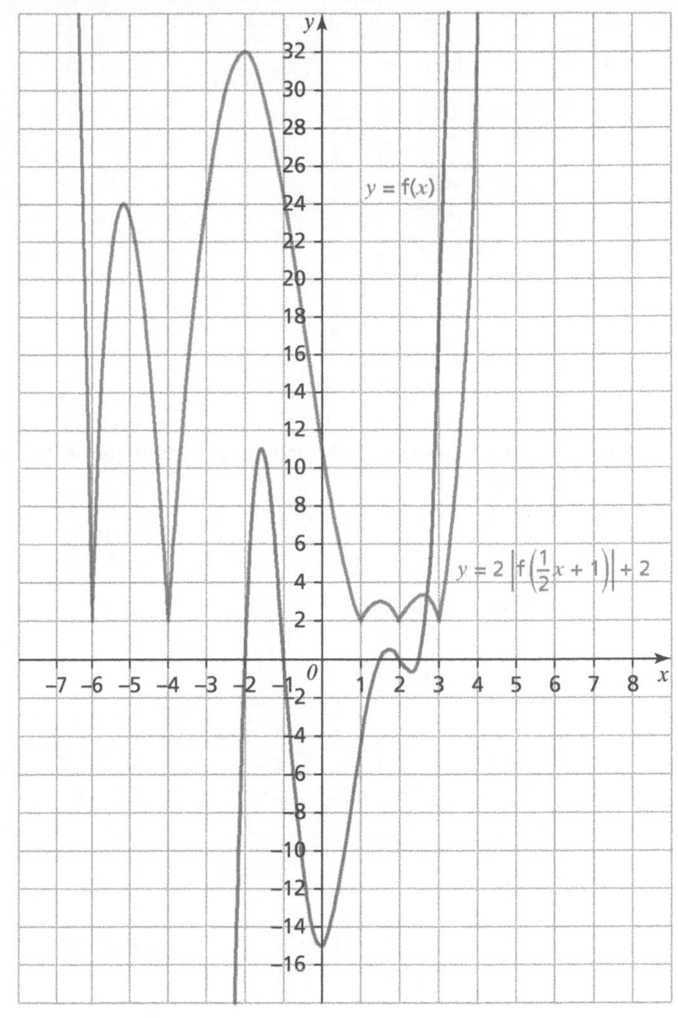

Page 47: Algebraic Fractions

1. $\frac{2x^2 + x - 3}{3x^2 - 13x + 4} \div \frac{2x^2 - x - 6}{x^2 - 5x + 4} = \frac{(x - 1)(2x + 3)}{(x - 4)(3x - 1)} \div \frac{(x - 2)(2x + 3)}{(x - 4)(x - 1)}$

Factorise.

[2 for all factorisations correct; 1 for at least two correct]

$= \frac{(x - 1)\cancel{(2x + 3)}}{\cancel{(x - 4)}(3x - 1)} \times \frac{\cancel{(x - 4)}(x - 1)}{(x - 2)\cancel{(2x + 3)}}$ **[1]** $= \frac{(x - 1)^2}{(x - 2)(3x - 1)}$ **[1]**

2. $\frac{7x^2 + 20x + 5}{(x + 3)(x^2 - 1)} = \frac{7x^2 + 20x + 5}{(x + 3)(x + 1)(x - 1)}$ **[1]**

Factorise the denominator.

$\frac{7x^2 + 20x + 5}{(x + 3)(x + 1)(x - 1)} \equiv \frac{A}{x + 3} + \frac{B}{x + 1} + \frac{C}{x - 1}$

$7x^2 + 20x + 5 \equiv A(x + 1)(x - 1) + B(x + 3)(x - 1) + C(x + 3)(x + 1)$ **[1]**

Multiply each term by $(x + 3)(x + 1)(x - 1)$.

To find A, substitute $x = -3$.

$7 \times (-3)^2 + (20 \times -3) + 5$
$\equiv A(-3 + 1)(-3 - 1) + B(-3 + 3)(-3 - 1) + C(-3 + 3)(-3 + 1)$
$8 = 8A$, so $A = 1$ **[1]**

To find B, substitute $x = -1$.

$7 \times (-1)^2 + (20 \times -1) + 5$
$\equiv A(-1 + 1)(-1 - 1) + B(-1 + 3)(-1 - 1) + C(-1 + 3)(-1 + 1)$
$-8 = -4B$, so $B = 2$ **[1]**

To find C, substitute $x = 1$.

$7 \times (1)^2 + (20 \times 1) + 5$
$\equiv A(1 + 1)(1 - 1) + B(1 + 3)(1 - 1) + C(1 + 3)(1 + 1)$
$32 = 8C$, so $C = 4$ **[1]**

$\frac{7x^2 + 20x + 5}{(x + 3)(x + 1)(x - 1)} \equiv \frac{1}{x + 3} + \frac{2}{x + 1} + \frac{4}{x - 1}$ **[1]**

3. $(2x^4 + x^3 + x^2 - 2x + 3) \div (x + 1)$

Do the algebraic division.

$$\begin{array}{r} 2x^3 - x^2 + 2x - 4 \\ x+1 \overline{) \, 2x^4 + x^3 + x^2 - 2x + 3} \\ -(2x^4 + 2x^3) \\ -x^3 + x^2 \\ -(x^3 - x^2) \\ 2x^2 - 2x \\ -(2x^2 + 2x) \\ -4x + 3 \\ -(-4x - 4) \\ 7 \end{array}$$

[1 for method of algebraic division; 1 for correct quotient; 1 for correct remainder]

$\frac{2x^4 + x^3 + x^2 - 2x + 3}{x + 1} = 2x^3 - x^2 + 2x - 4 + \frac{7}{x + 1}$ **[1]**

$A = 2, B = -1, C = 2, D = -4, E = 7$ **[1]**

Page 48: Parametric Equations

1. $x = 5t + 1 \Rightarrow t = \frac{x - 1}{5}$ **[1]**

$y = \frac{2}{\frac{x-1}{5}} + 3\left(\frac{x - 1}{5}\right) - 1$

$y = \frac{10}{x - 1} + \left(\frac{3(x - 1)}{5}\right) - 1$

$y = \frac{50 + 3(x - 1)^2 - 5(x - 1)}{5(x - 1)}$ **[1]**

$y = \frac{3x^2 - 11x + 58}{5x - 5}$ **[1]**

The range of $x = 5t + 1$ $\{t \in \mathbb{R}: t > 0\}$ and hence the domain of $y = \frac{3x^2 - 11x + 58}{5x - 5}$ is $\{x \in \mathbb{R}: x > 1\}$. **[1]**

2. Convert the parametric equations to a Cartesian equation.

$x = t + 3 \Rightarrow t = x - 3$

$y = 2(x - 3)^2 - 1$ **[1]**

The graph $y = f(x)$ is a transformation of $y = x^2$ by a translation of $\begin{pmatrix} 3 \\ -1 \end{pmatrix}$ and a vertical stretch by a factor of 2. The turning point is at (3, –1).

The range of $x = t + 3$ $\{t \in \mathbb{R}: -2 \leqslant t \leqslant 2\}$ and hence the domain of $y = 2(x - 3)^2 - 1$ is $\{x \in \mathbb{R}: 1 \leqslant x \leqslant 5\}$. **[1]**

The range of $y = 2t^2 - 1$ $\{t \in \mathbb{R}: -2 \leqslant t \leqslant 2\}$ is $\{y \in \mathbb{R}: y \leqslant 7\}$. **[1]**

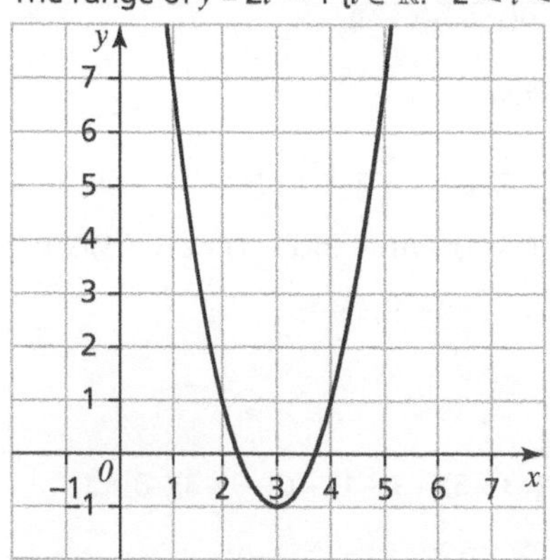

[1 for correct general shape; 1 for correct minimum; 1 for correct domain and range]

3. Substitute values of t to find the x- and y-coordinates.

t	$\frac{-\pi}{2}$	$\frac{-\pi}{4}$	0	$\frac{\pi}{4}$	$\frac{\pi}{2}$
$x = 2\cos(t)$	0	1.41	2	1.41	0
$y = 3\sin(t)$	–3	–2.12	0	2.12	3

[2 for all values correct; 1 for at least five values correct]

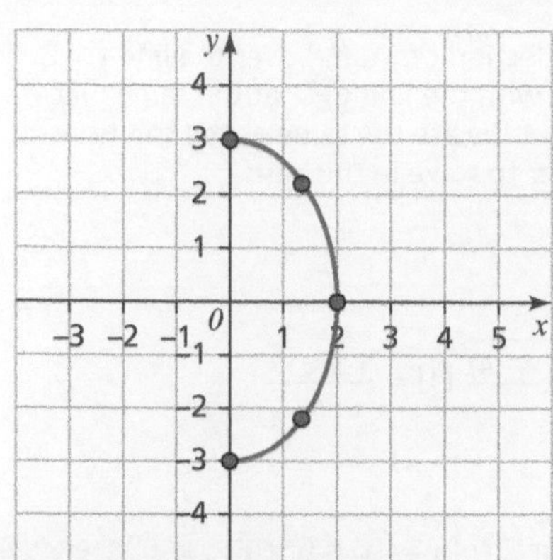

[1 for points plotted correctly (follow through); 1 for general shape; 1 for correct domain and range]

Page 48: Problems Involving Parametric Equations

1. a) Parabolic motion can be modelled by $x = t(v\cos a)$, $y = t(v\sin a) - 4.9t^2 + y_0$, then:
$x = t(15\cos(30))$ **[1]**, $y = t(15\sin(30)) - 4.9t^2 + 2$ **[1]**

b) Find the value of t when the curve intersects the x-axis.

$0 = t(15\sin(30)) - 4.9t^2 + 2$

$0 = -4.9t^2 + 7.5t + 2$ **[1]**

$t = \frac{-7.5 \pm \sqrt{(7.5)^2 - (4 \times -4.9 \times 2)}}{(2 \times -4.9)}$ **[1]**

$t = -0.23$ and $t = 1.76$ (to 2 d.p.)

Since t represents time, $t = 1.76$ is the only solution. **[1]**

c) From **b)**, the skier lands at $t = 1.76$. Find the value of x when $t = 1.76$.

$x = 1.76(15\cos(30)) = 22.86\,\text{m}$ **[1]**

2. When $x = 0$, $t^3 - at = 0$

$t(t^2 - a) = 0$

$t = 0$ or $t^2 - a = 0 \Rightarrow t = \pm\sqrt{a}$ **[1]**

When $y = 4$, $t^2 - 4 = 4$

$\left(\sqrt{a}\right)^2 - 4 = 4$ **[1]**

Substitute $t = \pm\sqrt{a}$

$a - 4 = 4$, so $a = 8$ **[1]**

Page 49: Types of Sequences

1. a) Increasing **[1]**

b) 2, 0, –4, 12, 140, 19596 **[1]**

c) $\sum_{1}^{6} u_n = 2 + 0 - 4 + 12 + 140 + 19596 = 19746$ **[1]**

2. $\sum_{1}^{4} 3n^2 - 2n = 1 + 8 + 21 + 40 = 70$ **[1]**

3. $\sum_{1}^{6} n^3 + 1$ **[1 for sigma notation; 1 for nth term formula]**

Page 50: Arithmetic Sequences and Series

1. a) $u_n = 23 + 7(n - 1)$ **[1]**

b) $S_n = \frac{n}{2}(23 + 23 + 7(n - 1)) = \frac{n}{2}(39 + 7n)$

$\frac{n}{2}(39 + 7n) > 400$ **[1]**

$7n^2 + 39n > 800$

Solving for n: $7n^2 + 39n - 800 = 0$

$n = \frac{-39 \pm \sqrt{39^2 - (4 \times 7 \times -800)}}{2 \times 7}$ **[1]** ≈ 8.26 **[1]**

Ignore the negative solution.

After nine terms, the sum will be greater than 400.

2. a) It is an arithmetic sequence because she is adding on nine sticks with each pattern. **[1]**

b) $u_n = 6 + 9(n - 1)$ **[1]**

$u_{20} = 6 + 9(20 - 1) = 177$ **[1]**

c) $S_{20} = \frac{20}{2}(6 + 177) = 1830$ **[1]**

3. $S_{100} = 2 + 4 + 6 + \ldots + 196 + 198 + 200$

$+\,S_{100} = 200 + 198 + 196 + \ldots + 6 + 4 + 2$

$2S_{100} = 202 + 202 + 202 + \ldots + 202 + 202 + 202$

[1 for setting up equations]

$2S_{100} = 202 \times 100$ [1]
$S_{100} = 10\,100$ [1]

Page 50: Geometric Sequences and Series

1. a) Substitute $n = 1$, $n = 2$, $n = 3$, $n = 4$ and $n = 5$ to get 5, 1, $\frac{1}{5}, \frac{1}{25}, \frac{1}{125}$ [1]

b) $S_{10} = \dfrac{5\left(\left(\frac{1}{5}\right)^{10} - 1\right)}{\left(\frac{1}{5} - 1\right)} = \dfrac{2\,441\,406}{390\,625}$ **[1 for correct substitution into correct formula; 1 for answer]**

c) Convergent ($|r| < 1$) [1]

$S_\infty = \dfrac{a}{1-r} = \dfrac{5}{1 - \frac{1}{5}} = \dfrac{25}{4}$ (= 6.25) [1]

2. a) $a = 100$, to find r, $u_1 = 100$, substitute into $\frac{25}{4} = 100 \times r^4$

$\frac{1}{16} = r^4 \Rightarrow r = \frac{1}{2}$ [1]

$u_n = 100 \times \left(\frac{1}{2}\right)^{n-1}$ [1]

b) $100 \times \left(\frac{1}{2}\right)^{n-1} < \frac{1}{4}$

$\left(\frac{1}{2}\right)^{n-1} < \frac{1}{400}$ [1]

$(n-1)\left(\log_{10}\left(\frac{1}{2}\right)\right) < \log_{10}\left(\frac{1}{400}\right)$

$n > \dfrac{\log_{10}\left(\frac{1}{400}\right)}{\log_{10}\left(\frac{1}{2}\right)} + 1$ [1]

$n > 9.64$

The first term that is less than $\frac{1}{4}$ will be the 10th term. [1]

c) $S_n = \dfrac{a(r^n - 1)}{(r-1)} = \dfrac{100\left(\left(\frac{1}{2}\right)^3 - 1\right)}{\left(\frac{1}{2} - 1\right)} = 175$ **[1 for correct substitution into correct formula; 1 for answer]**

d) $S_\infty = \dfrac{100}{1 - \frac{1}{2}} = 200$ [1]

3. a) $a = 180\,000$

$85\,000 = 180\,000 \times 0.92^{n-1}$ [1]

Remember to multiply by 92% to find a decrease of 8%.

$\frac{17}{36} = 0.92^{n-1}$

$\log_{10}\left(\frac{17}{36}\right) = (n-1)(\log_{10}(0.92))$

$n = \dfrac{\log_{10}(0.472)}{\log_{10}(0.92)} + 1$ **[1]** = 10 years [1]

b) $180\,000 \times 0.92^{n-1} < 1$ [1]

Since log(0) is undefined, find when the population will be less than 1.

$0.92^{n-1} < \dfrac{1}{180\,000}$

$((n-1)\log_{10}(0.92)) < \log_{10}\left(\dfrac{1}{180\,000}\right)$

$n > \dfrac{\log_{10}\left(\frac{1}{180\,000}\right)}{\log_{10}(0.92)} + 1$ [1]

$n > 146.12$ [1]

After 147 years, the population will be extinct.

Page 51: Binomial Sequences

1. Rewriting in index form: $\dfrac{1}{\sqrt{x+4}} = (x+4)^{-\frac{1}{2}}$

$(x+4)^{-\frac{1}{2}} = 4^{-\frac{1}{2}}\left(1 + \dfrac{-\frac{1}{2} \times \frac{1}{4}x}{1!} + \dfrac{-\frac{1}{2}\left(-\frac{1}{2} - 1\right)\left(\frac{1}{4}x\right)^2}{2!} + \dfrac{-\frac{1}{2}\left(-\frac{1}{2} - 1\right)\left(-\frac{1}{2} - 2\right)\left(\frac{1}{4}x\right)^3}{3!} + \ldots\right)$ [1]

$= \frac{1}{2}\left(1 - \frac{x}{8} + \frac{3x^2}{128} - \frac{5x^3}{1024} + \cdots\right)$ **[1]** $= \frac{1}{2} - \frac{x}{16} + \frac{3x^2}{256} - \frac{5x^3}{2048} + \cdots$ [1]

Valid for $|x| < 4$ [1]

2. Rewriting in index form: $(kx - 5)^{-1}$

$(-5 + kx)^{-1} = -5^{-1}\left(1 + \ldots + \dfrac{-1(-1-1)(-1-2)\left(\frac{-k}{5}x\right)^3}{3!} + \ldots\right)$ [1]

The x^3 term in the expansion of $(-5 + kx)^{-1}$ is

$\left(-\frac{1}{5}\right)\left(\dfrac{-1(-1-1)(-1-2)\left(\frac{-k}{5}x\right)^3}{3!}\right) = -\dfrac{k^3x^3}{625}$ [1]

To find the value of k: $-\dfrac{k^3}{625} = -\dfrac{8}{625} \Rightarrow k = 2$ [1]

3. Rewriting in index form: $\sqrt[4]{1+3x} = (1+3x)^{\frac{1}{4}}$

$(1+3x)^{\frac{1}{4}} = 1 + \dfrac{\frac{1}{4} \times 3x}{1!} + \dfrac{\frac{1}{4}\left(\frac{1}{4} - 1\right)(3x)^2}{2!} + \dfrac{\frac{1}{4}\left(\frac{1}{4} - 1\right)\left(\frac{1}{4} - 2\right)(3x)^3}{3!} + \cdots$ [1]

$= 1 + \dfrac{3x}{4} - \dfrac{27x^2}{32} + \dfrac{189x^3}{128}$ [1]

Valid for $|x| < \frac{1}{3}$ [1]

Pages 52–65 Revise Questions

Page 53 Quick Test

1. $-6\sin(6x - 2)$
2. $20e^{4x}$
3. $\dfrac{3}{2x}$
4. -54

Page 55 Quick Test

1. $\dfrac{dy}{dx} = \dfrac{10x - 7y}{7x - 12y}$
2. $26(\ln 2)y - 260(\ln 2) + x - 1 = 0$
3. $\dfrac{dy}{dx} = 24x^2(x^2 + 2) + 6(x^2 + 2)^2$

Page 57 Quick Test

1. $y = 3x - 1$
2. $\dfrac{dr}{dt} = \dfrac{k}{r}$

Page 59 Quick Test

1. $\frac{1}{3}\ln(x) + c$
2. $\frac{1}{3}e^{3x} + \frac{1}{5}\sin 5x + c$
3. $\frac{1}{32}(1 + 4x)^8 + c$
4. $\frac{-1}{16}\cos 8x + c$

Page 61 Quick Test

1. $\frac{1}{12}(x^3 + 3)^4 + c$
2. 1
3. $\frac{x^3}{3}\ln x - \frac{x^3}{9} + c$

Page 63 Quick Test

1. $\frac{1}{16}(2x + 6)^8 + c$
2. $\ln(2x^3 - 6) + c$
3. $\frac{7}{6}\ln(3x - 3) + \frac{4}{3}\ln(x + 3) + c$

Page 65 Quick Test

1. $y = \ln(2e^x + A)$
2. $\tan y = \frac{x^2}{2} + c$
3. $-\cot y = \tan x - 1 - \frac{1}{\sqrt{3}}$

Pages 66–69 Practice Questions

Page 66: Differentiating Exponentials, Logarithms and Trigonometric Ratios

1. a) $2\cos 2x$ [1]

b) $3 \times (1 + 6x^2)^2 \times 12x$ **[1]** $= 36x(1 + 6x^2)^2$ **[1]**

c) $5 \times \frac{1}{2x} \times 2$ **[1]** $= \frac{5}{x}$ **[1]**

2. $\dfrac{dy}{dx} = 3x^2 + 4x + 5$ [1]

$3x^2 + 4x + 5 = 0$ **[1]** $(3x + 1)(x + 1) = 0$, $x = \frac{-1}{3}$, $x = -1$ **[1]**

$\dfrac{d^2y}{dx^2} = 6x + 4$ **[1]**; at $x = \frac{-1}{3}$ +ve ∴ minimum; at $x = -1$ –ve ∴ maximum **[1]**

$\left(\frac{-1}{3}, \frac{-4}{27}\right)$ minimum, (–1, 0) maximum **[1]**

Remember to use the original equation to find the y-coordinate.

Page 66: Product Rule, Quotient Rule and Implicit Differentiation

1. a) $u = x^2$, $v = \ln x$

$\frac{du}{dx} = 2x$, $\frac{dv}{dx} = \frac{1}{x}$

$x^2 \times \frac{1}{x} + 2x \times \ln x$ **[1]** $= x + 2x\ln x$ **[1]**

b) $u = (1 + 2x)^2$, $v = x^3$

$\frac{du}{dx} = 4(1 + 2x)$, $\frac{dv}{dx} = 3x^2$

$\frac{x^3 \times 4(1 + 2x) - 3x^2 \times (1 + 2x)^2}{x^6}$ **[1]**

$\frac{-(2x + 1)(2x + 3)}{x^4}$ **[1]**

c) $y = \arcsin x$, $\sin y = x$ **[1]**

$\cos y \times \frac{dy}{dx} = 1$ **[1]**

$\frac{dy}{dx} = \frac{1}{\cos y} = \frac{1}{\sqrt{1 - x^2}}$ **[1]**

Remember $\sin^2 x + \cos^2 x = 1$

2. $y = 3^t$

$\ln y = \ln(3^t)$ **[1]**

$\ln y = t\ln 3$

$\frac{1}{y} \times \frac{dy}{dt} = \ln 3$, $\frac{dy}{dt} = y(\ln 3)$ **[1]**

$\frac{dy}{dt} = 3^t \ln 3$ **[1]**

3. a) $u = e^{3x}$, $v = \sin x$

$\frac{du}{dx} = 3e^{3x}$, $\frac{dv}{dx} = \cos x$ **[1]**

$\frac{dy}{dx} = e^{3x} \times \cos x + 3e^{3x} \times \sin x$ **[1]**

$\frac{dy}{dx} = e^{3x} (\cos x + 3\sin x)$ **[1]**

b) $e^{3x}(\cos x + 3\sin x) = 0$ **[1]**

$\cos x + 3\sin x = 0$ **[1]** $e^{3x} \neq 0$ **[1]**

$\tan x = -3$ ∴ shown **[1]**

c) $e^{3\left(\frac{\pi}{2}\right)}\left(\cos\frac{\pi}{2} + 3\sin\frac{\pi}{2}\right) = 3e^{\frac{3\pi}{2}}$ $m = \frac{-1}{3e^{\frac{3\pi}{2}}}$ **[1]**

$y - e^{\frac{3\pi}{2}} = \frac{-1}{3e^{\frac{3\pi}{2}}}\left(x - \frac{\pi}{2}\right)$ **[1]**

$y - e^{\frac{3\pi}{2}} = \frac{-1}{3}e^{\frac{-3\pi}{2}}\left(x - \frac{\pi}{2}\right)$ **[1]**

4. $18x^2$ **[1]** $3x \times \frac{dy}{dx} + 3y$ **[1]**

Remember to use the product rule.

$18x^2 + 3x\frac{dy}{dx} + 3y + 2x = 2$ **[1]**

$\frac{dy}{dx} = \frac{2 - 2x - 3y - 18x^2}{3x}$ **[1]**

Page 67: Differentiating Parametric Equations and Differential Equations

1. a) $t^3 + 8 = 0$, $t = \sqrt[3]{-8}$ **[1]**

$x = (1 + (-2))^2 = 1$ **[1]**

b) $\frac{dx}{dt} = 2(1 + t)$ $\frac{dy}{dt} = 3t^2$ **[1]**

$\frac{dy}{dx} = \frac{3t^2}{2(1 + t)}$ **[2]**

c) At A, $t = -2$, ∴ $m = -6$ **[1]**

$y - 0 = -6(x - 1)$ **[1]**

$y + 6x - 6 = 0$ **[1]**

2. a) $\frac{dx}{dy} = \ln(2y) \times 3y^2 + y^3 \times \frac{1}{y}$ **[1]**

$\frac{dx}{dy} = 3y^2 \ln(2y) + y^2$ **[1]**

b) $3(e^2)^2 \ln(2e^2) + (e^2)^2$ **[1]**

$3e^4 \ln(2e^2) + e^4$ **[1]** $e^4\left(3\ln\left(2e^2\right) + 1\right)$ **[1]**

Page 68: Standard Integrals and Definite Integrals

1. a) $\frac{\left(x^2 + 7\right)^4}{8}$ **[1]** $\frac{\left(6^2 + 7\right)^4}{8} - \frac{\left(3^2 + 7\right)^4}{8}$ **[1]** $= \frac{3353265}{8}$ **[1]**

b) $\sin^2 x = \frac{1}{2} - \frac{1}{2}\cos 2x$ **[1]**

Use double angle rule.

$\frac{1}{2}x + \frac{1}{4}\cos 2x$ **[1]**

$\left(\frac{1}{2}\pi + \frac{1}{4}\cos 2\pi\right) - \left(\frac{1}{4}\cos 0\right)$ **[1]**

$\frac{\pi}{2}$ **[1]**

c) $\frac{1}{2}\ln(2x + 1)$ **[1]** $\frac{1}{2}\ln(2(5) + 1) - \frac{1}{2}\ln(2(2) + 1)$ **[1]**

$= \frac{1}{2}(\ln 11 - \ln 5)$ **[1]**

d) $\tan x$ **[1]** $\tan\frac{\pi}{3} - \tan\frac{\pi}{6}$ **[1]** $= \frac{2\sqrt{3}}{3}$ **[1]**

Page 68: Further Integration 1 & 2

1. a) $A(x - 1)^2 + Bx(x - 1) + cx = 5x^2 - 8x + 1$ **[1]**

When $x = 1$, $0 + 0 + c = -2$, $c = -2$

Equating terms:

$A + B = 5$ and $A = 1$, therefore $B = 4$ **[1]**

$\frac{1}{x} + \frac{4}{x - 1} - \frac{2}{(x - 1)^2}$ **[1]**

b) $\ln(x)$ **[1]** $+ 4\ln(x - 1) + 2(x - 1)^{-1} + c$ **[1]**

$\frac{\ln(x(x - 1)^4 + 2}{(x - 1)} + c$ **[1]**

2. $\frac{du}{dx} = -2x$ **[1]**

When $x = 3$, $u = 7$, and when $x = 0$, $u = 16$ **[1]**

$\int_{16}^{7} \frac{x}{\sqrt{u}} \times \frac{du}{-2x} = \int_{16}^{7} \frac{-1}{2\sqrt{u}}\, du$ **[1]**

$\left[-u^{\frac{1}{2}}\right]_{16}^{7}$ **[1]**

$4 - \sqrt{7}$ **[1]**

3. $xe^x(3 - x) = 0$

$x = 0$, $x = 3$ $e^x = 0$ has no solutions. **[1]**

$\int_1^3 3xe^x - x^2e^x\, dx$

$= (3xe^x - 3e^x)$ **[1]** $- (x^2e^x - 2xe^x + 2e^x)$ **[1]** Use integration by parts.

$e^x(-x^2 + 5x - 5)$ **[1]**

$(e^3(-(3)^2 + 15 - 5)) - e^0(-(0)^2 + 0 - 5))$ **[1]**

$e^3 + 5$ **[1]**

4. a) $u = x$, $\frac{dv}{dx} = \sin x$

$\frac{du}{dx} = 1$, $v = -\cos x$ **[1]**

$-x\cos x - \int -\cos x\, dx$

$-x\cos x + \sin x$ **[1]**

$(-\pi\cos\pi + \sin\pi) - (-0\cos 0 + \sin 0)$ **[1]** $= \pi$ **[1]**

b) $u = x^2$, $\frac{dv}{dx} = \cos x$

$\frac{du}{dx} = 2x$, $v = \sin x$ **[1]**

$-x^2\sin x - \int 2x\sin x\, dx$ **[2]**

$-\pi^2\sin\pi - 0 - 2\pi = -2\pi$ **[1]**

Page 69: Differential Equations

1. a) $u = x$, $\frac{dv}{dx} = e^{-2x}$

$\frac{du}{dx} = 1, v = -\frac{1}{2}e^{-2x}$ **[1]**

$-\frac{1}{2}xe^{-2x} - \int -\frac{1}{2}e^{-2x}dx$ **[2]**

$-\frac{1}{2}xe^{-2x} - \frac{1}{4}e^{-2x} + c$ **[1]**

b) $\int \sec^2 y\, dy = \int xe^{-2x}\, dx$ **[1]**

$\tan y = -\frac{1}{2}xe^{-2x} - \frac{1}{4}e^{-2x} + c$ **[1]**

$\tan\frac{\pi}{4} = -\frac{1}{2}(0)e^{-2(0)} - \frac{1}{4}e^{-2(0)} + c$ [1]

$1 = -\frac{1}{4} + c,\ c = \frac{5}{4}$ [1]

Pages 70–73 Review Questions

Page 70: Radians, Arc Lengths and Areas of Sectors

1. $A = \frac{1}{2}r^2\theta \Rightarrow 100 = \frac{1}{2} \times r^2 \times \frac{2\pi}{3} \Rightarrow r = 9.772$ [2]

   $P = 2r + r\theta = 2 \times 9.772 + 9.772 \times \frac{2\pi}{3} = 40.0\text{cm}$ [2]

2. $\frac{1}{2}r^2 \times \frac{\pi}{6} = 75 \Rightarrow r = 16.93\text{cm}$ [2]

   Using cosine rule: $AB^2 = 16.93^2 + 16.93^2 - 2 \times 16.93^2 \times \cos\frac{\pi}{6}$ [1]
   AB = 8.76 cm [1]

3. $P = 2A \Rightarrow 2r + r\theta = r^2\theta$ [1]
   $2 + \theta = r\theta$

   $\Rightarrow r = \frac{2+\theta}{\theta} = \frac{2 + \frac{5\pi}{6}}{\frac{5\pi}{6}} = 1.764$ [2]

   $\Rightarrow 2r + r\theta = 2 \times 1.764 + 1.764 \times \frac{5\pi}{6} = 8.15\text{cm}$ [2]

Page 70: Vectors in 3D

1. $\sqrt{(4-2)^2 + (10-4)^2 + (6-8)^2} = \sqrt{2^2 + 6^2 + 2^2}$ [2]

   $= \sqrt{44} = 6.63$ [1]

2. $3\mathbf{a} - 2\mathbf{b} = 3\begin{pmatrix}2\\3\\-8\end{pmatrix} - 2\begin{pmatrix}4\\0\\2\end{pmatrix} = \begin{pmatrix}6\\9\\-24\end{pmatrix} - \begin{pmatrix}8\\0\\4\end{pmatrix} = \begin{pmatrix}-2\\9\\-28\end{pmatrix}$ [2]

   $|3\mathbf{a} - 2\mathbf{b}| = \sqrt{(-2)^2 + 9^2 + (-28)^2} = \sqrt{869} = 29.5$ [2]

3. Use of $\overline{OR} = \frac{n\mathbf{p} + m\mathbf{q}}{m+n}$ [1]

   $\overline{OR} = \frac{5\mathbf{p} + 3\mathbf{q}}{8}$ [1]

   $= \frac{5\begin{pmatrix}4\\16\\20\end{pmatrix} + 3\begin{pmatrix}-4\\-8\\-12\end{pmatrix}}{8} = \frac{\begin{pmatrix}20\\80\\100\end{pmatrix} + \begin{pmatrix}-12\\-24\\-36\end{pmatrix}}{8} = \frac{\begin{pmatrix}8\\56\\64\end{pmatrix}}{8}$ [1]

   $= \begin{pmatrix}1\\7\\8\end{pmatrix}$ [1]

Page 71: Trigonometric Ratios

1. a) $\cot\frac{11\pi}{6} = \cot\left(2\pi - \frac{\pi}{6}\right) = \frac{1}{\tan\left(2\pi - \frac{\pi}{6}\right)} = -\frac{1}{\tan\left(\frac{\pi}{6}\right)}$

   $= -\frac{1}{\frac{1}{\sqrt{3}}} = -\sqrt{3}$ [2]

   b) $\sec\frac{11\pi}{4} = \sec\left(2\pi + \pi - \frac{\pi}{4}\right) = \sec\left(\pi - \frac{\pi}{4}\right)$

   $= \frac{1}{\cos\left(\pi - \frac{\pi}{4}\right)} = -\frac{1}{\cos\frac{\pi}{4}} = -\frac{1}{\frac{1}{\sqrt{2}}} = -\sqrt{2}$ [2]

   c) $\operatorname{cosec}\frac{5\pi}{3} = \operatorname{cosec}\left(2\pi - \frac{\pi}{3}\right) = \frac{1}{\sin\left(2\pi - \frac{\pi}{3}\right)}$

   $= -\frac{1}{\sin\frac{\pi}{3}} = -\frac{1}{\frac{\sqrt{3}}{2}} = -\frac{2}{\sqrt{3}} = -\frac{2\sqrt{3}}{3}$ [2]

2. $\frac{x - x\cos 2x}{\sin^3 x} \approx \frac{x - x\left(1 - \frac{(2x)^2}{2}\right)}{x^3}$ [1] $= \frac{x - x + 2x^3}{x^3}$ [1] $= 2$ [1]

3. a) Domain is $x \in \mathbb{R},\ x \neq \frac{n\pi}{2}$ [2]
   Range is $y \in \mathbb{R}$ [1]

   b) Domain $-2 \leqslant x \leqslant 2$ [1]

   Range $-\frac{\pi}{2} \leqslant y \leqslant \frac{\pi}{2}$ [1]

   c) Domain is $x \in \mathbb{R},\ x \neq (2n+1)\frac{\pi}{2}$ [2]
   Range is $y \in \mathbb{R},\ y \leqslant -1$ or $y \geqslant 1$ [2]

Page 72: Further Trigonometric Equations and Identities

1. $\frac{(\sin x + \cos x)^2}{\cos x} = \frac{\sin^2 x + \cos^2 x + 2\sin x\cos x}{\cos x}$ [1]

   $= \frac{1 + 2\sin x\cos x}{\cos x}$ [1]

   $= \frac{1}{\cos x} + \frac{2\sin x\cos x}{\cos x}$ [1]

   $= \sec x + 2\sin x$ [1]

2. $\cot x(\operatorname{cosec} x - 2) = 0$ [1]

   $\cot x = 0 \Rightarrow \tan x$ undefined $\Rightarrow x = \frac{\pi}{2}$ or $x = \frac{3\pi}{2}$ [2]

   $\operatorname{cosec} x = 2 \Rightarrow \sin x = \frac{1}{2}$

   $\Rightarrow x = \frac{\pi}{6}$ or $x = \frac{5\pi}{6}$ [2]

3. $(\cot x + \operatorname{cosec} x)^2 = \cot^2 x + \operatorname{cosec}^2 x + 2\cot x\operatorname{cosec} x$ [1]

   $= \frac{\cos^2 x}{\sin^2 x} + \frac{1}{\sin^2 x} + 2\left(\frac{\cos x}{\sin x}\right)\left(\frac{1}{\sin x}\right)$ [1]

   $= \frac{\cos^2 x + 1 + 2\cos x}{\sin^2 x}$ [1]

   $= \frac{\cos^2 x + 1 + 2\cos x}{1 - \cos^2 x}$ [1]

   $= \frac{(1 + \cos x)^2}{1 - \cos^2 x}$ [1]

   $= \frac{(1 + \cos x)^2}{(1 + \cos x)(1 - \cos x)}$ [1]

   $= \frac{1 + \cos x}{1 - \cos x}$

4. $\sec 2x = -3 \Rightarrow \cos 2x = -\frac{1}{3}$ [1]

   $\arccos\left(\frac{1}{3}\right) = 1.231$ [1]

   $2x = \pi - 1.231 \Rightarrow x = 0.955$ [1]
   $2x = \pi + 1.231 \Rightarrow x = 2.186$ [1]
   $2x = 3\pi - 1.231 \Rightarrow x = 4.10$ [1]
   $2x = 3\pi + 1.231 \Rightarrow x = 5.33$ [1]

Page 73: Using Addition Formulae and Double Angle Formulae

1. $\cos 210° = -\frac{\sqrt{3}}{2}$ [1]

   $\cos x = 2\cos^2\left(\frac{x}{2}\right) - 1$ [1]

   $\pm\sqrt{\frac{1 + \cos x}{2}} = \cos\left(\frac{x}{2}\right)$

   Substitute $x = 210°$

   $\cos 105° = \pm\sqrt{\frac{1 - \frac{\sqrt{3}}{2}}{2}}$ [1]

   $= \pm\sqrt{\frac{2 - \sqrt{3}}{4}}$ [1]

   $= \frac{\pm\sqrt{2 - \sqrt{3}}}{2}$ [1]

   $= -\frac{\sqrt{2 - \sqrt{3}}}{2}$ since cos105° lies in the second quadrant. [1]

2. a) $\cos x = 1 - 2\sin^2\left(\frac{x}{2}\right) \Rightarrow \sin\left(\frac{x}{2}\right) = \pm\sqrt{\frac{1 - \cos x}{2}}$

   $\sin\left(\frac{x}{2}\right) = \pm\sqrt{\frac{1 - \left(-\frac{1}{3}\right)}{2}} = \pm\sqrt{\frac{2}{3}}$ [2]

   $\pi < x < \frac{3\pi}{2} \Rightarrow \sin\left(\frac{x}{2}\right) = -\sqrt{\frac{2}{3}}$ [1]

   b) $\cos x = 2\cos^2\left(\frac{x}{2}\right) - 1 \Rightarrow \cos\left(\frac{x}{2}\right) = \pm\sqrt{\frac{1 + \cos x}{2}}$

   $\cos\left(\frac{x}{2}\right) = \pm\sqrt{\frac{1 + \left(-\frac{1}{3}\right)}{2}} = \pm\frac{1}{\sqrt{3}}$ [2]

   $\pi < x < \frac{3\pi}{2} \Rightarrow \cos\left(\frac{x}{2}\right) = -\frac{1}{\sqrt{3}} = -\frac{\sqrt{3}}{3}$ [1]

3. $\sin\left(\frac{x}{2}\right) - 2\sin\left(\frac{x}{2}\right)\cos\left(\frac{x}{2}\right) = 0$ [1]

$\sin\left(\frac{x}{2}\right)\left(1 - 2\cos\left(\frac{x}{2}\right)\right) = 0$ [1]

$\sin\left(\frac{x}{2}\right) = 0 \Rightarrow \frac{x}{2} = 0 \Rightarrow x = 0$ [2]

$\cos\left(\frac{x}{2}\right) = \frac{1}{2} \Rightarrow \frac{x}{2} = \frac{\pi}{3} \Rightarrow x = \frac{2\pi}{3}$ [2]

Page 73: Problem Solving Using Trigonometry

1. $6\sin x - 4\cos x = R\sin(x - \alpha)$
$= R\sin x\cos\alpha - R\cos x\sin\alpha$
$6 = R\cos\alpha$ [1]
$4 = R\sin\alpha$ [1]
$R^2 = 6^2 + 4^2 = 52 \Rightarrow R = \sqrt{52} = 2\sqrt{13}$ [2]
$\frac{R\sin\alpha}{R\cos\alpha} = \tan\alpha = \frac{2}{3} \Rightarrow \alpha = 0.588^c$ [2]
$6\sin x - 4\cos x = 2\sqrt{13}\sin\left(x - 0.588^c\right)$

2. $\cos x - 2\sin x = R\cos(x + \alpha)$
$= R\cos x\cos\alpha - R\sin x\sin\alpha$
$1 = R\cos\alpha$ [1]
$2 = R\sin\alpha$ [1]
$R^2 = 1^2 + 2^2 = 5 \Rightarrow R = \sqrt{5}$ [2]
$\frac{R\sin\alpha}{R\cos\alpha} = \tan\alpha = 2 \Rightarrow \alpha = 1.107^c$ [2]
$\cos x - 2\sin x = \sqrt{5}\cos\left(x - 1.107^c\right)$
$\cos x - 2\sin x = \sqrt{2}$
$\sqrt{5}\cos\left(x + 1.107^c\right) = \sqrt{2}$
$\cos\left(x + 1.107^c\right) = \sqrt{\frac{2}{5}}$ [1]
$x + 1.107 = 2\pi - 0.886 \Rightarrow x = 4.29$ [2]
$x + 1.107 = 2\pi + 0.886 \Rightarrow x = 6.06$ [2]

3. From question 2, $f(x) = 3 - (\cos x - 2\sin x)$
$= 3 - \sqrt{5}\cos\left(x - 1.107^c\right)$ [1]
Minimum value is $3 - \sqrt{5}(= 0.764)$ [1]
Maximum value is $3 - \left(-\sqrt{5}\right)(= 5.24)$ [1]
Maximum occurs when $\cos(x + 1.107^c) = -1$,
i.e. $x + 1.107^c = \pi \Rightarrow x = 2.03^c$ [2]

Pages 74–79 Revise Questions

Page 75 Quick Test

1. Using $x_{n+1} = e^{-x_n}$, $\alpha = 0.567$
2. $-3 < x < -2$, $0 < x < 1$, $2 < x < 3$
3. 0.90

Page 77 Quick Test

1. 3.46
2. $x = 0$ is a stationary (minimum) point on the curve
3. 3.98. Underestimate

Page 79 Quick Test

1. $\operatorname{cosec}\left(\theta + \frac{\pi}{2}\right) = \frac{1}{\sin\left(\theta + \frac{\pi}{2}\right)} = \frac{1}{\sin\theta\cos\frac{\pi}{2} + \cos\theta\sin\frac{\pi}{2}}$
$= \frac{1}{\sin\theta \times 0 + \cos\theta \times 1} = \frac{1}{\cos\theta} = \sec\theta$
2. $2\cot\theta\sin^2\theta = 2\left(\frac{\cos\theta}{\sin\theta}\right)\sin^2\theta = 2\sin\theta\cos\theta = \sin 2\theta$
3. Supposing p^2 is even implies p is odd. So you can write p in the form $p = 2r + 1$.
Now $p^2 = (2r + 1)^2 = 4r^2 + 2r + 1 = 2(2r^2 + r) + 1$, which is odd, thus contradicting the fact that p^2 is even. Therefore the initial assumption is incorrect, and so p must be even.

Pages 80–81 Practice Questions

Page 80: Iterative Methods and Finding Roots

1. $e^{\sin 3} - \ln 3 = 0.0530 > 0$ [2]
$e^{\sin 3.1} - \ln 3.1 = -0.0889 < 0$ [2]
The function is continuous in the interval $3.0 < \alpha < 3.1$ and there is a sign change, so there exists a root α such that $3.0 < \alpha < 3.1$ [1]

2. $f(1.5) = -2.44 < 0$ [2]
$f(2) = 5 > 0$ [2]
f is continuous in the interval $1.5 < \alpha < 2$ and there is a sign change, so there exists a root α such that $1.5 < \alpha < 2$ [1]
$x_1 = 1.821$ [1]
$x_2 = 1.767$
$x_3 = 1.750$
$x_4 = 1.744$ [1]
$\alpha = 1.74$ [1]

3. $f(1) = -5 < 0$ [2]
$f(1.5) = 7.59 > 0$ [2]
f is continuous in the interval $1 < \alpha < 1.5$ and there is a sign change, so there exists a root α such that $1 < \alpha < 1.5$. [1]
$\alpha = 1.358$ [2]

Page 80: Numerical Integration and Solving Problems in Context

1. $h = \frac{\frac{\pi}{2} - 0}{4} = \frac{\pi}{8}$ [1]

x	0	$\frac{\pi}{8}$	$\frac{\pi}{4}$	$\frac{3\pi}{8}$	$\frac{\pi}{2}$
$e^x \sin x$	0	0.5667	1.551	3.001	0

[2]

$\int_0^{\frac{\pi}{2}} e^x \sin x \, dx \approx \frac{\left(\frac{\pi}{8}\right)}{2}[0 + 2(0.5667 + 1.551 + 3.001) + 0]$ [2]
$= 2.01$ units² [1]

2. $f'(x) = 2x$ [1]
$x_{n+1} = x_n - \left(\frac{x^2 - 3}{2x}\right)$ [1]
$x_1 = 1.73235...$ [1]
$x_2 = 1.732...$ [1]
$\sqrt{3} = 1.732$ [1]

3. $h = \frac{8 - 2}{6} = 1$ [1]

x	2	3	4	5	6	7	8
$(\ln x)^2$	0.480	1.207	1.922	2.590	3.210	3.787	4.324

[2]

$\int_2^8 (\ln x)^2 \, dx \approx$
$\frac{1}{2}[0.480 + 2(1.207 + 1.922 + 2.590 + 3.210 + 3.787) + 4.324]$ [2]
$= 15.2$ units² [1]

Page 81: Proof by Contradiction and Further Trigonometric Identities

1. $\sin\left(x + \frac{\pi}{6}\right)\cos\left(x + \frac{\pi}{6}\right)$
$= \left(\sin x\cos\frac{\pi}{6} + \cos x\sin\frac{\pi}{6}\right)\left(\cos x\cos\frac{\pi}{6} - \sin x\sin\frac{\pi}{6}\right)$ [1]
$= \left(\sin x\left(\frac{\sqrt{3}}{2}\right) + \cos x\left(\frac{1}{2}\right)\right)\left(\cos x\left(\frac{\sqrt{3}}{2}\right) + \sin x\left(\frac{1}{2}\right)\right)$ [1]
$= \sin x\cos x\left(\frac{3}{4}\right) - \sin^2 x\left(\frac{\sqrt{3}}{4}\right) + \cos^2 x\left(\frac{\sqrt{3}}{4}\right) - \sin x\cos x\left(\frac{1}{4}\right)$ [1]
$= \frac{1}{2}\sin x\cos x + \frac{\sqrt{3}}{4}(\cos^2 x - \sin^2 x)$ [1]
$= \frac{1}{4}\sin 2x + \frac{\sqrt{3}}{4}\cos 2x$ [2]
$= \frac{\sin 2x + \sqrt{3}\cos 2x}{4}$

2. $\sin x \sin 2x = \sin x (2\sin x \cos x)$ [1]
$= 2\sin^2 x \cos x$ [1]
$= 2\cos x(1 - \cos^2 x)$ [1]
$= 2\cos x - 2\cos^3 x$ [1]

3. Assume that $a^2 + b^2 < 2ab$ [1]
Then $a^2 - 2ab + b^2 < 0$ [1]
Or $(a - b)^2 < 0$ [1]
This is a contradiction, since any squared number is equal to or greater than zero. [1]
Therefore $a^2 + b^2 \geqslant 2ab$.

Pages 82–85 Review Questions

Page 82: Differentiating Exponentials, Logarithms and Trigonometric Ratios

1. a) $u = x$, $v = \ln(x)$

$\frac{du}{dx} = 1 \qquad \frac{dv}{dx} = \frac{1}{x}$ [1]

$\frac{dy}{dx} = x \times \frac{1}{x} + \ln x$ [1]; $1 + \ln x$ [1]

b) $1 + \ln x = 0$ [1]; $\ln x = -1$ [1]; $x = e^{-1}$, $y = -e^{-1}$ [1]

c) $\frac{d^2y}{dx^2} = \frac{1}{x}$ [1]; $\frac{1}{e^{-1}}$ positive, therefore minimum [1]

Page 82: Product Rule, Quotient Rule and Implicit Differentiation

1. a) $u = x^2$, $v = \ln(3x)$ [1]

$\frac{du}{dx} = 2x \qquad \frac{dv}{dx} = \frac{1}{x}$ [1]

$\frac{dy}{dx} = x^2 \times \frac{1}{x} + 2x \times \ln 3x$ [1]

$x + 2x\ln 3x$ [1]

b) $u = \sin 4x$, $v = x^3$ [1]

$\frac{du}{dx} = 4\cos 4x \qquad \frac{dv}{dx} = 3x^2$ [1]

$\frac{x^3 \times 4\cos 4x - 3x^2 \times \sin 4x}{x^6}$ [1]

$\frac{4x\cos 4x - 3\sin 4x}{x^4}$ [1]

2. a) $u = 3 + \sin 2x$, $v = 2 + \cos 2x$ [1]

$\frac{du}{dx} = 2\cos 2x \qquad \frac{dv}{dx} = -2\sin 2x$ [1]

$\frac{2\cos 2x(2 + \cos 2x) + 2\sin 2x(3 + \sin 2x)}{(2 + 2\cos 2x)^2}$ [1]

$\frac{6\sin 2x + 4\cos 2x + 2}{(2 + \cos 2x)^2}$ [1]

Use $\sin^2 x + \cos^2 x = 1$.

b) Where $x = \frac{\pi}{2}, \frac{dy}{dx} = -2$ [1]

Where $x = \frac{\pi}{2}$, $y = 3$ [1]

$y - 3 = -2\left(x - \frac{\pi}{2}\right)$ [1]

$y = -2x + (3 + \pi)$ [1]

3. a) $u = \ln(x^2 + 1)$, $v = x$ [1]

$\frac{du}{dx} = \frac{2x}{x^2 + 1}; \frac{dv}{dx} = 1$ [1]

$\frac{\frac{2x^2}{x^2+1} - \ln\left(x^2 + 1\right)}{x^2}$ [1]

$\frac{2x^2 - \left(x^2 + 1\right)\ln\left(x^2 + 1\right)}{x^2\left(x^2 + 1\right)}$ [1]

b) $\frac{dx}{dy} = \sec^2 y$ [1]

$\frac{dy}{dx} = \frac{1}{\sec^2 y}$ [1]

$1 + \tan^2 y = \sec^2 y$ [1]

$1 + x^2 = \sec^2 y$ [1]

$\frac{dy}{dx} = \frac{1}{1 + x^2}$ [1]

Page 83: Differentiating Parametric Equations and Differential Equations

1. a) 30 [1]

b) $500 = 30e^{\frac{t}{5}}, \qquad e^{\frac{t}{5}} = \frac{50}{3}$ [1]

$t = 5\ln\left(\frac{50}{3}\right) = 14.067....$ 15 years [1]

c) $\frac{dF}{dt} = 6e^{\frac{t}{5}}$ [2 if fully correct; 1 for $e^{\frac{t}{5}}$]

d) $6e^{\frac{t}{5}} = 50$ [1]; $t = 5\ln(\frac{50}{6})$ [1]

$F = 250$ [1]

Page 84: Standard Integrals and Definite Integrals

1. a) $\frac{-1}{9}e^{-9x-3} + c$ [2]

b) $\ln(x^2 + 1) + c$ [2]

c) $\frac{1}{3}\ln(\tan 3x) + c$ [2]

2. $\int_0^{\pi} x + \sin x \, dx$ [1]

$\frac{x^2}{2} - \cos x$ [2]

$(\frac{\pi^2}{2} - \cos\pi) - (0 - \cos 0)$ [1]

$\frac{\pi^2}{2}$ [1]

Page 84: Further Integration 1 & 2

1. a) $A(2x-1) + B(x+2) = 1$ [1]

$A = \frac{-1}{5}; B = \frac{2}{5}$ [1]

$\int \frac{2}{5(2x-1)} - \frac{1}{5(x+2)} dx$ [1]

$\frac{1}{5}\ln(2x - 1) - \frac{1}{5}\ln(x + 2) + c$ [1]

b) $u = x, \frac{dv}{dx} = \operatorname{cosec}^2 x$ [1]

$\frac{du}{dx} = 1, v = -\cot x$ [1]

$-x\cot x - \int -\cot x \, dx$ [1]

$-x\cot x + \ln(\sin x) + c$ [1]

2. a) $A(3 + x) + B = 2x$ [1]

$B = -6, A = 2$ [1]

$\frac{2}{3+x} - \frac{6}{(3+x)^2}$ [1]

b) $\int \frac{2}{3+x} - \frac{6}{(3+x)^2} = 2\ln(3 + x) + 6(3 + x)^{-1}$ [1]

$(2\ln(3 + 3) + 6(3 + 3)^{-1}) - (2\ln(3) + 6(3)^{-1})$ [1]

$\ln 4 - 1$ [1]

Use your log rules to simplify.

3. $\frac{du}{dx} = -\sin x$ [1]

$\frac{du}{-\sin x} = dx$ [1]

$\int e^{\cos x}\sin x \, dx = \int e^u \sin x \times \frac{du}{-\sin x}$ [1]

$\int -e^u du$ [1]

$= -e^u$ [1]

$= -e^{\cos x} + c$ [1]

4. $\frac{-x^2 - 2}{x(x+1)^2} = \frac{A}{x} + \frac{B}{x+1} + \frac{C}{(x+1)^2}$

$A(x + 1)^2 + Bx(x + 1) + Cx = -x^2 - 2$ [1]

$A + B = 1, A = -2, 2A + B + C = 0$ [1] Equate coefficients.

$A = -2, B = 1, C = 3$ [1]

$\int \frac{-2}{x} + \frac{1}{x+1} + \frac{3}{(x+1)^2} dx = -2\ln x + \ln(x + 1) - 3(x + 1)^{-1} + c$ [1]

$\ln\left(\frac{x+1}{x^2}\right) - \frac{3}{x+1} + c$ [1]

Page 85: Differential Equations

1. a) $y\,dy = e^{2x}\,dx$ [1]

$\int y\,dy = \int e^{2x}dx$ [1]

$\frac{y^2}{2} = \frac{1}{2}e^{2x} + c$ [1]

$y^2 = e^{2x} + c$ [1]

b) $9 = 1 + c$ [1] $\Rightarrow c = 8$ [1]

2. $\frac{1}{y(y+2)}dy = 3dx$ [1]

$\int \frac{1}{y(y+2)}dy = \int 3\,dx$ [1]

$\frac{1}{y(y+2)} = \frac{1}{2y} - \frac{1}{2(y+2)}$ **[1]**

Use partial fractions.

$\frac{1}{2}\ln(y) - \frac{1}{2}\ln(y+2) = 3x + c$ **[1]**

Pages 86–97 Revise Questions

Page 87 Quick Test

1. 0.12 2. $\frac{1}{13}$
3. $0.7 \times 0.4 = 0.28$
   $0.7 + 0.4 - 0.82 = 0.28$, therefore shown

Page 89 Quick Test

1. **a)** 74.5 **b)** Reliable as interpolation

Page 91 Quick Test

1. **a)** 0.8413 **b)** 0.9088 2. 86.76 3. 11.5

Page 93 Quick Test

1. $P(X < 14.5)$ 2. B(40, 0.5) 3. 0.162

Page 95 Quick Test

1. Strength of linear correlation between two variables
2. 0.5923 is critical value so evidence of positive correlation
3. −0.3783 is critical value so evidence of negative correlation

Page 97 Quick Test

1. $P(Z > 0.596) = 0.2756$; $0.2756 > 0.025$; greater than significance level so not enough evidence to reject null hypothesis
2. $P(Z < -5.478) = 0$; less than significance level so enough evidence to reject null hypothesis
3. $X > 86.6$ and $X < 83.4$

Pages 98–101 Practice Questions

Page 98: Set Notation and Conditional Probability

1. $P(A \mid B) = \frac{P(A \cap B)}{P(B)} \therefore 0.6 = \frac{P(A \cap B)}{0.7}$; $P(A \cap B) = 0.42$

   By same method $P(A \cap B') = 0.06$

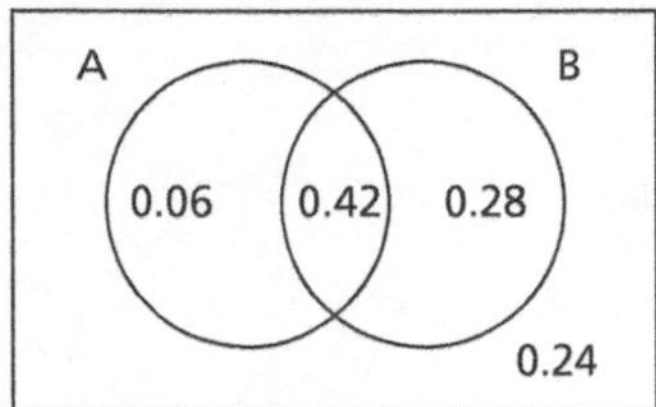

   a) $0.42 + 0.06$ **[1]** $= 0.48$ **[1]**
   b) $0.06 + 0.42 + 0.28$ **[1]** $= 0.76$ **[1]**
   c) $0.06 + 0.42 + 0.24$ **[1]** $= 0.72$ **[1]**
   d) $\frac{0.42}{0.48}$ **[1]** $= 0.875$ **[1]**
   e) $\frac{0.24}{0.3}$ **[1]** $= 0.8$ **[1]**

   Remember $P(A' \mid B') = \frac{P(A' \cap B')}{P(B')}$

2. **a)** $\frac{37}{60}$ **[1]**
   b) $\frac{20}{60} = \frac{1}{3}$ **[1]**
   c) $\frac{12}{12+8}$ **[1]** $= \frac{3}{5}$ **[1]**
   d) $\frac{12}{25+12+8+15}$ **[1]** $= \frac{12}{60} = \frac{1}{5}$ **[1]**

Page 99: Modelling and Data Presentation

1. **a)** $0.0146 \times 47 - 0.0829 = 0.6033$ cm **[1]**
   b) $0.0146 \times 60 - 0.0829 = 0.7931$ cm **[1]**
   c) The first estimate is more reliable **[1]**. This estimate has been interpolated so is inside the data range given. 60 is outside the data range given **[1]**.
2. **a)** $X \sim B(300, \frac{1}{6})$ **[1]**; binomial **[1]**

   $n = 300$ **[1]**; $P = \frac{1}{6}$ **[1]**

   b) The dice is fair **[1]**. The rolls are independent of each other **[1]**.
3. $\ln y = \ln (ka^x)$
   $\ln y = \ln k + x\ln a$ **[1 for both statements]**
   $\ln 18 = \ln k + \ln a$ and $\ln 54 = \ln k + 2\ln a$ **[1]**
   $2\ln 18 = 2\ln k + 2\ln a$
   $2\ln 18 - \ln 54 = \ln k$ **[1]**

   These are simultaneous equations.

   Use log rules.

   $\ln 6 = \ln k$; $k = 6$ **[1]**

Page 100: Statistical Distributions 1 & 2

1. **a)** $P(X < 62) = P\left(Z < \frac{62-60}{4}\right) = P(Z < 0.5)$ **[2]**
   0.6915 **[1]**
   b) $P(X > 54) = P\left(Z > \frac{54-48}{3}\right) = P(Z > 2)$ **[2]**
   0.0228 **[1]**
   c) $P(X < 23) = P\left(Z < \frac{23-30}{5}\right) = P(Z < -1.4)$ **[2]**
   0.0808 **[1]**
2. **a)** $H \sim N(165, 5^2)$ **[1]**
   $(H > 170) = P\left(Z > \frac{170-165}{5}\right) = P(Z > 1)$ **[2]**
   0.1587 **[1]**
   b) $P(H > b) < 0.05$ **[1]**
   From percentage points table $Z = 1.6449$ **[1]**
   $1.6449 = \frac{b-165}{5}$ **[1]**; 173.2 cm **[1]**
3.

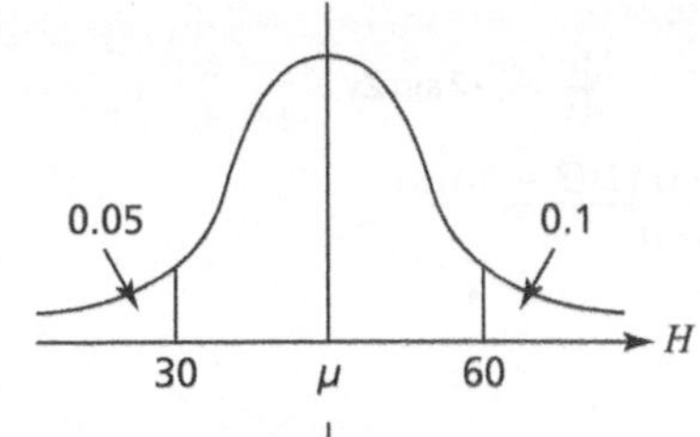

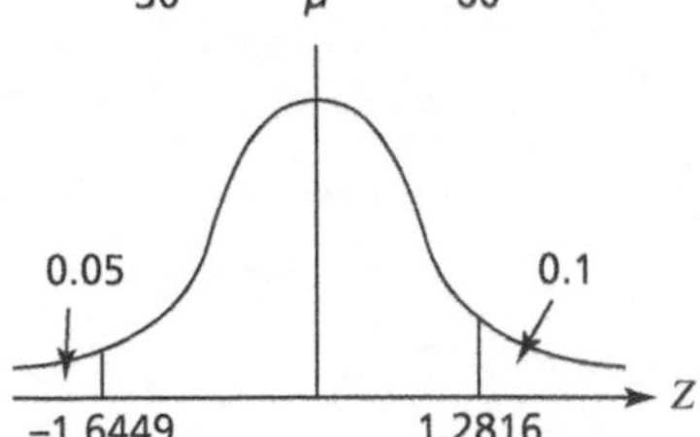

   $Z = -1.6449$ and $Z = 1.2816$ **[1]**
   $-1.6449 = \frac{30-\mu}{\sigma}$ and $1.2816 = \frac{60-\mu}{\sigma}$ **[1]**
   $-1.6449\sigma = 30 - \mu$ and $1.2816\sigma = 60 - \mu$ **[1]**
   $2.9265\sigma = 30$ **[1]**

   These are simultaneous equations.

   $\sigma = 10.25$ cm **[1]** $\mu = 46.86$ cm **[1]**
4. **a)** From percentage points table $Z = 3.2905$ **[1]**
   $3.2905 = \frac{20-\mu}{4.5}$ **[1]**
   5.19 **[1]**
   b) $P(T < 10) = P\left(Z < \frac{10-5.19}{4.5}\right) = P(Z < 1.069)$ **[1]**
   0.8575 **[1]**

Page 101: Hypothesis Testing 1 & 2

1. $\bar{X} \sim N\left(26, \frac{9}{10}\right)$ **[1]** $P(\bar{X} > 28)$ **[1]**

   $P\left(Z > \frac{28-26}{\frac{3}{\sqrt{10}}}\right)$ **[1]** $= 0.0175$ **[1]**

   $0.0175 < 0.025$ Therefore sufficient evidence to reject the null hypothesis; the sample did not come from a population with mean 26. **[1]**

   This is a two-tailed test.

2. $H_0: \mu = 65 \quad H_1: \mu > 65$

$\bar{X} \sim N\left(65, \frac{100}{8}\right)$ **[1]** $P(\bar{X} > 70)$ **[1]**

$P\left(Z > \frac{70-65}{\frac{10}{\sqrt{8}}}\right)$ **[1]** = 0.0787 **[1]**

0.0787 > 0.05 Therefore insufficient evidence to reject the null hypothesis; there is no evidence the results have improved. **[1]**

This is one-tailed.

3. $H_0: \mu = 7.5 \quad H_1: \mu > 7.5$

$\bar{X} \sim N\left(7.5, \frac{1.2}{30}\right)$ **[1]** $P(\bar{X} > 7.6)$ **[1]**

$P\left(Z > \frac{7.6-7.5}{\frac{1.2}{\sqrt{30}}}\right)$ **[1]** = 0.324 **[1]**

0.324 > 0.05 Therefore insufficient evidence to reject the null hypothesis; there is no evidence the strengths have improved. **[1]**

This is one-tailed.

4. a) $H_0: p = 0 \quad H_1: p < 0$ **[1]**
   Critical value –0.3365 **[1]**
   –0.56 < –0.3365
   Therefore sufficient evidence to reject the null hypothesis; there is evidence of negative correlation. **[1]**

   b) $H_0: p = 0 \quad H_1: p \neq 0$ **[1]**
   Critical value –0.3961 **[1]**

   This is two-tailed.

   –0.56 < –0.3961 Therefore sufficient evidence to reject the null hypothesis; there is evidence of correlation. **[1]**

5. $H_0: p = 0$ **[1]** $H_1: p > 0$ **[1]**
   Critical value = 0.4259 **[1]**

   This is one-tailed.

   0.69 > 0.4259 **[1]** Therefore sufficient evidence to reject the null hypothesis **[1]**; there is evidence of positive correlation. **[1]**

Pages 102–103 Review Questions

Page 102: Iterative Methods and Finding Roots

1. $0.9^2 - \sin 1.8 = -0.164 < 0$ **[2]**
   $1.0^2 - \sin 1.0 = 0.159 > 0$ **[2]**
   The function is continuous in the interval $0.9 < \alpha < 1.0$ and there is a sign change, so there exists a root α such that $0.9 < \alpha < 1.0$. **[1]**
   0.159 is nearer to 0 than –0.164, so the root is likely to be closer to 1.0. **[1]**

2. $f(0.5) = 0.35 > 0$ **[2]**
   $f(1) = -0.16 < 0$ **[2]**
   f is continuous in the interval $0.5 < \alpha < 1$ and there is a sign change, so there exists a root α such that $0.5 < \alpha < 1$. **[1]**
   $x_1 = 0.944$ **[1]**
   $x_2 = 0.932$
   $x_3 = 0.929$
   $x_4 = 0.929$ **[1]**
   $\alpha = 0.929$ **[1]**

3. Use iteration formula $x_{n+1} = \sqrt{1 - x_n}$ **[2]**
   Evidence of a significant number of iterations used: **[2]**
   $\alpha = 0.62$ **[1]**

Page 102: Numerical Integration and Solving Problems in Context

1. $h = \frac{\frac{\pi}{4} - 0}{4} = \frac{\pi}{16}$ **[1]**

x	0	$\frac{\pi}{16}$	$\frac{\pi}{8}$	$\frac{3\pi}{16}$	$\frac{\pi}{4}$
$\sqrt{1+\sec x}$	$\sqrt{2}$	1.421	1.443	1.484	1.336

[2]

$$\int_0^{\frac{\pi}{4}} \sqrt{1+\sec x}\, dx \approx \frac{\left(\frac{\pi}{16}\right)}{2}\left[\sqrt{2} + 2(1.421 + 1.443 + 1.484) + 1.336\right]$$ **[2]**

= 1.12 units² **[1]**

2. $f'(x) = 2e^{2x} - 2x$ **[2]**
   $x_{n+1} = x_n - \left(\frac{e^{2x} - x^2 - 5}{2e^{2x} - 2x}\right)$ **[1]**
   $x_1 = 0.891$ **[1]**
   $x_2 = 0.876$ **[1]**
   $\alpha = 0.88$ **[1]**

3. $f(1.5) = 1.5^3 + 4 \times 1.5 - 12 = -1.63 < 0$ **[2]**
   $f(2) = 2^3 + 4 \times 2 - 12 = 4 > 0$ **[2]**
   There is a sign change and f(x) is continuous in the interval $1.5 < x < 2$, therefore there exists a root α such that $1.5 < \alpha < 2$. **[1]**
   $f'(x) = 3x^2 + 4$ **[1]**
   $x_{n+1} = x_n - \left(\frac{x^3 + 4x - 12}{3x^2 + 4}\right)$ **[1]**
   $x_1 = 1.723$ **[1]**
   $\alpha = 1.72$ **[1]**

Page 103: Proof by Contradiction and Further Trigonometric Identities

1. $\sin 3x = \sin(2x + x)$ **[1]**
   $= \sin 2x \cos x + \cos 2x \sin x$ **[1]**
   $= 2\sin x \cos^2 x + (1 - 2\sin^2 x)\sin x$ **[1]**
   $= 2\sin x(1 - \sin^2 x) + (1 - 2\sin^2 x)\sin x$ **[1]**
   $= 2\sin x - 2\sin^3 x + \sin x - 2\sin^3 x$ **[1]**
   $= 3\sin x - 4\sin^3 x$

2. Assume both a and b are odd. **[1]**
   Then $a^2 + b^2$ is even, so c^2 is even, so c is even. **[1]**
   Write $a = 2p + 1$, $b = 2q + 1$ **[2]**
   and $c = 2r$ **[1]**
   So $(2p + 1)^2 + (2q + 1)^2 = (2r)^2$ **[1]**
   $4p^2 + 4p + 1 + 4q^2 + 4q + 1 = 4r^2$
   $4(p^2 + p + q^2 + q) + 2 = 4r^2$ **[1]**
   Now the right-hand side is divisible by 4, but the left-hand side is not, since 2 is not divisible by 4. **[1]**
   So you have a contradiction. Therefore at least one of a or b must be even.

3. $\sec 2x + \tan 2x = \frac{1}{\cos 2x} + \frac{\sin 2x}{\cos 2x}$ **[1]**
   $= \frac{1 + \sin 2x}{\cos 2x}$ **[1]**
   $= \frac{1 + 2\sin x \cos x}{\cos^2 x - \sin^2 x}$ **[1]**
   $= \frac{(\cos^2 x + \sin^2 x) + 2\sin x \cos x}{\cos^2 x - \sin^2 x}$ **[1]**
   $= \frac{(\cos x + \sin x)^2}{(\cos x + \sin x)(\cos x + \sin x)}$ **[2]**
   $= \frac{\cos x + \sin x}{\cos x - \sin x}$
   $= \frac{\frac{\cos x}{\sin x} + \frac{\sin x}{\sin x}}{\frac{\cos x}{\sin x} - \frac{\sin x}{\sin x}}$ **[1]**
   $= \frac{\cot x + 1}{\cot x - 1}$
   Substituting $x = 15°$: **[1]**
   $\sec 30° + \tan 30° = \frac{\cot 15° + 1}{\cot 15° - 1}$
   $\frac{2}{\sqrt{3}} + \frac{1}{\sqrt{3}} = \frac{\cot 15° + 1}{\cot 15° - 1}$ **[2]**
   $\frac{3}{\sqrt{3}} = \frac{\cot 15° + 1}{\cot 15° - 1}$
   $\sqrt{3} = \frac{\cot 15° + 1}{\cot 15° - 1}$ **[1]**
   $\sqrt{3}\cot 15° - \sqrt{3} = \cot 15° + 1$
   $\sqrt{3}\cot 15° - \cot 15° = \sqrt{3} + 1$
   $(\sqrt{3} - 1)\cot 15° = \sqrt{3} + 1$
   $\cot 15° = \frac{\sqrt{3} + 1}{\sqrt{3} - 1}$ **[2]**
   $= \left(\frac{\sqrt{3} + 1}{\sqrt{3} - 1}\right)\left(\frac{\sqrt{3} + 1}{\sqrt{3} + 1}\right)$ **[1]**
   $= \frac{4 + 2\sqrt{3}}{2}$ **[1]**
   $= 2 + \sqrt{3}$

Pages 104–113 Revise Questions

Page 105 Quick Test

1. Speed = $\sqrt{13}$ ms^{-1}. Acceleration = **i** ms^{-2}, so constant.
2. $\sqrt{10}$ ms^{-1}
3. arctan2 = 63.4°

Page 107 Quick Test

1. 2.34 s
2. 14.4 m
3. 63.4°

Page 109 Quick Test

1. 10.7 N, 36.1°
2. a) 90.6 N b) 42.3 N

Page 111 Quick Test

1. $\mu = 0.223$
2. 0.332 ms^{-2}

Page 113 Quick Test

1. 4.78 Nm
2. 3.15 m
3. $\frac{\sqrt{3}}{6}$

Pages 114–117 Practice Questions

Page 114: Using Vectors in Kinematics

1. a) $\mathbf{v} = \frac{d\mathbf{r}}{dt} = (-2t)\mathbf{i} + (3t^2 - 1)\mathbf{j}$ [2]

$t = 3 \Rightarrow \mathbf{v} = -6\mathbf{i} + 26\mathbf{j}$ [1]

Speed $\sqrt{(-6)^2 + 26^2} = 26.7\,\text{ms}^{-1}$ [1]

b) $\mathbf{a} = \frac{d\mathbf{v}}{dt} = -2\mathbf{i} + 6t\mathbf{j}$ [2]

$t = 1 \Rightarrow \mathbf{a} = -2\mathbf{i} + 6\mathbf{j}$ [1]

$|\mathbf{a}| = \sqrt{(-2)^2 + 6^2} = \sqrt{40} = 2\sqrt{10} = 6.32\,\text{ms}^{-2}$ [1]

2. a) When $t = 2$, $\mathbf{v} = 5\mathbf{i} + 4\mathbf{j}$ [1]

Its direction of motion is therefore at an angle of $\arctan\left(\frac{4}{5}\right) = 38.7°$ to the x-axis. [1]

b) $\mathbf{a} = \frac{d\mathbf{v}}{dt} = 3\mathbf{i} + \mathbf{j}$ [2]

$|\mathbf{a}| = \sqrt{3^2 + 1^2} = \sqrt{10} = 3.16\,\text{ms}^{-2}$ [1]

3. a) $t = 1$, $\mathbf{v} = 3\mathbf{i} + (3 + c)\mathbf{j}$ [1]

$|\mathbf{v}|^2 = 3^2 + (3 + c)^2 = 34$ [1]

$c^2 + 6c + 18 = 34$

$c^2 + 6c - 16 = 0$

$(c - 2)(c + 8) = 0$ [1]

$\Rightarrow c = 2$ or $c = -8$ [2]

b) $\mathbf{r} = \int \mathbf{v}\,dt = \int \left((4t - 1)\mathbf{i} + (t^2 + 2t + 2)\mathbf{j}\right) dt$

$= (2t^2 - t)\mathbf{i} + \left(\frac{t^3}{3} + t^2 + 2t\right)\mathbf{j} + \mathbf{c}$ [2]

Substituting $t = 0$, $\mathbf{r} = \mathbf{i} - \mathbf{j}$ [1]

$\Rightarrow \mathbf{i} - \mathbf{j} = \mathbf{c}$ [1]

$\Rightarrow \mathbf{r} = (2t^2 - t + 1)\mathbf{i} + \left(\frac{t^3}{3} + t^2 + 2t - 1\right)\mathbf{j} + \mathbf{c}$

Page 115: Projectiles

1. At maximum range, the vertical displacement is zero.

Therefore $ut\sin\alpha - \frac{1}{2}gt^2 = 0$ [2]

$\Rightarrow t = \frac{2u\sin\alpha}{g}$ [1]

The horizontal range is $x = u\cos\alpha\left(\frac{2u\sin\alpha}{g}\right) = \frac{2u^2\sin\alpha\cos\alpha}{g}$

$= \frac{u^2\sin 2\alpha}{g}$ [1]

$\frac{65^2 \times \sin 60°}{9.8} = 373\,\text{m}$ [1]

2. Horizontal motion: $x = ut\cos\alpha \Rightarrow t = \frac{x}{u\cos\alpha}$ (1) [1]

Vertical motion: $y = ut\sin\alpha - \frac{1}{2}gt^2$ (2) [1]

Substituting (1) into (2) (and simplifying) to give

$y = x\tan\alpha - \frac{gx^2}{2u^2}(1 + \tan^2\alpha)$ [2]

Substituting $x = 18$, $y = -3$, $u = 20$ to give

$-3 = 18\tan\alpha - 3.969(1 + \tan^2\alpha)$ [1]

$\Rightarrow 3.969\tan^2\alpha - 18\tan\alpha + 0.969 = 0$ [1]

Using quadratic formula to give $\tan\alpha = 0.0545$ or $\tan\alpha = 4.48$ [1]

So $\alpha = 3.1°$ or $\alpha = 77.4°$ [2]

3. Considering ball at point P:

Horizontal motion: $x = ut\cos\alpha$ [1]

$x = 10T\cos 30 = 10T\frac{\sqrt{3}}{2} = 5\sqrt{3}T$ [1]

Vertical motion: $y = ut\sin\alpha - \frac{1}{2}gt^2$ [1]

$y = 10T\sin 30 - \frac{1}{2} \times 4.9 \times T^2$

$y = 5T - 4.9T^2$ [1]

$\tan 60° = \frac{-(5T - 4.9T^2)}{5\sqrt{3}T}$ [2]

$\sqrt{3} = \frac{4.9T^2 - 5T}{5\sqrt{3}T}$

Solving to give: $T = \frac{20}{4.9} = 4.08$ seconds [1]

Page 115: Resolving Forces and Forces in Equilibrium

1. Resolving forces horizontally: $P\cos\theta = 6\cos60° + 4\cos15°$ [2]

Resolving forces vertically: $P\sin\theta + 6\sin60° = 4\sin15° + 5$ [2]

$\Rightarrow P\sin\theta = 4\sin15° + 5 - 6\sin60°$

$\frac{P\sin\theta}{P\cos\theta} = \frac{4\sin15° + 5 - 6\sin60°}{6\cos60° + 4\cos15}$ [1]

$\tan\theta = 0.122$

$\theta = 7.0°$ [1]

$P = \frac{6\cos60° + 4\cos15°}{\cos6.97°}$

$P = 6.91\,\text{N}$ [1]

2. a) Resolving parallel to the plane: $5g\sin 30° = T\cos20°$ [2]

$T = \frac{5g\sin30}{\cos20} = 26.1\,\text{N}$ [1]

b) Resolving perpendicular to plane: $R + T\sin20° = 5g\cos30°$ [2]

$R = 5g\cos30° - T\sin20°$ N

$= 5g\cos30° - 26.07\sin20°$ [1]

$= 33.5$ N [1]

3. Let the tension in the first string be T_1 and the tension in the second be T_2.

Resolving horizontally: $T_1\cos55° = T_2\cos70° + 5$ (1) [2]

Resolving vertically: $T_1\sin55° + T_2\sin70° = 2g$ (2) [2]

From equation (1): $T_1 = \frac{T_2\cos70°}{\cos55°} + \frac{5}{\cos55°}$

$\left(\frac{T_2\cos70°}{\cos55°} + \frac{5}{\cos55°}\right)\sin55° + T_2\sin70° = 2g$ [2]

Substitute in (2).

$\frac{T_2\cos70°\sin55°}{\cos55°} + T_2\sin70° = 2g - \frac{5\sin55°}{\cos55°}$

$T_2\left(\frac{\cos70°\sin55°}{\cos55°} + \sin70°\right) = 2g - \frac{5\sin55°}{\cos55°}$

$\Rightarrow T_2 = 8.72\,\text{N}$ [1]

$T_1 = \frac{8.724\cos70°}{\cos55°} + \frac{5}{\cos55°} = 13.9\,\text{N}$ [2]

Substitute in (1).

Page 116: Friction and Motion on an Inclined Plane

1. Resolving perpendicular to plane: $R = 1.5g\cos40°$ (1) [1]

Friction limiting: $F = \mu R$ (2) [1]

Equation of motion down plane: $1.5g\sin40° - F = 1.5a$ (3) [2]

Solving (1), (2), (3) simultaneously $\Rightarrow a = 2.92\,\text{ms}^{-2}$ [2]

2. Resolving perpendicular to the plane:

$R = 3g\cos25° + P\sin25°$ (1) [2]

Resolving parallel to the plane: $F + P\cos25° = 3g\sin25°$ (2) [2]

Friction limiting: $F = 0.3R$ (3) [1]

Solving (1), (2), (3) simultaneously $\Rightarrow P = 4.29\,\text{N}$ [2]

3. Resolving perpendicular to the plane: $R = mg\cos\theta$ (1) [1]

Resolving parallel to the plane: $F = mg\sin\theta$ (2) [1]

Laws of friction: $F \leqslant \mu R$ (3) [1]

Equation (3) gives $\mu \geqslant \frac{F}{R} = \frac{mg\sin\theta}{mg\cos\theta} = \tan\theta$ [2]

Page 117: Moments

1. a) $-(3 \times 0.5) + (1 \times 1) - (7 \times 2.25) + (4 \times 3) = -4.25\,\text{Nm}$ (or 4.25 Nm clockwise) [2]

b) $(3 \times 1) + (4 \times 1.5) - (1 \times 0.5) - (7 \times 0.75) = 3.25\,\text{Nm}$ [2]

c) $(7 \times 0.75) + (3 \times 2.5) - (1 \times 2) = 10.75\,\text{Nm}$ [2]

2. Taking moments about O: $5gl\sin 75° = 2lF$ **[3]**

$$F = \frac{5g\sin 75°}{2} = 23.7\text{N}$$ **[2]**

3.

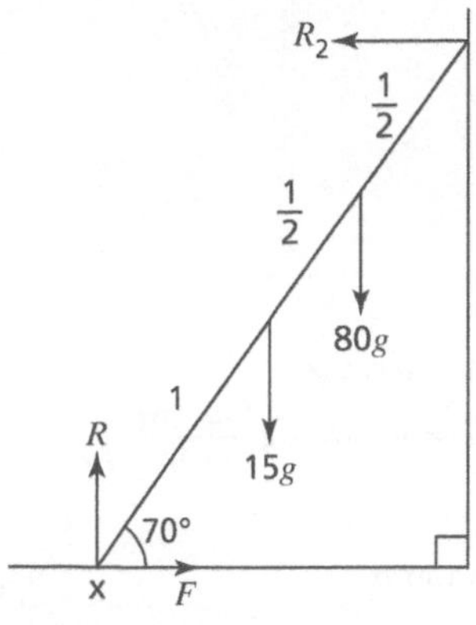

Resolving vertically: $R = 95g$ **[1]**
Resolving horizontally: $R_2 = F$ **[1]**
Friction limiting: $F = \mu R$ **[1]**

Moments about X: $15gl\cos 70° + 80g\left(\frac{3l}{2}\right)\cos 70° = R_2(2l\sin 70°)$ **[3]**

$R_2 = 240.766$

Solving simultaneously: $\mu = \frac{F}{R} = \frac{240.766}{95g} = 0.259$ **[2]**

Pages 118–121 Review Questions

Page 118: Set Notation and Conditional Probability

1. a) $P(A \mid B) = \frac{P(A \cap B)}{P(B)}$ as $P(B) = \frac{1}{2} = P(A' \cap B) = \frac{1}{2}$, $P(A \cap B) = 0$ **[1]**

$P(A \mid B) = \frac{0}{0.5}$ **[1]** $= 0$ **[1]**

b) Mutually exclusive **[1]**

c) $1 - \frac{1}{3} - \frac{1}{2}$ **[1]** $= \frac{1}{6}$ **[1]** Probability of neither A nor B occurring. **[1]**

2. a)

	Chicken	Beef	Lamb	Total
Small	14	26	23	63
Medium	10	12	17	39
Total	24	38	40	102

[3 if fully correct; 2 if two errors or fewer; 1 if three cells correct]

b) $\frac{24}{102}$ **[1]** $= \frac{4}{17}$ **[1]**

c) $\frac{23}{102}$ **[2]**

d) $\frac{12}{39}$ **[1]** $= \frac{4}{13}$ **[1]**

Page 119: Modelling and Data Presentation

1. a)

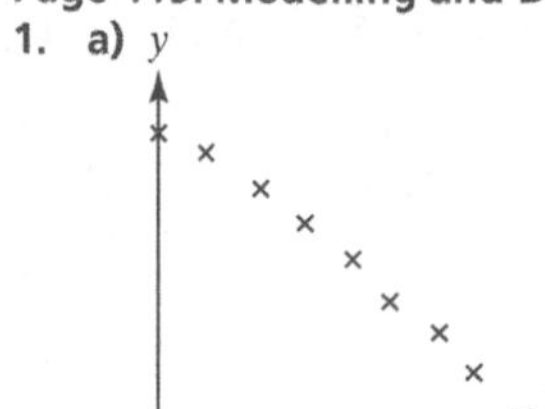

[1]

b)

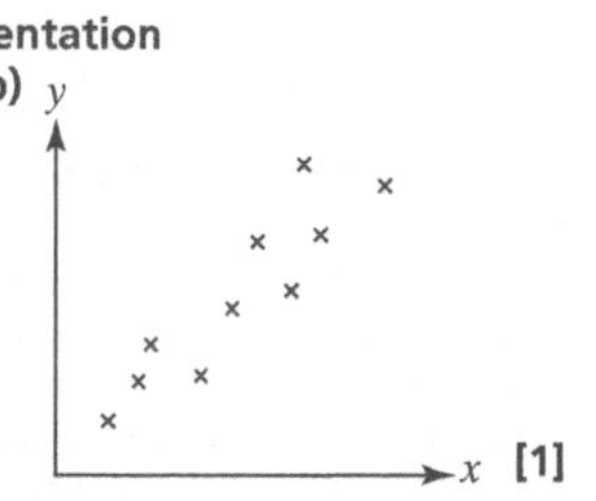

[1]

c)

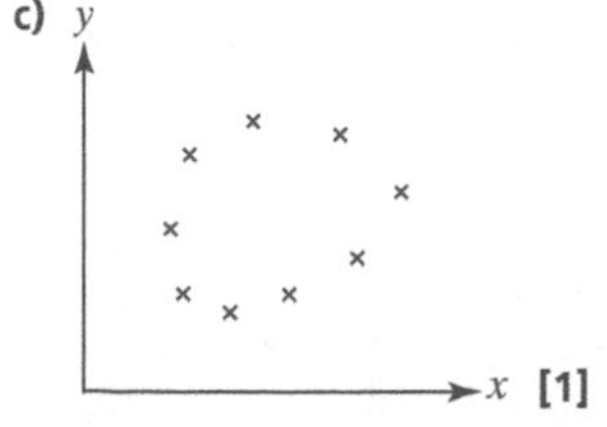

[1]

2. a) Binomial distribution **[1]** $X \sim B(100, \frac{1}{2})$ **[1]**

b) Coin is fair **[1]**; tosses are independent **[1]**

3. $\ln y = \ln(ka^x)$
$\ln y = \ln k + x\ln a$ **[1 for both statements]**
$\ln 8 = \ln k + \ln a$ and $\ln 32 = \ln k + 2\ln a$ **[1]**
$2\ln 8 = 2\ln k + 2\ln a$
$2\ln 8 - \ln 32 = \ln k$ **[1]**

These are simultaneous equations.

Use log rules.

$\ln 2 = \ln k$; $k = 2$ **[1]**

4. a) Binomial **[1]** $X \sim B(120, \frac{1}{3})$ **[1]**

b) Probability of buying an iced finger is constant **[1]**

Page 120: Statistical Distributions 1 & 2

1. a) $P(X < 52) = P\left(Z < \frac{52-50}{4}\right) = P(Z < 0.5)$ **[2]**

0.6915 **[1]**

b) $P(X > 20) = P\left(Z > \frac{20-24}{3}\right) = P(Z > \frac{-4}{3})$ **[2]**

0.9088 **[1]**

c) $P(14 < X < 23) = P\left(\frac{14-20}{6} < Z < \frac{23-20}{6}\right)$

$= P(-1 < Z < 0.5)$ **[2]**

0.5328 **[1]**

2. a) $np = 80$ $\quad np(1-p) = 48$ **[1]**

$P(X < 65) = P\left(Z < \frac{64.5-80}{\sqrt{48}}\right) = P(Z < -2.24)$ **[1]**

0.0125 **[1]**

b) $P(X > 90) = P\left(Z > \frac{90.5-80}{\sqrt{48}}\right) = P(Z > 1.52)$ **[1]**

0.0643 **[1]**

c) $P(75 < X < 85) = P\left(\frac{75.5-80}{\sqrt{48}} < Z < \frac{84.5-80}{\sqrt{48}}\right)$

$= P(-0.65 < Z < 0.65)$ **[1]**

0.4843 **[1]**

3. $X \sim B(80, 0.25)$ **[1]**
$np = 20$ **[1]**; $np(1-p) = 15$ **[1]**

$P(X > 30) = P\left(Z > \frac{30.5-20}{\sqrt{15}}\right)$ **[1]**

$P(Z > 2.71)$ **[1]**
0.0034 **[1]**

4. From percentage points table $Z = 1.6449$ **[1]**

$Z = \frac{34-30}{\sigma} = 1.6449$ **[2]**
$4 = 1.6449\sigma$ **[1]**
$\sigma = 2.432$ **[1]**

Page 121: Hypothesis Testing 1 & 2

1. $\bar{X} \sim N\left(19, \frac{4}{100}\right)$ **[1]** $P(\bar{X} > 20)$ **[1]**

$P\left(Z > \frac{20-19}{\frac{2}{\sqrt{100}}}\right)$ **[1]**; 0 **[1]**

$0 < 0.005$ Therefore sufficient evidence to reject the null hypothesis; the sample did not come from a population with mean 19. **[1]**

This is a two-tailed test.

2. H_0: $\mu = 8.7$ $\quad H_1$: $\mu > 8.7$ **[1]**

$\bar{X} \sim N\left(8.7, \frac{1.44}{100}\right)$ **[1]** $P(\bar{X} > 9.1)$ **[1]**

$P\left(Z > \frac{9.1-8.7}{\frac{1.2}{10}}\right)$ **[1]**; 0.0004 **[1]**

$0.0004 < 0.05$ **[1]** Therefore sufficient evidence to reject the null hypothesis; there is evidence the results have improved. **[1]**

This is one-tailed.

3. H_0: $\mu = 250$ $\quad H_1$: $\mu < 250$ **[1]**

$\bar{X} \sim N\left(250, \frac{36}{20}\right)$ **[1]** $P(\bar{X} < 248)$ **[1]**

$P\left(Z < \frac{248-250}{\frac{6}{\sqrt{20}}}\right)$ **[1]**; 0.0681 **[1]**

$0.0681 > 0.05$ **[1]** Therefore insufficient evidence to reject the null hypothesis; there is no evidence to support the customer's complaint. **[1]**

This is one-tailed.

4. a) H_0: $p = 0$ $\quad H_1$: $p > 0$ **[1]**
Critical value 0.4259 **[1]**
$0.61 > 0.4259$ Therefore sufficient evidence to reject the null hypothesis; there is evidence of positive correlation. **[1]**

b) H_0: $p = 0$ $\quad H_1$: $p \neq 0$ **[1]**
Critical value 0.4973 **[1]**

0.61 > 0.4973 Therefore sufficient evidence to reject the null hypothesis; there is evidence of correlation. [1]

Pages 122–125 Review Questions

Page 122: Using Vectors in Kinematics

1. a) $\mathbf{v} = \frac{d\mathbf{r}}{dt} = (2t+1)\mathbf{i} + (4-3t)\mathbf{j}$ [2]

North-easterly direction, so equate coefficients of **i** and **j** [1]

$2t + 1 = 4 - 3t$ [1]

$\Rightarrow t = \frac{3}{5}$ [1]

b) $\mathbf{r} = \left(\left(\frac{3}{5}\right)^2 + \frac{3}{5} + 1\right)\mathbf{i} + \left(4\left(\frac{3}{5}\right) - \frac{3}{2}\left(\frac{3}{5}\right)^2\right)\mathbf{j}$

$\mathbf{r} = 1.96\mathbf{i} + 1.86\mathbf{j}$ [1]

Distance $= \sqrt{1.96^2 + 1.86^2} = 2.70\,\text{m}$ [1]

2. Using $\mathbf{s} = \mathbf{v}t - \frac{1}{2}\mathbf{a}t^2$ [1]

$\mathbf{s} = (12\mathbf{i} + 20\mathbf{j}) \times 3 - \frac{1}{2} \times (\mathbf{i} - 2\mathbf{j}) \times 3^2$ [1]

$\mathbf{s} = (36\mathbf{i} + 60\mathbf{j}) - \left(\frac{9}{2}\mathbf{i} - 9\mathbf{j}\right)$

$\mathbf{s} = \frac{63}{2}\mathbf{i} + 69\mathbf{j}$ [1]

3. a) $\mathbf{v}_P = \begin{pmatrix} t-2 \\ 2-2t \end{pmatrix}$ [2]

$\mathbf{v}_Q = \begin{pmatrix} 5-t \\ 2t-11 \end{pmatrix}$ [2]

b) $\mathbf{v}_P = k\mathbf{v}_Q \Rightarrow \begin{pmatrix} t-2 \\ 2-2t \end{pmatrix} = k\begin{pmatrix} 5-t \\ 2t-11 \end{pmatrix}$ [1]

$t - 2 = k(5-t)$ and $2 - 2t = k(2t - 11)$ [1]

Solving simultaneously: [1]

$\frac{t-2}{5-t} = \frac{2-2t}{2t-11}$ [1]

$(t-2)(2t-11) = (2-2t)(5-t)$

$2t^2 - 15t + 22 = 2t^2 - 12t + 10$

$\Rightarrow t = 4$ [1]

Page 122: Projectiles

1. a) Considering vertical motion and applying $v = u + at$: [1]

$0 = 20 - gt$ [1]

$t = \frac{20}{9.8} = 2.04$ seconds [1]

b) When $t = 4$, horizontal displacement $= ut = 12 \times 4 = 48\,\text{m}$ [1]

When $t = 4$, vertical displacement

$= 20 \times 4 - \frac{1}{2} \times 9.8 \times 4^2 = 1.6\,\text{m}$ [1]

Distance $= \sqrt{48^2 + 1.6^2} = 48.02\text{m}$ [1]

2. Considering maximum height $\frac{u^2\sin^2\alpha}{2g} = 86$ (1) [1]

Considering horizontal range $\frac{u^2\sin 2\alpha}{g} = 150$ (2) [1]

Attempt to solve simultaneously: [1]

Equation (2) divided by equation (1) gives $\frac{2\sin 2\alpha}{\sin^2\alpha} = \frac{150}{86}$ or

$\frac{\sin 2\alpha}{\sin^2\alpha} = \frac{75}{86}$ [1]

$2\sin\alpha\cos\alpha = \frac{75\sin^2\alpha}{86}$

$\sin\alpha\cos\alpha = \frac{75\sin^2\alpha}{172}$

$\sin\alpha\cos\alpha - \frac{75\sin^2\alpha}{172} = 0$

$\sin\alpha\left(\cos\alpha - \frac{75\sin\alpha}{172}\right) = 0$ [1]

$\sin\alpha \neq 0$, so $\frac{\sin\alpha}{\cos\alpha} = \frac{172}{75} \Rightarrow \tan\alpha = \frac{172}{75} \Rightarrow \alpha = 66.44°$ [2]

$u^2 = \frac{2g \times 86}{\sin^2 66.44°}$ Substitute in equation (1). [1]

$\Rightarrow u = 44.8\,\text{ms}^{-1}$ [1]

3. a) Vertical motion: $y = ut\sin\alpha - \frac{1}{2}gt^2$

At B, the vertical displacement is zero.

Therefore $ut\sin\alpha - \frac{1}{2}gt^2 = 0$ [1]

$t\left(u\sin\alpha - \frac{1}{2}gt\right) = 0$ [1]

$(t = 0)$ or $u\sin\alpha - \frac{1}{2}gt = 0$

$\Rightarrow u\sin\alpha = \frac{1}{2}gt$ [1]

$\Rightarrow T = \frac{2u\sin\alpha}{g}$

b) $t = \frac{2T}{3} \Rightarrow t = \frac{4u\sin\alpha}{3g}$ [1]

At this time, horizontal velocity $= u\cos\alpha$ [1]

Vertical velocity $= u\sin\alpha - gT = u\sin\alpha - g\left(\frac{4u\sin\alpha}{3g}\right)$

$= -\frac{u\sin\alpha}{3}$ [2]

Initial direction of particle is $\tan\alpha$

Using $m_1m_2 = -1$ [1]

$\tan\alpha\left(\frac{\frac{-u\sin\alpha}{3}}{u\cos\alpha}\right) = -1$ [1]

$-\frac{\tan^2\alpha}{3} = -1$

$\tan^2\alpha = 3$ [1]

$\tan\alpha = \sqrt{3}$

$\alpha = 60°$ [1]

Page 123: Resolving Forces and Forces in Equilibrium

1. a) Resolving forces horizontally: $P = 12\cos 55° = 6.88$ N [2]

b) Resolving forces vertically: $mg = 12\sin 55°$ [2]

$m = \frac{12\sin 55°}{g} = 1.00\,\text{kg}$ [1]

2. a) Resolving parallel to the plane: $X\cos 25° = 12g\sin 25°$ [2]

$X = \frac{12g\sin 25°}{\cos 25°} = 12g\tan 25° = 54.9$ N [1]

b) Resolving perpendicular to plane: $R = 12g\cos 25° + X\sin 25°$ [2]

$R = 12g\cos 25° + (12g\tan 25°)\sin 25° = 130$ N [2]

3. Resolving horizontally: $12\cos 50° = T\cos 70°$ [2]

$T = \frac{12\cos 50°}{\cos 70°} = 22.6$N [1]

Resolving vertically: $12\sin 50° + T\sin 70° = mg$ [2]

$12\sin 50° + 22.55\sin 70° = mg$ [1]

$m = \frac{12\sin 50° + 22.55\sin 70°}{g} = 3.10\,\text{kg}$ [1]

Page 124: Friction and Motion on an Inclined Plane

1. Resolving horizontally: $F = 8$ (1) [1]

Friction limiting: $F = \mu R$ (2) [1]

Resolving vertically: $R = 2.5g$ (3) [1]

Solving (1), (2), (3) simultaneously gives $\mu = \frac{8}{2.5g} = 0.327$ [2]

2. Resolving perpendicular to the plane: $R = mg\cos 30°$ (1) [1]

Friction limiting: $F = 0.2R$ (2) [1]

Equation of motion up the plane: $-F - mg\sin 30° = ma$ (3) [2]

Solving (1), (2), (3) simultaneously: $a = -6.60\,\text{ms}^{-2}$ [2]

$\Rightarrow$ deceleration $= 6.60\,\text{ms}^{-2}$ [1]

When particle comes to instantaneous rest:

$\mu R = 0.2 \times mg\cos 30° = 0.173\,mg$ [1]

If particle in equilibrium then (resolving parallel to plane):

$F = mg\sin 30° = 0.5mg$ [1]

$F > \mu R$ so particle must slide. [1]

3. a) Equation of motion for P: $T - F = 4a$ (1) [2]

Resolving P vertically: $R = 4g$ (2) [1]

Friction limiting: $F = 0.2R$ (3) [1]

Solving (1), (2), (3) simultaneously: $T - 0.8g = 4a$ (4) [2]

Equation of motion for Q: $1.5g - T = 1.5a$ (5) [2]

$0.7g = 5.5a \Rightarrow a = \frac{0.7g}{5.5} = 1.25\,\text{ms}^{-2}$ [2]

Adding equations (4) and (5).

b) $T = 0.8g + 4 \times 1.247 = 7.84$ N [1]

Substitute into equation (4).

Page 125: Moments

1. Resolving vertically: $R_B + R_D = 70g + 20g$ [2]

$3R_B = 90g$

$R_B = 30g$ and $R_D = 60g$ [2]

Let x be distance of the centre of mass of the rod from B.

Moments about B: $70gx + 20g \times 0.7 = R_D \times 0.9$ [2]

$70gx + 20g \times 0.7 = 60g \times 0.9$

$x = \frac{4}{7} = 0.571$ [1]

Therefore distance of centre of mass of the rod from A is $0.5 + 0.571 = 1.07\,\text{m}$ [1]

2. Resolving vertically: $12 + 10 = 15 + P$ [2]

$P = 7\text{ N}$ [1]

Let distance CD $= x$

Moments about D: $7x + 15(x + 6) = 12(x + 10)$ [2]

$22x + 90 = 12x + 120$

$10x = 30$, so $x = 3\text{m}$ [1]

3. Resolving vertically: $T_1 \sin 30° = T_2 \sin\theta = 5g$ (1) [2]

Resolving horizontally: $T_1 \cos 30° = T_2 \cos\theta$ (2) [2]

Moments about A: $5gl = 3lT_2 \sin\theta \Rightarrow T_2 \sin\theta = \frac{5g}{3}$ (3) [2]

Substituting (3) in (1): $\frac{T_1}{2} + \frac{5g}{3} = 5g$ [1]

$T_1 = \frac{20g}{3} = 65.3\text{ N}$ [1]

Substituting T_1 in (2): $\frac{20g\cos 30°}{3} = T_2\cos\theta$ [1]

$T_2\cos\theta = \frac{10g\sqrt{3}}{3}$ (4)

Substituting T_1 in (1): $T_2\sin\theta = 5g - T_1\sin 30°$ [1]

$T_2\sin\theta = 5g - \frac{20g}{3}\left(\frac{1}{2}\right) = \frac{5g}{3}$ (5)

Solving (4) and (5) simultaneously: $\frac{\sin\theta}{\cos\theta} = \frac{\frac{5g}{3}}{\frac{10g\sqrt{3}}{3}}$ [1]

$\tan\theta = \frac{1}{2\sqrt{3}}$ [1] $\Rightarrow \theta = 16.1°$ [1]

Substituting in (2): $T_2 = \frac{T_1\cos 30°}{\cos\theta} = \frac{\frac{20g\cos 30°}{3}}{0.961} = 58.9\text{ N}$ [1]

Pages 126–139 Mixed Questions

Pure Mathematics

1. a) $2\sin(3x)$ **[1]** $\times \cos(3x) \times 3$ **[1]** $6\sin(3x)\cos(3x)$ **[1]**

b) $2\tan(2x)$ **[1]** $\times \sec^2(2x) \times 2$ **[1]** $4\tan(2x)\sec^2(2x)$ **[1]**

2. a) $\int x^2 + 2x = \frac{x^3}{3} + x^2 + c$ [1]

$2 = \frac{1^3}{3} + 1^3 + c$ **[1]**, $c = \frac{2}{3}$

$y = \frac{x^3}{3} + x^2 + \frac{2}{3}$ [1]

b) $\int_{-2}^{0} \frac{x^3}{3} + x^2 + \frac{2}{3} = \left[\frac{x^4}{12} + \frac{x^3}{3} + \frac{2x}{3}\right]_{-2}^{0}$ [1]

$= (0) - \left(\frac{16}{12} - \frac{8}{3} - \frac{4}{3}\right) = -\left(-\frac{8}{3}\right)$ **[2]** $= \frac{8}{3}$ units^2 **[1]**

3. $O\hat{A}B$ and $O\hat{C}B$ are right angles [1]

Area triangle OAB $= \frac{1}{2} \times 5 \times 12 = 30\text{ cm}^2$ [1]

So area OABC $= 30 \times 2 = 60\text{ cm}^2$ [1]

Area sector OAC $= \frac{1}{2} \times 5^2 \times \theta = \frac{25\theta}{2}$ [1]

$\Rightarrow \frac{25\theta}{2} \times 2 = 60$ [1]

$\Rightarrow \theta = \frac{60}{25} = \left(\frac{12}{5}\right)^c$ [1]

4. a) Factorising $x^2 + x - 2 = (x + 2)(x - 1)$.

Multiply so that each term has the common denominator of $(x + 2)(x - 1)$ and write as a single fraction.

$x - \frac{3}{x+2} + 1 = \frac{x(x+2)(x-1)}{(x+2)(x-1)} - \frac{3(x-1)}{(x+2)(x-1)} + \frac{(x+2)(x-1)}{(x+2)(x-1)}$ [1]

$= \frac{x(x+2)(x-1) - 3(x-1) + (x+2)(x-1)}{(x+2)(x-1)}$ [1]

Expanding and simplifying the numerator:

$\frac{x(x+2)(x-1) - 3(x-1) + (x+2)(x-1)}{(x+2)(x-1)} =$

$\frac{(x^3 + x^2 - 2x) - (3x - 1) + (x^2 + x - 2)}{(x+2)(x-1)}$ **[1]** $= \frac{x^3 + 2x^2 - 4x + 1}{x^2 + x - 2}$ **[1]**, as required.

b) Use the Factor Theorem to show that $(x - 1)$ is a factor of $x^3 + 2x^2 - 4x + 1$.

$1^3 + (2 \times 1^2) - (4 \times 1) + 1 = 0$ **[1]**, so $x^3 + 2x^2 - 4x + 1 = (x - 1) \times p(x)$ for some polynomial $p(x)$, therefore $\frac{x^3 + 2x^2 - 4x + 1}{(x+2)(x-1)} = \frac{(x-1) \times p(x)}{(x+2)(x-1)}$

Divide the numerator by $(x - 1)$ using algebraic division.

$$\begin{array}{r} x^2 + 3x - 1 \\ x - 1 \overline{)\, x^3 + 2x^2 - 4x + 1} \\ -(x^3 - x^2) \\ \hline 3x^2 - 4x \\ -(3x^2 - 3x) \\ \hline -x + 1 \\ -(-x + 1) \\ \hline 0 \end{array}$$

$(x^3 + 2x^2 - 4x + 1) \div (x - 1) = x^2 + 3x - 1$

[1 for method; 1 for correct answer]

So, $\frac{x^3 + 2x^2 - 4x + 1}{(x-1)(x+2)} = \frac{\cancel{(x-1)}(x^2 + 3x - 1)}{\cancel{(x-1)}(x+2)} = \frac{x^2 + 3x - 1}{x + 2}$ [1]

5. a) Finding $f^{-1}(x)$ is equivalent to changing the subject of a formula.

Let $y = 3x + 2$ [1]

$y - 2 = 3x \Rightarrow \frac{y-2}{3} = x$

Then $f^{-1}(x) = \frac{x-2}{3}$ [1]

The domain is $\mathbb{R}$ and the range is $\mathbb{R}$. **[1 for both]**

b) To find gf(−3), either find gf(x) and substitute $x = -3$ or find the value of f(−3) and substitute it into g(x).

$f(-3) = (3 \times -3) + 2 = -7$ [1]

$g(-7) = \frac{3}{-7 + 4} = -1$ [1]

c) $3x + 2 + \frac{3}{x+4} = -1$

$3x(x + 4) + 2(x + 4) + 3 = -x - 4$ [1]

$3x^2 + 12x + 2x + 8 + 3 + x + 4 = 0$

$3x^2 + 15x + 15 = 0$

$x^2 + 5x + 5 = 0$ [1]

$x = \frac{-5 \pm \sqrt{5^2 - 4 \times 5}}{2}$ **[1]** $= \frac{-5 \pm \sqrt{5}}{2}$

$x = \frac{-5 + \sqrt{5}}{2}$, $x = \frac{-5 - \sqrt{5}}{2}$ [1]

6. $2y\frac{dy}{dx} + 1\frac{dy}{dx} = -2x$ [1]

$(2y + 1)\frac{dy}{dx} = -2x$ [1]

$x = -2, y = 2 \Rightarrow \frac{dy}{dx} = \frac{4}{(4 + 1)} = \frac{4}{5}$

$y - 2 = \frac{4}{5}(x + 2) \Rightarrow 4x - 5y + 18 = 0$ [1]

7.

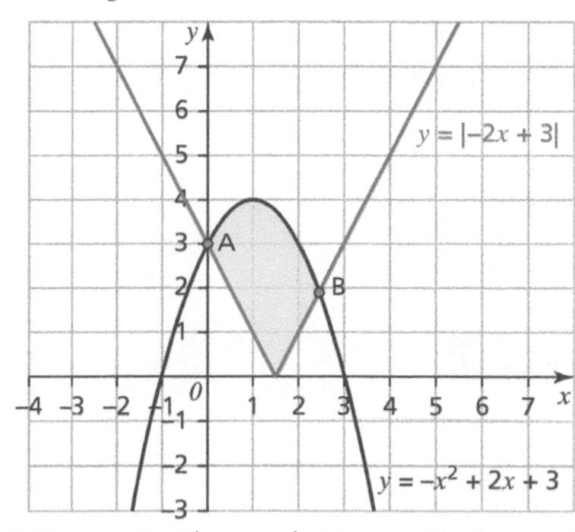

[1 for graph of $|-2x + 3|$; 1 for graph of $y = -x^2 + 2x + 3$]

$y = |-2x + 3|$ and $y = -x^2 + 2x + 3$ intersect at two points, A and B.

Point A: $-2x + 3 = -x^2 + 2x + 3$ **[1]**
$0 = x^2 - 4x$
$x = 0$ and $x = 4$
The graph shows the solution is $x = 0$. **[1]**
Point B: $-(-2x + 3) = -x^2 + 2x + 3$ **[1]**
$0 = x^2 - 6$
$x = \pm\sqrt{6}$
The graph shows the solution is $x = \sqrt{6}$ **[1]**

The solution to the inequality $|-2x + 3| \leqslant -x^2 + 2x + 3$ is $\{x: 0 \leqslant x \leqslant \sqrt{6}\}$ **[1]**

8. $\dfrac{3x - 11}{(x-1)(x+3)} = \dfrac{A}{x-1} + \dfrac{B}{x+3}$

$3x - 11 = A(x+3) + B(x-1)$ **[1]**
When $x = -3$: $-20 = -4B \Rightarrow B = 5$
When $x = 1$: $-8 = 4A \Rightarrow A = -2$ **[1 for both]**
Rewrite in index form:

$$\frac{3x - 11}{(x-1)(x+3)} = \frac{-2}{x-1} + \frac{5}{x+3} = -2(x-1)^{-1} + 5(x+3)^{-1}$$

$$-2(x-1)^{-1} = -2\left(-1^{-1}\left(1 + \frac{-1\times -1x}{1!} + \frac{-1(-1-1)(-1x)^2}{2!} + \frac{-1(-1-1)(-1-2)(-1x)^3}{3!} + \ldots\right)\right) \quad \textbf{[1]}$$

$= 2(1 + x + x^2 + x^3 + \ldots) = 2 + 2x + 2x^2 + 2x^3 + \ldots$ **[1]**

$$5(x+3)^{-1} = 5\left(3^{-1}\left((1 + \frac{-1\times\frac{1}{3}x}{1!} + \frac{-1(-1-1)(\frac{1}{3}x)^2}{2!} + \frac{-1(-1-1)(-1-2)\left(\frac{1}{3}x\right)^3}{3!} + \ldots\right)\right) \quad \textbf{[1]}$$

$$= \left(\frac{5}{3}\right)\left(1 - \frac{x}{3} + \frac{x^2}{9} - \frac{x^3}{27} + \ldots\right) = \frac{5}{3} - \frac{5x}{9} + \frac{5x^2}{27} - \frac{5x^3}{81} + \ldots \quad \textbf{[1]}$$

$$\frac{3x - 11}{(x-1)(x+3)} = 2 + 2x + 2x^2 + 2x^3 + \frac{5}{3} - \frac{5x}{9} + \frac{5x^2}{27} - \frac{5x^3}{81} + \ldots$$

$$= \frac{11}{3} + \frac{13x}{9} + \frac{59x^2}{27} + \frac{157x^3}{81} + \ldots \quad \textbf{[1]}$$

The expansion of $-2(x-1)^{-1}$ is valid for $|x| < 1$ and the expansion of $5(x+3)^{-1}$ is valid for $|x| < 3$, so the expansion of $\dfrac{3x-11}{(x-1)(x+3)}$ is valid for $|x| < 1$. **[1]**

9. $\dfrac{dv}{dx} = x, \quad u = \ln x$

$v = \dfrac{x^2}{2}$; $\dfrac{du}{dx} = \dfrac{1}{x}$ **[1]**

$\dfrac{x^2}{2}\ln x - \displaystyle\int \frac{x^2}{2} \times \frac{1}{x}\,dx$ **[1]** $= \dfrac{x^2}{2}\ln x - \dfrac{x^2}{4}$ **[1]**

$= (8\ln(4) - 4) - (2\ln(2) - 1)$
$= 8\ln(4) - 4 - \ln(4) + 1$ **[1]**
$= 7\ln(4) - 3$

10. $\cot\left(x - \frac{\pi}{4}\right) = \dfrac{1}{\tan\left(x - \frac{\pi}{4}\right)}$ **[1]**

$= \dfrac{1}{\dfrac{\tan x - \tan\frac{\pi}{4}}{1 + \tan x \tan\frac{\pi}{4}}}$ **[1]**

$= \dfrac{1 + \tan x \tan\frac{\pi}{4}}{\tan x - \tan\frac{\pi}{4}}$ **[1]**

$= \dfrac{1 + \tan x}{\tan x - 1}$ **[1]**

$= \dfrac{1 + \frac{\sin x}{\cos x}}{\frac{\sin x}{\cos x} - 1}$ **[1]**

$= \dfrac{\sin x + \cos x}{\sin x - \cos x}$

11. $\dfrac{dx}{dt} = 2t$ **[1]**

$\dfrac{dy}{dt} = 3t^2$ **[1]**, $\dfrac{dy}{dx} = \dfrac{3t}{2}$ **[1]**

At $t = 1$, $\dfrac{dy}{dx} = \dfrac{3}{2}$ **[1]**

$y - 1 = \dfrac{-2}{3}(x - 1)$ **[1]**

$3y = -2x + 5$ **[1]**

12. Suppose A has the coordinate (0, y, 0):
Then $AB^2 = (-1)^2 + (y-2)^2 + 0^2$ **[1]**
and $AC^2 = (-2)^2 + (y+1)^2 + (-2)^2$ **[1]**
So $(-1)^2 + (y-2)^2 + 0^2 = (-2)^2 + (y+1)^2 + (-2)^2$ **[1]**
$1 + y^2 - 4y + 4 = 4 + y^2 + 2y + 1 + 4$
$5 - 4y = 2y + 9$ **[1]**
$-4 = 6y \Rightarrow y = -\dfrac{2}{3}$ **[1]**

So coordinate is $A\left(0, -\frac{2}{3}, 0\right)$ **[1]**

13. a) Increasing **[1]**

b) $\dfrac{7}{2}, 4, \dfrac{9}{2}, 5, \dfrac{11}{2}$ **[1]**

c) $u_{n+1} = u_n + \dfrac{1}{2}$, $u_1 = \dfrac{7}{2}$ **[1]**

d) $\displaystyle\sum_{1}^{5} \frac{1}{2}n + 3 = \frac{7}{2} + 4 + \frac{9}{2} + 5 + \frac{11}{2} = \frac{45}{2}$ **[1]**

14. a) Substitute the information given for Fredrick's training into the nth term formula to find the number of weeks spent training.

$d = 1.5 - 1 = 0.5$ **[1]**
$10 = 1 + (n \times 0.5)$
$n = 9 \div 0.5 = 18$ weeks **[1]**

Substitute the information given for Amber's training to find d.

$10 = 2.8 + 18d$
$d = (10 - 2.8) \div 18 = 0.4$ **[1]**

b) Amber: $S_{26} = \dfrac{18}{2}(2.8 + 12) = 133.2$ km **[1 for finding the sum for both Amber and Fredrick]**
$133.2 \times 5.50 = 732.60$ raised **[1]**
Fredrick: $S_{26} = \dfrac{18}{2}(1 + 10) = 99$ km

$m = £732.60 \div 99 = £7.40$ **[1]**

15. a) Let $\dfrac{dv}{dx} = \cos x \Rightarrow v = \sin x$

Let $u = x \Rightarrow \dfrac{du}{dx} = 1$ **[1]**

$x\sin x - \int \sin x\,dx$ **[1]**
$\int x\cos x\,dx = x\sin x + \cos x + c$ **[1]**

b) $\int \cos^2 y\,dy = \int \frac{1}{2}(\cos 2y + 1)\,dy$ **[1]** $= \frac{1}{2}\left(\frac{1}{2}\sin 2y + y\right) + c$ **[1]**

$= \frac{1}{4}\sin 2y + \frac{1}{2}y + c$ **[1]**

c) $\int \cos^2 y\,dy = \int x\cos x\,dx$ **[1]**

$\frac{1}{4}\sin 2y + \frac{1}{2}y = x\sin x + \cos x + c$ **[1]**

16. a) $x = 2\sec(t)$

$\sec^{-1}\left(\dfrac{x}{2}\right) = t$ **[1]**

$y = 3\cos(t)$

$y = 3 \times \dfrac{1}{\sec\left(\sec^{-1}\left(\frac{x}{2}\right)\right)}$

Recall $\cos(\theta) = \dfrac{1}{\sec(\theta)}$

$y = 3 \times \dfrac{1}{\frac{x}{2}} \Rightarrow y = \dfrac{6}{x}$ **[1]**

To find the range, look at the equation $x = 2\sec(t)$, $0 \leqslant t < \dfrac{\pi}{2}$

$\sec(t)$ will have values $\sec(t) \geqslant 1$, so the range of $x = 2\sec(t)$ and hence the domain of $y = \dfrac{6}{x}$ is $\{x \in \mathbb{R}: x \geqslant 2\}$. **[1]**
Similarly, $y = 3\cos(t)$, $\cos(t)$ will have values $\{0 < \cos(t) \leqslant 1\}$, so the range of $y = \dfrac{6}{x}$ is $\{y \in \mathbb{R}: 0 < y \leqslant 3\}$. **[1]**

b)

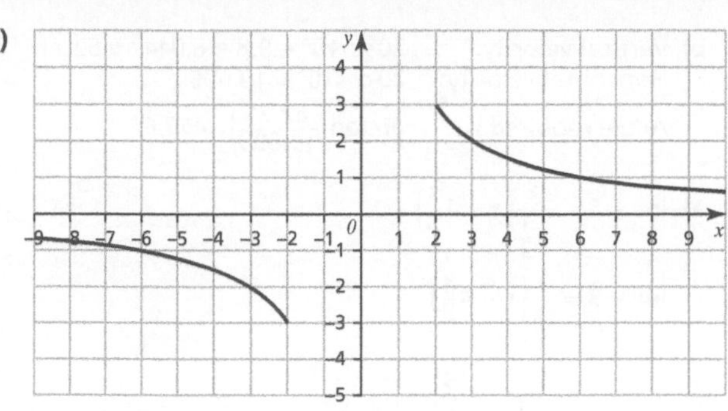

[1 for general shape and asymptote at $y = 0$; 1 for correct domain and range]

17. a) 6, 33, 1086, 1 179 393 [1]
b) 0, −3, 6, 33, 1086 [1]
c) The sequence is defined recursively so the successive values are the same after the term value of 6. [1]

18. $\frac{1}{\cos^2 x} + \frac{\sin^2 x}{\cos^2 x} = 3$ [1]

$\sec^2 x + \tan^2 x = 3$ [1]
$(1 + \tan^2 x) + \tan^2 x = 3$ [1]
$1 + 2\tan^2 x = 3$
$2\tan^2 x = 2 \therefore \tan^2 x = 1$ [1]
$\tan x = \pm 1$ [1]

So solutions are $x = \frac{\pi}{4}$, $x = \frac{3\pi}{4}$, $x = \frac{5\pi}{4}$, $x = \frac{7\pi}{4}$ [2]

Other methods can be used, e.g. converting to sin.

19. $f(0.8) = -0.073 < 0$ [2]
$f(0.9) = 0.101 > 0$ [2]
f is continuous in the interval $0.8 < x < 0.9$ and there is a sign change, so there exists a root α such that $0.8 < \alpha < 0.9$. [1]
$3^{-x} + x^3 - 1 = 0$
$x^3 = 1 - 3^{-x}$
$x = \sqrt[3]{1 - 3^{-x}}$
So use iteration $x_{n+1} = \sqrt[3]{1 - 3^{-x_n}}$ with $x_0 = 0.9$:
$x_1 = 0.8563$ [1]
$x_2 = 0.8479$
$x_3 = 0.8462$ [1]
$\alpha = 0.846$ [1]

20. a) The first year, they are charged interest on £200 000, so they owe £200 000 × 1.04 = £208 000. [1]
Each year they pay £1000 × 12 = £12 000, so the second year they are charged interest on £208 000 − £12 000 = £196 000; the balance owed is then (£196 000 × 1.04) − £12 000 = £191 840 [1]
The third year, they are charged interest on £191 840; the balance owed is then (£191 840 × 1.04) − £12 000 = £187 513.60 [1]
b) 1st year: £200 000 × 1.04
2nd year: (£200 000 × 1.04 − 12 000) × 1.04 − 12 000
$= 200\,000 \times 1.04^2 - 12\,000 \times 1.04 - 12\,000$ [1]
$= 200\,000 \times 1.04^2 - 12\,000(1.04 - 1)$ [1]
3rd year: $(200\,000 \times 1.04^2 - 12\,000(1.04 - 1) \times 1.04 - 12\,000$
$= 200\,000 \times 1.04^3 - 12\,000 \times 1.04(1.04 - 1) - 12\,000$ [1]
$= 200\,000 \times 1.04^3 - 12\,000 \times 1.04^2 - 12\,000 \times 1.04 - 12\,000$
$= 200\,000 \times 1.04^3 - 12\,000(1.04^2 + 1.04 + 1)$ [1]
c) $D_n = 200\,000 \times 1.04^n - 12\,000(1.04^{n-1} + 1.04^{n-2} + \ldots + 1.04 + 1)$ [1]

$$D_n = 200000 \times 1.04^n - \frac{12000(1.04^n - 1)}{(1.04 - 1)}$$

$D_n = 200\,000 \times 1.04^n - 300\,000(1.04^n - 1)$ [1]

21. a) $\arctan\left(\sin\frac{3\pi}{2}\right) = \arctan(-1) = -\frac{\pi}{4}$ [2]

b) $\sec\left(\arctan\sqrt{3}\right) = \sec\frac{\pi}{3} = 2$ [2]

c) $\sec\left(\arcsin\left(-\frac{1}{2}\right)\right) = \sec\left(-\frac{\pi}{6}\right) = \sec\left(\frac{\pi}{6}\right) = \frac{2\sqrt{3}}{3}$ [2]

22. $h = \frac{\frac{\pi}{3}}{3} = \frac{\pi}{9}$ [1]

x	0	$\frac{\pi}{9}$	$\frac{2\pi}{9}$	$\frac{\pi}{3}$
$\cos\left(\sqrt{\sin x}\right)$	1	0.8338	0.6955	0.5973

[2]

$\int_0^{\frac{\pi}{3}} \cos\left(\sqrt{\sin x}\right)\,dx \approx \frac{\left(\frac{\pi}{9}\right)}{2}\left[1 + 2(0.8338 + 0.6955) + 0.5973\right]$ [2]

$= 0.813$ units^2 [1]

23. The performer lands when $y = 0$.
$0 = -4.9t^2 + t(30\sin(\theta))$
$4.9t^2 = t(30\sin(\theta))$
$4.9t = 30\sin(\theta)$
$t = \frac{30\sin(\theta)}{4.9}$ [1]
Substituting t into $x = t(30\cos(\theta))$
$x = \frac{30\sin(\theta)}{4.9} \times (30\cos(\theta))$
$x = \frac{30^2}{4.9}(\sin(\theta)\cos(\theta))$ [1]
$x = \frac{30^2}{4.9}\left(\frac{1}{2}\sin(2\theta)\right)$ [1]

Use the double angle formula.

The net starts 75 m away, so $\frac{30^2}{4.9}\left(\frac{1}{2}\sin(2\theta)\right) \geqslant 75$ [1]

$\frac{30^2}{9.8}(\sin(2\theta)) \geqslant 75$

$\sin(2\theta) \geqslant 0.8167$
$2\theta \geqslant \sin^{-1}(0.8167) \Rightarrow 2\theta \geqslant 54.75°$
$\theta \geqslant 27.37°$ (to 2 d.p.) [1]
The net ends 90 m away, so $\frac{30^2}{4.9}\left(\frac{1}{2}\sin(2\theta)\right) \leqslant 90$ [1]
$\sin(2\theta) \leqslant 0.98$
$2\theta \leqslant \sin^{-1}(0.98) \Rightarrow 2\theta \leqslant 78.52°$
$\theta \leqslant 39.26°$ (to 2 d.p.) [1]

24. a) $\sin 70°\cos 10° - \sin 10°\cos 70° = \sin(70° - 10°) = \sin 60° = \frac{\sqrt{3}}{2}$ [2]

b) $\frac{\sqrt{3} + \tan 75°}{1 - \sqrt{3}\tan 75°} = \frac{\tan 60° + \tan 75°}{1 - \tan 60°\tan 75°}$ [1]
$= \tan(60° + 75°) = \tan 135° = \tan(180° - 45°) = -\tan 45°$ [1]
$= -1$ [1]

c) $\sin 75°\cos 75° = \frac{1}{2}(2\sin 75°\cos 75°)$ [1]
$= \frac{1}{2}\sin 150°$ [1]
$= \frac{1}{2}\sin(180° - 30°) = \frac{1}{2} \times \frac{1}{2}$ [1] $= \frac{1}{4}$ [1]

25. $3\cos 2x - \sin 2x = R\cos(2x + \alpha)$
$= R\cos 2x\cos\alpha - R\sin 2x\sin\alpha$
$3 = R\cos\alpha$ [1]
$1 = R\sin\alpha$ [1]
$R^2 = 3^2 + 1^2 = 10 \Rightarrow R = \sqrt{10}$ [2]
$\frac{R\sin\alpha}{R\cos\alpha} = \tan\alpha = \frac{1}{3} \Rightarrow \alpha = 0.3218^c$ [2]
$3\cos 2x - \sin 2x = \sqrt{10}\cos(2x + 0.3218)$
$\sqrt{10}\cos(2x + 0.3218) = 1$
$\cos(2x + 0.3218) = \frac{1}{\sqrt{10}}$ [1]
$2x + 0.3218 = 1.249 \Rightarrow x = 0.464$ [2]
$2x + 0.3218 = 2\pi - 1.249 \Rightarrow x = 2.36$ [2]

Statistics and Mechanics

1.

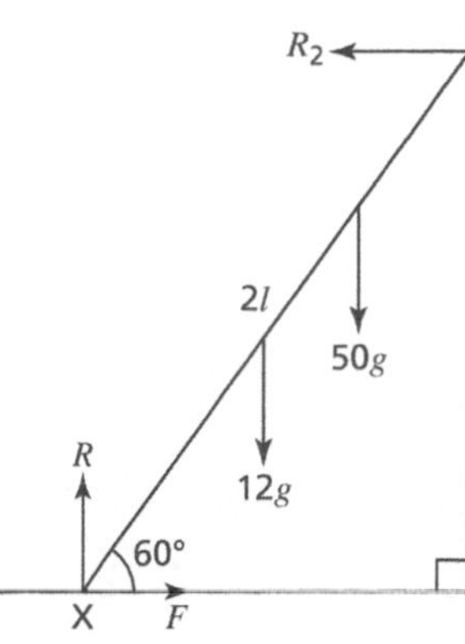

Suppose the child climbs a distance x up the ladder before it begins to slip.

Resolving vertically: $R = 62g$ [1]
Resolving horizontally: $R_2 = F$ [1]
Friction limiting: $F = 0.3R$ [1]
So $R_2 = F = 0.3R = 0.3 \times 62g = 18.6g$
Moments about X:
$12gl\cos60° + 50gx\cos60° = 18.6g\,(2l\sin60°)$ [3]
$50x\cos60° = 18.6\,(2l\sin60°) - 12l\cos60°$

$x = \dfrac{18.6(2l\sin60°) - 12l\cos60°}{50\cos60°}$ [1] $\Rightarrow x = 1.05l$ [1]

2. **a)** $3t - 2 = 0 \Rightarrow t = \frac{2}{3}$ [2]

b) $6 - 2t = 0 \Rightarrow t = 3$ [2]

c) $\mathbf{r} = \int_{\frac{2}{3}}^{3}((3t-2)\mathbf{i} + (6-2t)\mathbf{j})\,dt = \left[\left(\frac{3t^2}{2} - 2t\right)\mathbf{i} + (6t - t^2)\mathbf{j}\right]_{\frac{2}{3}}^{3}$ [2]

$= \frac{15}{2}\mathbf{i} + 9\mathbf{j} - \left(-\frac{2}{3}\mathbf{i} + \frac{32}{9}\mathbf{j}\right) = \frac{49}{6}\mathbf{i} + \frac{49}{9}\mathbf{j}$ [1]

3. **a)** $\bar{M} \sim N\left(85, \frac{6.76}{10}\right)$ [1]

b) $P(\bar{M} < 83.5) = P\left(Z < \dfrac{83.5 - 85}{\frac{2.6}{\sqrt{10}}}\right)$ [1]

$P(Z < -1.82)$ [1]
0.0344 [1]

c) $H_0: \mu = 85 \quad H_1: \mu > 85$
$\bar{M} \sim N\left(85, \frac{6.76}{10}\right) \quad P(\bar{M} > 87)$ [1]

$P\left(Z > \dfrac{87 - 85}{\frac{2.6}{\sqrt{10}}}\right)$ [1]

0.0075 [1]
$0.0075 < 0.05$ Therefore sufficient evidence to reject the null hypothesis; there is evidence the mean has increased. [1]

This is a one-tailed test.

4. **a)** $0.6 \times 0.55 + 0.4 \times 0.35$ [1] $= 0.47$ [1]

b) $\dfrac{0.6 \times 0.55}{0.6 \times 0.55 + 0.4 \times 0.35}$ [1] $= \dfrac{0.33}{0.47}$ [1] $= 0.702$ [1]

5. **a)** Vertical motion downwards: $200 = (20\sin10°)t + 4.9t^2$ [2]
$4.9t^2 + (20\sin10°)t - 200 = 0$
Solving using quadratic formula gives: $t = 6.04$ s [2]

b) Vertical velocity: $20\sin10° + 9.8 \times 6.0442 = 62.71$ [2]
Horizontal velocity: $20\cos10° = 19.696$ [2]
Angle required is $\arctan\left(\frac{62.71}{19.696}\right) = 72.6°$ [1]

6. **a)** $\ln y - 3 = \dfrac{3}{4\frac{1}{3}}\left(x - 4\frac{1}{3}\right)$

$\ln y - 3 = \dfrac{3}{\frac{13}{3}}\left(x - 4\frac{1}{3}\right)$ [1]

$\ln y - 3 = \frac{9}{13}\left(x - \frac{13}{3}\right)$

$13\ln y - 39 = 9x - 3$

$13\ln y = 9x + 36$ [1]

b) $\ln y = \ln(ab^x)$
$\ln y = \ln a + \ln b^x$
$\ln y = x\ln b + \ln a$ [1]

$\ln b = \frac{9}{13} \quad \ln a = \frac{36}{13}$

$b = e^{\frac{9}{13}}$ [1] $\quad a = e^{\frac{36}{13}}$ [1]

7. **a)** $P\left(Z < \dfrac{180 - 200}{14}\right)$ [1]
$P(Z < -1.43)$ [1]
0.0764 [1]

b) $0.516 \div 2 + 0.5 = 0.758$ [1]
$Z = 0.7$ [1]
$\dfrac{200 + k - 200}{14} = 0.7$ [1]
$k = 14 \times 0.7$ [1] $\Rightarrow k = 9.8$ [1]

8. Resolving perpendicular to the plane:
$R = 1.5g\cos30° + P\sin30°$ (1) [2]
Resolving parallel to the plane: $F + 1.5g\sin30° = P\cos30°$ (2) [2]
Friction limiting: $F = 0.45R$ (3) [1]
Solving (1), (2), (3) simultaneously $\Rightarrow P = 20.4$ N [2]

9. Resolving horizontally:
$P\cos\theta° + 13\cos55° = 10\cos40° + 4\cos52°$ (1) [2]
Resolving vertically:
$P\sin\theta° + 10\sin40° = 13\sin55° + 4\sin52°$ (2) [2]
Solving simultaneously:
$\dfrac{P\sin\theta°}{P\cos\theta°} = \tan\theta° = \dfrac{13\sin55° + 4\sin52° - 10\sin40°}{10\cos40° + 4\cos52° - 13\cos55°}$ [2]
$\theta = 70.1°$ [1]
Substituting into (1) or (2): $P = 7.84$ N [1]

Glossary and Index

A

acceleration vector the derivative of the velocity vector **104–105**

addition formulae in trigonometry, the identities for $\sin(A + B)$, $\cos(A + B)$ and $\tan(A + B)$ **38, 78–79**

adjacent next to **39, 108–109**

algebraic division also called polynomial division; the process of dividing a polynomial by another polynomial **10**

algebraic fraction a fraction in which the numerator, the denominator, or both are polynomials **10**

alternate hypothesis the statement which contradicts the null hypothesis **95, 96–97**

angle of elevation the angle measured from the horizontal line of sight and an object **14–15, 106–107**

approximation a value that is nearly, but not exactly correct **34, 75, 76–77, 92–93**

arc length the distance along an arc (part of the circumference of a circle) **30–31, 34**

arithmetic progression the sum of an arithmetic sequence; also the arithmetic series **19**

arithmetic sequence a sequence in which the difference between each consecutive term is constant **18–19**

assumption a mathematical condition that is assumed to hold true **78, 88–89**

asymptote a line on a graph that a function approaches but never meets **9, 12, 74**

B

binomial expansion also known as the Binomial Theorem; shows the algebraic expansion of a binomial when n is a rational number $(a+bx)^n = a^n\left(1+\frac{nbx}{1!}+\frac{n(n-1)(bx)^2}{2!}+\frac{n(n-1)(n-2)(bx)^3}{3!}+\ldots\right)$ **22–23, 34**

boundary conditions conditions for the given problem to determine the particular solution from the general solution **64**

C

Cartesian equation an equation of a curve in which the variables are x and y, where y is expressed as a function of x **12–13, 14**

chain rule the rule used to differentiate composite functions **52–53, 54, 56–57, 59, 62**

chord a line joining any two points on the circumference of a circle **31**

cobweb diagram a type of diagram obtained when finding approximations to roots. The sequence of approximations oscillates around the actual root **74–75**

common difference the amount by which an arithmetic sequence increases or decreases between each term **18–19**

common ratio the ratio by which each term is multiplied to generate a geometric sequence **20**

comparing coefficients in trigonometric identities, for example, if $A\cos x + B\sin x \equiv C\cos x + D\sin x$, then $A = C$ and $B = D$ **40**

components separate parts of a vector **104–105, 108–109**

composite function a combination of functions in which the output of one function is the input of another function; these are denoted as fg(x), which means do the function g first, then the function f **6, 52**

conditional probability the probability of an event occurring given that another event has already occurred **86**

constant vector a vector that does not change **104**

continuity correction the correction required when using a continuous distribution to estimate probabilities from a discrete distribution **92–93**

continuous (function) a function is said to be continuous if, for sufficiently small changes in the input, there are arbitrarily small changes in the output **74–75, 76, 90, 92–93**

contradiction a result obtained that indicates a previous mathematics assumption must be false **78**

converge when a sequence or function tends toward a finite number as the variable tends to infinity **20–21, 74–75, 76**

converse the converse to a proposition is obtained when the hypothesis and conclusion are swapped (e.g. the converse of 'Gruff is a goat, therefore he has four legs' is 'Gruff has four legs, therefore he is a goat') **78**

critical value the boundary used for determining if a result is statistically significant or not **94–95**

D

distance the length of space between two points **15, 32–33, 76, 105, 112**

direction of motion equivalent to the direction of the velocity (vector) **104, 110**

diverge when a sequence or function does not tend toward a finite number as the variable tends to infinity **21**

domain the set of all possible values of the input of a function **6–7, 12–13, 15, 35, 76**

double angle double-angle formulae are used to express trigonometric functions of $2x$ in terms of x (e.g. $\sin 2x = 2\sin x\cos x$) **38–39, 78–79**

E

equating coefficients comparing the coefficients of like terms in two expressions **11**

equation a mathematical statement involving an '=' sign **8, 11, 12–15, 18, 36–37, 39, 40–41, 53, 54–58, 64–65, 74–77, 78, 88, 91, 106–107, 109, 110–113**

equation of motion for a resultant force F, applied to a mass m, the equation of motion is $F = ma$, where a is the acceleration of the mass **110–111**

equilibrium a particle is said to be in equilibrium if the resultant force on it is zero; a rigid body is in equilibrium if the resultant force is zero and the total moment is zero **108–109, 110, 112**

exact trigonometric values trigonometric ratios expressed in surd form **38**

extrapolation an estimate taken outside the given data set **88**

F

finite a definite number, or something that can be measured or given a value **21, 58, 78**

friction a type of contact force opposing motion **110–111, 113**

function a mapping, in which each input maps to a unique output; functions can be one-to-one or many-to-one mappings **6–7, 8, 10–12, 22, 34–35, 37, 41, 52, 54–55, 59, 61, 62–63, 74, 76–77, 89, 107**

G

general solution the solution to a differential equation for which the constant of integration is yet to be found; this represents the family of solutions **64**

geometric progression (sequence) a sequence in which each successive term is found by multiplying the previous term by a common ratio **20–21**

greatest height the maximum height reached by a projectile, occurring when the vertical velocity is instantaneously zero **106**

H

half angle half-angle formulae are used to express trigonometric functions of x in terms of $\frac{x}{2}$ (e.g. $\sin x = 2\sin\frac{x}{2}\cos\frac{x}{2}$) **39, 79**

horizontal motion in projectile motion, equations referring to horizontal displacement and horizontal velocity **106**

horizontal range the greatest horizontal distance reached by a projectile, occurring when the vertical displacement is zero **106**

hypothesis test the method used to test if a result from an experiment or sample is considered statistically significant or not **94, 96**

I

identity an equation that is true for all values **36–39, 59, 63, 78–79**

implicit differentiation a special case of the chain rule **54–55**

independent events the probability of one event occurring is not changed by the fact that another event occurred **87**

inextensible (string) a string is said to be inextensible if its length remains unchanged **111**

infinite without an end (e.g. the sequence of natural numbers is infinite) **21, 22, 78, 89**

integration by inspection a method used to integrate an expression **62–63**

interpolation an estimate taken within a given data set **88–89**

interval a range of values **74–75**

inverse function given a one-to-one function f(x), its inverse function $f^{-1}(x)$ is a function that performs the opposite operations of f(x) in the reverse order; if f(a) = b, then $f^{-1}(b) = a$ **7, 35**

iteration formula a formula which generates successive improved approximations to a solution of an equation **75, 76**

L

limiting friction friction is 'limiting' if an object is moving, or on the point of moving; in this case, $F = \mu R$ **110**

linear motion motion following the path of a straight line **14**

M

magnitude the size of a number or vector **32, 108–109, 112**

mapping a transformation of values from one set into values in another set **6**

modulus the absolute size of a number **8**

moment the turning effect obtained due to a force applied to a rigid body **112–113**

mutually exclusive two events which cannot occur at the same time **86**

N

***n*th term formula** a formula that defines the nth term of a sequence **16–17, 18, 20–21**

null hypothesis the statement assumed to be true when carrying out a hypothesis test **95, 97**

P

parabolic motion motion following the shape of a parabola **14**

parameter the variable for which parametric equations are defined **12–13, 96**

parametric equations a pair of functions used to define the x- and y-coordinates on a graph; given parametric functions $x = p(t)$, $y = q(t)$, each point (x, y) on the curve is given by $(p(t), q(t))$ **12–15, 56**

partial fraction a fraction resulting from decomposing a fraction into a sum of two or more fractions **10–11, 22–23, 62–63**

particular solution the solution to a differential equation where the constant of integration has been found **64–65**

periodic sequence a sequence that repeats itself after a set interval, called the period **16–17**

perpendicular at 90° to **32, 108–112**

pivot the point about which a turning force acts **112**

point of inflection the point at which a curve changes from convex to concave, or vice-versa **53**

population parameter a quantity or measure for a given population **96**

position vector a vector that defines an object's location, or position **32, 104–105**

positive displacement a specified direction in which an object moves; displacement can be positive or negative **106, 111**

product moment correlation coefficient a value on a scale of –1 to 1 which measures the strength of correlation between two variables **94**

proposition a mathematical statement that can (sometimes) be categorised as being true or false **78**

R

radian one radian is the size of an angle subtended (made) at the centre of a circle by an arc equal in length to the radius; one radian is approximately 57.3° **30–31, 34, 37**

random variable a variable whose possible outcomes are randomly generated **90–93**

range the set of all possible values of the output of a function **6–7, 12–13, 22–23, 34–35, 37, 40, 88, 106**

Ratio Theorem $\overrightarrow{OA} = \mathbf{a}$ and $\overrightarrow{OB} = \mathbf{b}$
If C divides AB in the ratio $m:n$, then $\overrightarrow{OC} = \frac{n\mathbf{a}+m\mathbf{b}}{m+n}$ **33**

rational a number is rational if it can be written in the form $\frac{p}{q}$, where p and q are both integers **22, 38, 78**

rational expression another term for algebraic fraction; a fraction in which the numerator, denominator, or both are polynomials **10, 23**

rational function a function that can be expressed as a fraction with a polynomial in the numerator and/or denominator **11, 22**

recursive formula a formula that defines the $(n + 1)$th term of a sequence in relation to the nth term **16**

resolve consideration of the components of one or more vectors in one or more particular directions **108–109, 111**

resultant force the algebraic sum of two or more forces **108–112**

rigid body a shape in two or three dimensions which cannot change shape (when forces are applied) **112**

root a solution to the equation f(x) = 0 **39, 74–77**

rough (surface) a surface is said to be 'rough' if friction exists **110, 113**

S

sector the plane shape enclosed by two radii of a circle **30–31, 34**

segment the shape(s) formed when a circle is split into two parts by a chord **31**

sequence a set of numbers in a certain order that follow a particular pattern (e.g. the set of odd numbers 1, 3, 5, 7, ...) **16–18, 20, 36, 74–76**

series the sum of a sequence **16–21, 78**

sketching graphs an illustration of a graph that, while not necessarily accurate, shows the characteristics of the graph (i.e. the shape, asymptotes and coordinates of intersections with axes) **40**

small angles angles that are generally small enough so that approximations such as $\theta \approx \sin\theta$ may be used **34**

smooth (surface) a surface is said to be 'smooth' if there is no friction **110–111, 113**

solving equations finding the solutions to equations **36, 40, 74**

staircase diagram a type of diagram obtained when finding approximations to roots; the sequence of approximations is either an increasing sequence, or a decreasing sequence **74–75**

standard normal distribution the normal distribution with a mean of 0 and a standard deviation of 1 **90**

stationary point a point on a curve corresponding to a maximum point, a minimum point, or a point of inflection, where the gradient is equal to zero **53, 76**

T

tangent a tangent to a point on a curve is a straight line touching that point **55, 56, 76**

test statistic a standardised value calculated from sample data **96**

transformation changing the size or location of an object or graph of a function **8–9, 12, 40**

time of flight the time for which a projectile is moving **106**

U

uniform rod a (theoretical) one-dimensional object in mechanics which has an equally distributed mass **112**

unit vector a vector of magnitude 1 **32, 104**

V

vector form any quantity in mathematics expressed as a vector **104, 107, 108**

velocity vector the derivative of the position vector; the magnitude of the velocity vector gives you the speed **104–105**

vertical motion in projectile motion, equations referring to vertical displacement and horizontal velocity **106**

Collins

Edexcel A-Level Maths Year 2

Workbook

Phil Duxbury, Rebecca Evans
and Leisa Bovey

Revision Tips

Rethink Revision

Have you ever taken part in a quiz and thought *'I know this!'* but, despite frantically racking your brain, you just couldn't come up with the answer?

It's very frustrating when this happens but, in a fun situation, it doesn't really matter. However, in your A-level exams, it will be essential that you can recall the relevant information quickly when you need to.

Most students think that revision is about making sure you ***know** stuff*. Of course, this is important, but it is also about becoming confident that you can **retain** that *stuff* over time and **recall** it quickly when needed.

Revision That Really Works

Experts have discovered that there are two techniques that help with all of these things and consistently produce better results in exams compared to other revision techniques.

Applying these techniques to your A-level revision will ensure you get better results in your exams and will have all the relevant knowledge at your fingertips when you start studying for further qualifications or begin work.

It really isn't rocket science either – you simply need to:

- **test yourself** on each topic as many times as possible
- **leave a gap** between the test sessions.

Three Essential Revision Tips

1. Use Your Time Wisely

- Allow yourself plenty of time.
- Try to start revising at least six months before your exams – it's more effective and less stressful.
- Your revision time is precious so use it wisely – using the techniques described on this page will ensure you revise effectively and efficiently and get the best results.
- Don't waste time re-reading the same information over and over again – it's time-consuming and not effective!

2. Make a Plan

- Identify all the topics you need to revise (this Complete Revision & Practice book will help you).
- Plan at least five sessions for each topic.
- One hour should be ample time to test yourself on the key ideas for a topic.
- Spread out the practice sessions for each topic – the optimum time to leave between each session is about one month but, if this isn't possible, just make the gaps as big as realistically possible.

3. Test Yourself

- Methods for testing yourself include: quizzes, practice questions, flashcards, past papers, explaining a topic to someone else, etc.
- This Complete Revision & Practice book provides seven practice opportunities per topic.
- Don't worry if you get an answer wrong – provided you check what the correct answer is, you are more likely to get the same or similar questions right in future!

Visit our website to download your free flashcards, for more information about the benefits of these techniques, and for further guidance on how to plan ahead and make them work for you.

www.collins.co.uk/collinsalevelrevision

Contents

Algebra and Functions

1 f: $x \mapsto \dfrac{3x}{x-1}$ g: $x \mapsto x^2 - 3$

a) Find the function fg(x). [1]

b) Find the function $f^{-1}(x)$, stating the domain and the range. [5]

2 $f(x) = \sqrt{x+4}$

a) State the domain and the range of f(x). [2]

b) Find $f^{-1}(x)$, stating the domain and the range. [4]

c) Sketch a graph of $y = f(x)$ and $y = f^{-1}(x)$ on a separate piece of paper. [3]

3 $f(x) = 2x - 3$ $g(x) = x^2 + 4$

a) Find the function fg(x). [1]

b) Find the function gf(x). [1]

c) Find the function $f^2(x)$. [1]

d) Find the values of x for which $gf(x) = 29$. [2]

4 Sketch a graph on a separate piece of paper and use it to solve the equation $|2x+3|-1=\frac{1}{2}x+3$. [6]

5 Given the graph of $y = f(x)$,

sketch a graph of $y = 2(f(x-3)) - 1$. [4]

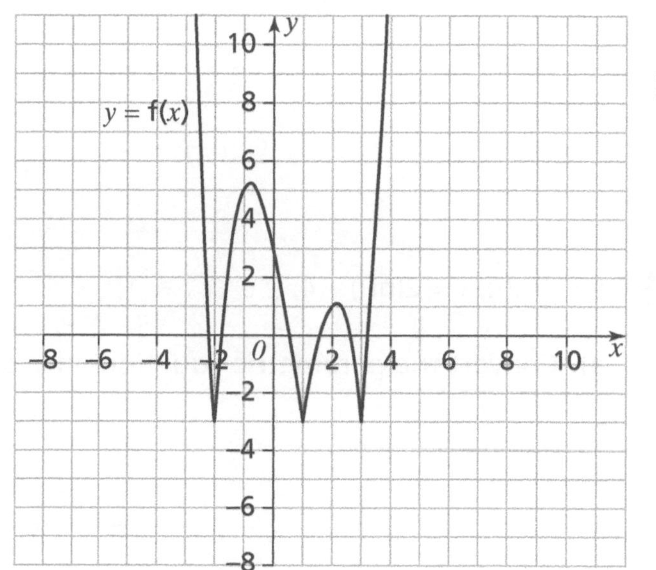

6 $f(x) = \frac{4}{x+2} + \frac{2x}{x-1} + 5$

a) Simplify $f(x)$. [2]

b) Show that $f(x) \div (x+2) = \frac{2}{3(x-1)} + \frac{19}{3(x+2)} + \frac{4}{(x+2)^2}$ [8]

7 Show that $\frac{2x^3+5x^2-4x}{x+3}$ can be written in the form $Ax^2 + Bx + C + \frac{D}{x+3}$

and hence find the values of A, B, C and D. [5]

Total Marks .... / 45

Coordinate Geometry

1 Show that the parametric equations

$x = 2 + 5\cos(t)$, $y = -4 + 5\sin(t)$, $\left\{t \in \mathbb{R}: \frac{\pi}{2} \leqslant t \leqslant \pi\right\}$

give this graph. [7]

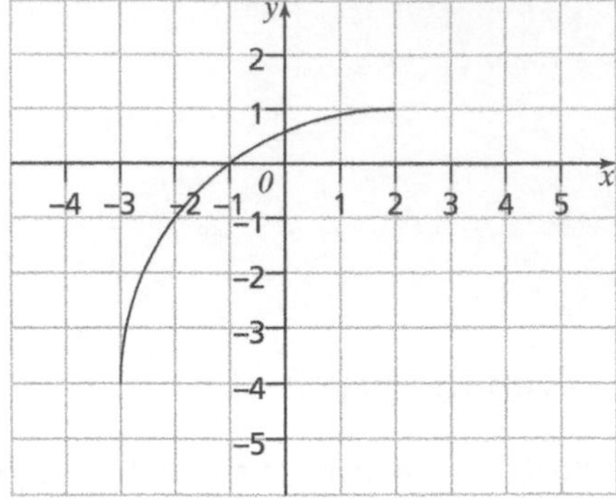

2 $x = 3\ln(t)$, $y = 2\ln(t) + 4$, $\{t \in \mathbb{R}: t \geqslant 1\}$

a) Convert the parametric equations into a Cartesian equation $y = \text{f}(x)$. [5]

b) Sketch a graph of $\text{f}(x)$ on a separate piece of paper. [2]

3 $x = 30\left(\frac{t}{1+t^3}\right)$, $y = 30\left(\frac{t^2}{1+t^3}\right)$, $\{t \in \mathbb{R}: 0 \leqslant t \leqslant 100\}$

a) Complete the table below. [2]

t	0	0.25	0.5	0.75	1	1.5	2	3	10	100
$x = 30\left(\frac{t}{1+t^3}\right)$										
$y = 30\left(\frac{t^2}{1+t^3}\right)$										

b) Sketch the graph of the parametric equations on a separate piece of paper. [2]

4 A boat is launched from a riverbank at a constant velocity of 1.5 ms^{-1}, at an angle of 30° relative to the bank, across an 80 m wide river. The river current flows with a velocity of –1 ms^{-1} relative to the bank.

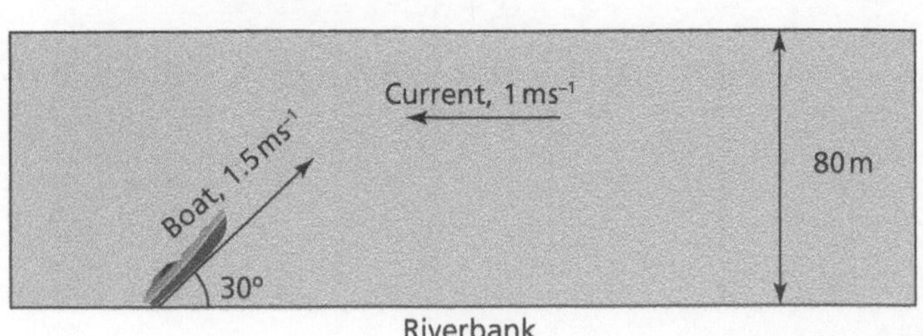

a) Find the parametric equations that can be used to model the boat's path as it crosses the river. [3]

b) Find the total time taken for the boat to cross the river. [3]

c) Find the horizontal distance travelled by the boat down the river. [1]

5 A projectile is launched off a 100 m high cliff at an angle of 30° with initial speed v. The path of the projectile can be modelled by the parametric equations
$x = (v\cos(30°))t$, $y = 100 + (v\sin(30°))t - 4.9t^2$.

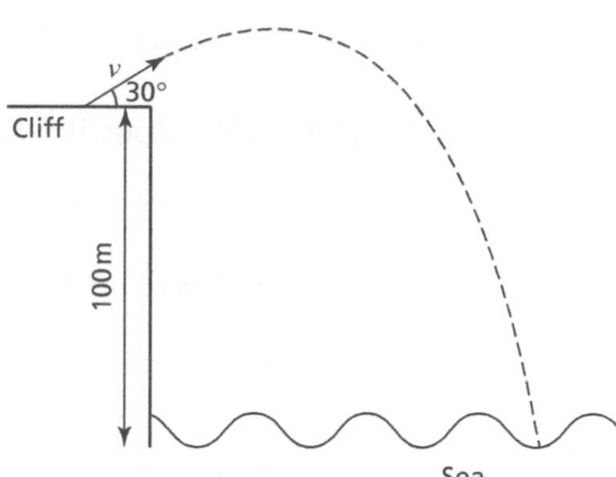

a) Find the range of values of v such that the projectile takes between five and six seconds to hit the sea. [4]

b) Find the total horizontal distance travelled by the projectile if it takes exactly five seconds to hit the sea. [1]

6 The path traced by a point, P, on the edge of a wheel rolling along a flat surface at a speed of one radian per second, can be modelled by the parametric equations $x = a(t - \sin(t))$, $y = a(1 - \cos(t))$ where P is touching the surface at $t = 0$.

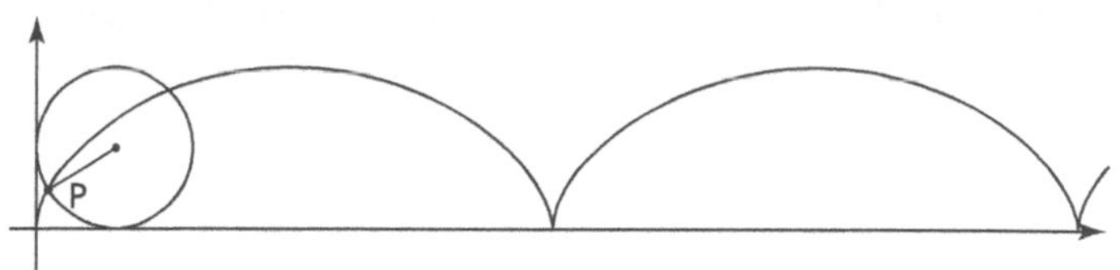

a) Given that point P is 0.25 cm above the surface at π seconds, find the value of a. [2]

b) Find the times at which point P is at the greatest height above the surface. [2]

c) Find the horizontal distance the wheel travels in one full revolution. [2]

Total Marks .... / 36

Sequences and Series

1 A sequence is given by $u_n = 2n - 8$.

a) Describe the sequence as increasing, decreasing or periodic. [1]

b) Write the first five terms of the sequence. [1]

c) Write a recursive formula to describe the sequence. [1]

d) Find $\sum_{1}^{5} u_n$ [1]

2 A runner trains for a 42.195 km marathon. She starts training by running 10 km the first day, then increases by 200 m each day.

a) How many days does it take her to train for the full distance? [2]

b) How many kilometres has she run in total by the end of her training? [1]

3 The first term of an arithmetic sequence is 1295 and the 15th term is 4795. Find the sum of the first 30 terms. [3]

4 Find the sum of the first 100 odd numbers and express it using sigma notation. [3]

5 A geometric sequence is defined by $u_n = 7 \times 2^{n-1}$

a) List the first five terms of the sequence. [1]

b) Find the sum of the first 10 terms of the sequence. [2]

c) Decide if the series is convergent or divergent and find the sum of the infinite series, if possible. [2]

6 Adelaide opens a savings account with £5000. The account pays 3% interest, compounded annually at the end of each year. She does not add any more money to the account.

a) How much money will be in the account after five years? [2]

b) After how many years will Adelaide have £7500 in the account? [4]

7 A 100-litre tank is full of dye. One-quarter of the dye is removed and replaced with water, then mixed. One-quarter of the new mixture is removed and replaced with water, then mixed, and so on.

a) Find the amount of dye in the first five dilutions. [1]

b) Write a formula for the amount of dye in the nth dilution. [1]

c) Find the number of dilutions after which the mixture is less than 1% dye. [4]

8 Find the first four terms in the expansion of $\frac{1}{(4x-3)^3}$ and state the values of x for which the expansion is valid. [4]

9 **a)** Find the first four terms in the expansion of $\frac{2}{\sqrt{x+1}}$ [3]

b) A student substitutes $x = 4$ to find an approximation to $\frac{2}{\sqrt{5}}$. Comment on the validity of this substitution. [1]

10 Given the coefficient of x^2 is –60 in the expansion of $\frac{1}{(kx-1)^5}$, find the value of k. [3]

Total Marks .... / 41

Trigonometry

1 A circle has radius 14 cm. Given that a chord AB, of length 25 cm, splits the circle into a minor segment and a major segment, find the area of the minor segment formed. [4]

2 The diagram shows a circle of radius 10 cm split into a major sector, an equilateral triangle and a minor segment.

Find the ratio of the minor segment to the major sector, expressing your answer in the form 1: n. [5]

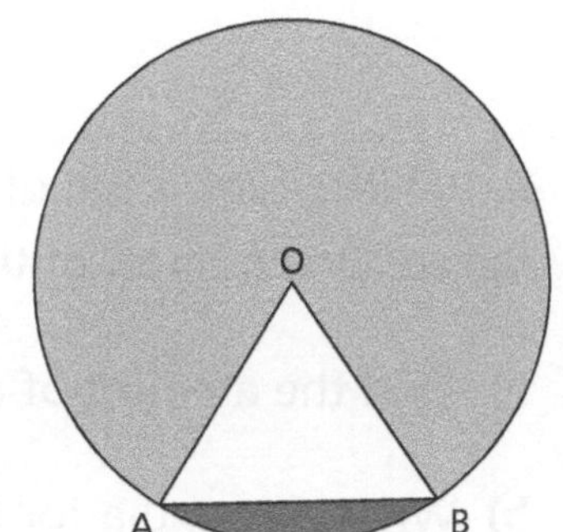

3 The diagram shows a semicircle, centre O and radius OA = OB = r.

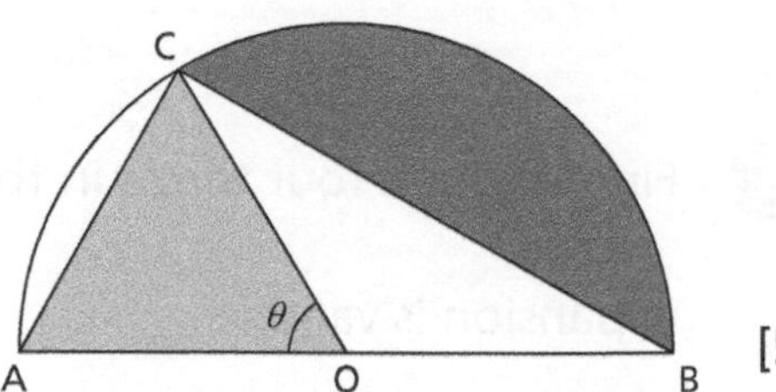

a) Given that the two shaded parts have the same area, show that $\theta + 2\sin\theta = \pi$. [5]

b) Hence show that a solution $\theta = \alpha$ exists to the above equation, where $1.2^c < \alpha < 1.3^c$. [5]

4 Given the vector $\overrightarrow{OA} = \begin{pmatrix} 2 \\ -3 \\ 1 \end{pmatrix}$ and $\overrightarrow{OB} = \begin{pmatrix} -1 \\ 1 \\ 3 \end{pmatrix}$, find the vector $\overrightarrow{AB}$.

Hence find the distance between the two points A and B. [4]

5 Given the position vector $\overrightarrow{OA} = \begin{pmatrix} 2t+4 \\ 2t \\ 1-4t \end{pmatrix}$, find the least possible value of $|\overrightarrow{OA}|$. [7]

6 Solve the equation $\tan 2x - \tan x = 0$ for $0 \leqslant x < 2\pi$. [6]

7 If x is small and in radians, show that $\dfrac{4\sin x - 4x\cos x}{\sin^2 x \tan x} \approx 2$. [5]

Trigonometry

8 Find the exact values of the following, without using a calculator:

a) $\operatorname{cosec}\left(-\frac{\pi}{4}\right)$ [2]

b) $\cot\left(-\frac{5\pi}{4}\right)$ [2]

c) $\sec\left(\frac{7\pi}{3}\right)$ [2]

9 Prove the identity $\cos^2 x - 1 \equiv (1 - \sec^2 x)(1 - \sin^2 x)$. [4]

10 Solve the equation $\sec^2\left(x - \frac{\pi}{6}\right) - 2\tan\left(x - \frac{\pi}{6}\right) - 2 = 0$ in the range $0 \leqslant x < 2\pi$. [9]

11 Solve the equation $\cot x \operatorname{cosec}^2 x - \cot x = 3\sqrt{3}$ in the range $0 \leqslant x < 2\pi$. [5]

12 Given $\sin x = -\frac{1}{4}$ and $\frac{3\pi}{2} < x < 2\pi$, find exact values for:

a) $\cos 2x$ [2]

b) $\tan 2x$ [4]

13 Using a half-angle identity, find the exact value of $\sin 195°$. [3]

14 Solve the equation $\sin x + \cos x = 1$ in the range $0 \leqslant x < 2\pi$. [5]

15 Express $4\cos x + 2\sin x$ in the form $R\cos(x - \alpha)$ where $R > 0$ and $0 \leqslant \alpha \leqslant \frac{\pi}{4}$

Hence solve the equation $4\cos x + 2\sin x = 2$ in the range $0 \leqslant x < 2\pi$. [11]

Total Marks .... / 90

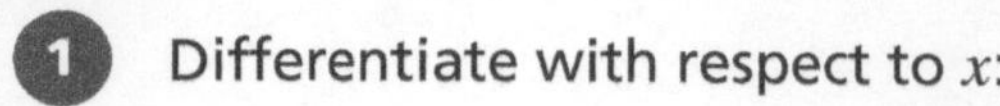

Differentiation

1 Differentiate with respect to x:

a) $x^2\sin 4x$ [3]

b) $\dfrac{\cos x}{x}$ [3]

c) $4\ln(3x) + \dfrac{1}{2x}$ [3]

2 Find the turning points in the graph with equation $y = \dfrac{e^x}{x}$ [7]

3 A curve has the parametric equation $x = \dfrac{2t}{1+t}$ $y = \dfrac{t^2}{1+t}$

a) Show that the normal to the curve at the point $A\left(1, \frac{1}{2}\right)$ has equation $6y = 7 - 4x$. [5]

b) The normal meets the curve again at the point B. Find the coordinates of B. [4]

4 A curve C satisfies the equation $x^3 + y^2 + 4x^2y = 2$.

a) Find $\dfrac{dy}{dx}$ in terms of x and y. [4]

b) When the x-coordinate of the curve is 1, show that the y-coordinate satisfies the equation $y^2 + 4y - 1 = 0$. [2]

c) The coordinate $(1, a)$ lies on C. Given that $y > 0$, find a and hence find the value of $\dfrac{dy}{dx}$ at the point $(1, a)$. [3]

Total Marks .... / 34

Integration

1 The diagram shows the curve with equation $y = x^3 - x^2 - 6x + 10$ and the line with equation $y = 3x + 1$.

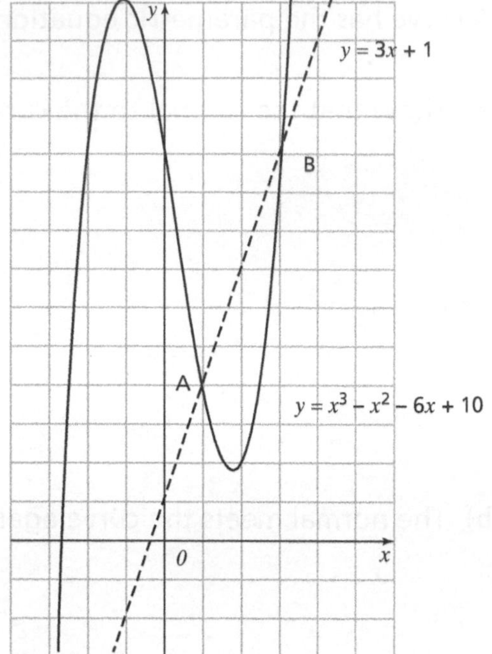

a) Find the coordinates of the points of intersection between the curve and the line. [3]

b) Find the area of the region enclosed by the curve, the x-axis and the lines $x = \text{A}$ and $x = \text{B}$. [5]

2 **a)** Find the general solution to the differential equation $\dfrac{dy}{dx} = xy^2$, $y > 0$. [5]

b) Given that $y = 1$ at $x = 1$, show that $y = \dfrac{2}{3 - x^2}, -\sqrt{3} < x < \sqrt{3}$. [3]

3 $y = \dfrac{5x^2 - 8x + 1}{2x(x - 1)^2}$

a) Write y in the form $\dfrac{A}{2x} + \dfrac{B}{x-1} + \dfrac{C}{(x-1)^2}$, where A, B and C are constants to be found. [6]

b) Hence find $\displaystyle\int \frac{5x^2 - 8x + 1}{2x(x - 1)^2}\,dx$ [4]

Total Marks .... / 26

Numerical Methods

1 Show that a root α of the equation $e^{-x} + \frac{1}{x} = 0$ exists in the interval $-0.6 < x < -0.5$
Find α to 2 significant figures by applying the Newton-Raphson iteration once, taking $x_0 = -0.6$ as your starting value. [9]

2 Show that the equation $f(x) = x^7 - \sin x = 0$ has a root α such that $0.9 < \alpha < 1.0$

By using a suitable iteration formula of the form $x_{n+1} = g(x_n)$ with $x_0 = 1$, write down x_1, x_2 and x_3, and hence determine α to 3 significant figures. [8]

3 Consider the function $f(x) = e^{-x} + x^3 - 1$.

a) Show that $f(-2) < 0$ and $f(2) > 0$.

What does this tell you about the number of roots of $f(x) = e^{-x} + x^3 - 1 = 0$? [3]

b) By calculating $f(-0.25)$ and $f(0.25)$, prove that there are at least three roots of $f(x) = 0$ in the interval $-2 < x < 2$. [5]

4 Show that a root α of the equation $f(x) = \ln(\tan x) - 1 = 0$ exists in the interval $1 < x < 1.5$

Find α to 4 significant figures by applying the Newton-Raphson iteration twice, taking $x_0 = 1.25$ as your starting value. [10]

5 Use the trapezium rule with four strips to estimate the value for $\int_0^2 \sqrt{e^x + x^2}\, dx$. [6]

6 Consider the function $f(x) = \frac{1}{x+1} - x^2$ for $x \geqslant 0$.

a) Calculate the value of f(0) and the value of f(1) and hence show there exists a root α such that $f(\alpha) = 0$ in the range $0 < x < 1$. [5]

b) Show that no further real roots exist. [3]

Total Marks .... / 49

Proof

1 Prove the trigonometric identity $\frac{1+\cos x}{\sin x}+\frac{\sin x}{1+\cos x}=2\operatorname{cosec} x$ [4]

2 Prove the trigonometric identity $\sin x+\cos x=\frac{\cos 2x}{\cos x-\sin x}$ [2]

3 Prove the trigonometric identity $\frac{\sin 3x-\sin x}{\cos x-\cos 3x}=\cot 2x$ [10]

4 Prove by contradiction that $n^2 + 2$ (where n is a positive integer) is not divisible by 4. [7]

5 Prove by contradiction that there are no integers a and b such that $14a + 21b = 1$. [4]

6 Consider a circle, centre O, radius r, with a tangent line L intersecting a point P on the circumference of the circle. Prove by contradiction that L is perpendicular to the line OP. [6]

7 Consider the function $f(x) = \frac{3x+1}{x+2}$, $x \neq -2$.

Prove by contradiction that for every x, $f(x) \neq 3$. [4]

Total Marks .... / 37

Probability

1 The events A and B are such that $P(A) = \frac{1}{4}$, $P(B) = \frac{1}{3}$ and $P(A \cup B) = \frac{5}{12}$

a) Find $P(A \cap B)$. [3]

b) Draw a Venn diagram to represent this information. [3]

c) Find $P(A|B)$. [3]

d) Write down $P(A \cap B')$. [1]

e) Write down $P(A' \cap B')$. [1]

f) Find $P(A \cup B')$. [2]

2 A production machine has three main components: F, G and H. All components are subject to failures and need to be replaced when failures occur.

An engineer has a box of 30 components, of which 20 are component F, 6 are component G and the remaining 4 are component H.

The engineer selects a component at random from the box and the component is **not** replaced. Further selections are then made, also without replacement.

Find the probability that:

a) the engineer selects a component F at first selection. [2]

b) the engineer does not select component G in either of her first two selections. [3]

c) the engineer selects at least one component H in either of her first two selections. [2]

d) Given that the first two selections were component F, find the probability that the engineer selects component G at third selection. [2]

Total Marks .... / 22

Statistical Distributions

1 The random variable X has a mean of 36 and a standard deviation of 2.5.

Find $P(X < 32)$. [2]

2 Thirty members of the Malmesbury Orchestra each recorded the amount of time they spent practising for an upcoming show. They measured the numbers of hours, x, spent practising in the first week of September and the results are recorded below:

$\sum x = 225$ $\sum x^2 = 1755$

a) Find the mean, μ, and the standard deviation, σ, of this data. [3]

b) Two new members of the orchestra joined and their practice times were $\mu + 2\sigma$ and $\mu - 2\sigma$.

State, giving a reason, the effect that these values will have on the mean and the standard deviation of the practice times. [4]

c) The five musicians who practised for the greatest amount of time had practice times of 10.5, 12.5, 14, 14 and 16.

State, giving a reason, if this data could be modelled by a normal distribution. [2]

3 A random variable $X \sim B(200, 0.45)$.

Using a suitable approximation, find $P(X < 80)$. [5]

4 **a)** Write down two conditions for $X \sim B(n, p)$ to be approximated by a normal distribution. [2]

b) Write down the mean and the variance of this normal approximation in terms of n and p. [2]

5 A company sells 100 skateboards every day. It is known that 40% of skateboards are type A.

Using a normal approximation, estimate the probability that at least 50 of type A skateboards are sold in one day. [5]

Total Marks .... / 25

Statistical Hypothesis Testing

1 A random sample of eight students sat examinations in biology and mathematics. The product moment correlation coefficient between their results was 0.572.

Using a 5% significance level, test whether or not the correlation is greater than zero. [7]

2 A company has a rule that employees may spend, on average, 60 minutes every working day on personal use of the Internet. The company takes a random sample of 30 employees and finds their mean personal Internet use is 65 minutes with a standard deviation of 10 minutes.

The company's managing director claims that his employees spend more time on average on personal use of the Internet than is allowed.

Test, at the 5% level of significance, the managing director's claim. State your hypotheses clearly. [9]

3 A company sells packets of crisps. The weight of crisps in the packet follows a normal distribution with a standard deviation of 1.5 g.

The company says that the weight of a packet of crisps is 30 g. A random sample of 10 packets is taken as part of the company's quality control process and the mean of the sample is found to be 31.4 g.

The quality control manager thinks there is enough evidence to suggest that the mean weight of the packets has gone above 30 g and production should be stopped.

Test, at a 5% significance level, whether or not there is enough evidence to support his argument. [8]

Total Marks .... / 24

Kinematics

1 A particle P moves in a plane such that at time t its velocity vector $\mathbf{v}$ is given by $\mathbf{v} = (2 - t^2)\mathbf{i} + (5t - 1)\mathbf{j}$.

a) Find the magnitude of the acceleration after two seconds. [3]

b) Given that at time $t = 0$, P has position vector $-2\mathbf{i}$, find its position vector when $t = 2$. [5]

2 A particle P moves in a plane such that at time t ($t \geqslant 0$), its position vector is given by $\mathbf{r} = \left(\frac{1}{2}\sin 2t\right)\mathbf{i} - (\sin t)\mathbf{j}$.

a) Find the smallest value of t for which the particle is travelling in a south-westerly direction. [8]

b) Find the magnitude of the acceleration at this time. [6]

3 A particle P moves in a plane such that at time t ($t \geqslant 0$), its position vector is given by $\mathbf{r} = \begin{pmatrix} a\sin kt \\ b\cos kt \end{pmatrix}$, where a, b and k are constants and not equal to zero.

a) Show that $\frac{d^2\mathbf{r}}{dt^2} \propto -\mathbf{r}$. [5]

b) Show that particle P is never stationary. [5]

4 A ball is projected from a point O at the top of a vertical cliff. The particle is projected with speed 25 ms^{-1} at an angle of 30° to the horizontal, as shown in the diagram. T seconds after projection, it hits the sea at B.

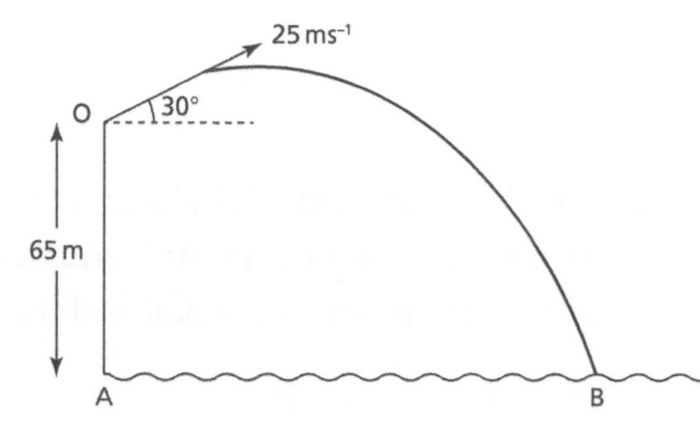

a) Find the horizontal distance AB. [6]

b) Find the speed of the particle at B. [3]

5 A ball is projected from a point on the ground, 21 m away from a vertical wall. If it is projected with speed 23 ms^{-1}, at an angle of 42° above the horizontal, determine the speed with which it hits the wall. [7]

Total Marks .... / 48

Forces

1 A particle of mass 2 kg slides down a rough plane inclined at 25° to the horizontal. Given that the acceleration of the particle is 2.65 ms^{-2}, find the coefficient of friction between the particle and the plane. [6]

2 A ball of mass 0.8 kg is held at A, on a smooth plane inclined at 20° to the horizontal. The plane meets the ground at point B and the distance AB = 2 m. BC is rough ground, and the coefficient of friction between the ball and the ground is 0.3. The ball is released and comes to rest at C.

Find the distance BC. [11]

3 A particle P of mass 3 kg is attached to one end of a light, inextensible string. P is held at rest on a rough, fixed plane inclined at an angle of 30° to the horizontal. The coefficient of friction between the particle and the plane is 0.2.

A particle Q of mass 2.5 kg is attached to the other end of the string, hanging over a smooth, fixed pulley. The string is taut, and the system then released from rest.

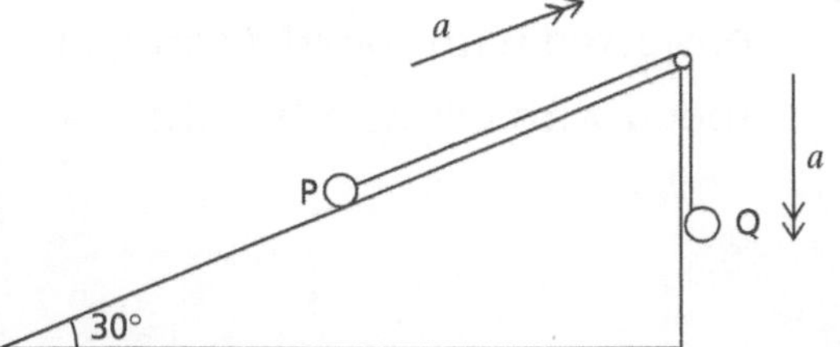

a) Find the acceleration of the system. [7]

b) Find the tension in the string. [1]

c) Find the magnitude of the force exerted on the pulley. [2]

4 A mass of 4 kg is held in place on a smooth, inclined plane by a force X, where X acts at an angle of 50° to the plane as shown in the diagram. The plane makes an angle of 40° with the horizontal.

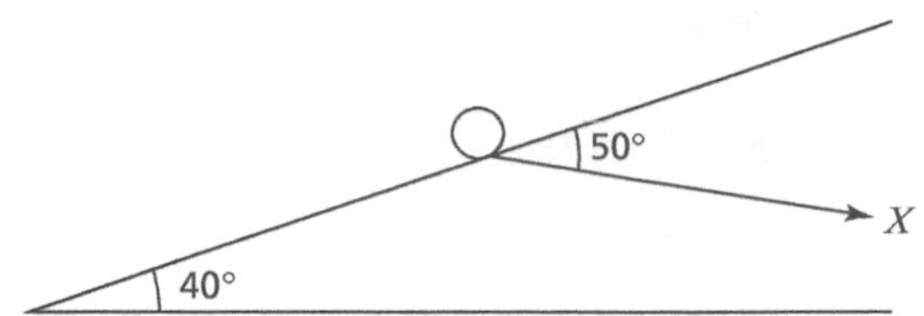

Given that the mass is in equilibrium, find:

a) the value of X [3]

b) the magnitude of the normal reaction of the plane on the mass. [3]

5 Two masses, A (5 kg) and B (10 kg), are resting on a smooth, inclined plane as shown in the diagram. A and B are joined by light, inelastic string and the system is in equilibrium due to a force P being exerted on B.

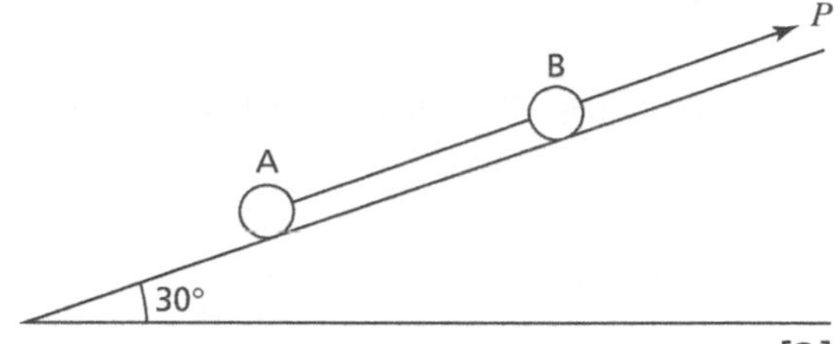

a) Find the magnitude of the tension in the string. [2]

b) Find the magnitude of P. [3]

6 A 15 kg mass is joined to the ends of two inelastic strings. The other two ends are attached to the ceiling at A and B respectively, as shown in the diagram. The system hangs in equilibrium.

Find the tensions in each string. [6]

Total Marks .... / 44

Moments

1 A non-uniform rod AB, of mass 2 kg and length 6 m, is held in horizontal equilibrium by two strings attached at A and B respectively.

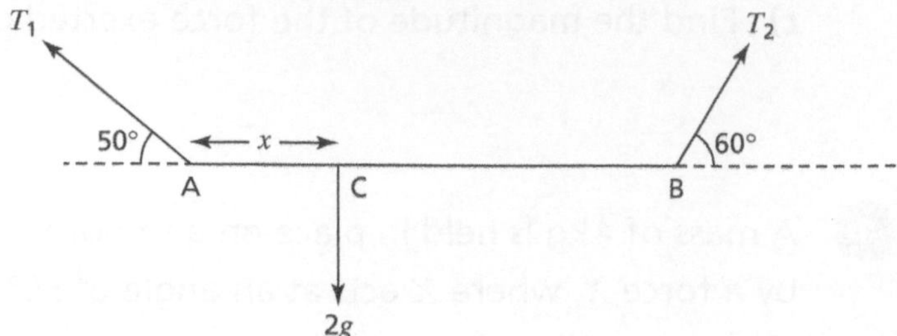

a) Find the value of T_1 and the value of T_2. [6]

b) Find x, the distance of the centre of mass of the rod from point A. [3]

2 A uniform rod AC, of mass 6 kg and length 7 m, rests on a smooth peg at B. The end A rests on rough, horizontal ground.

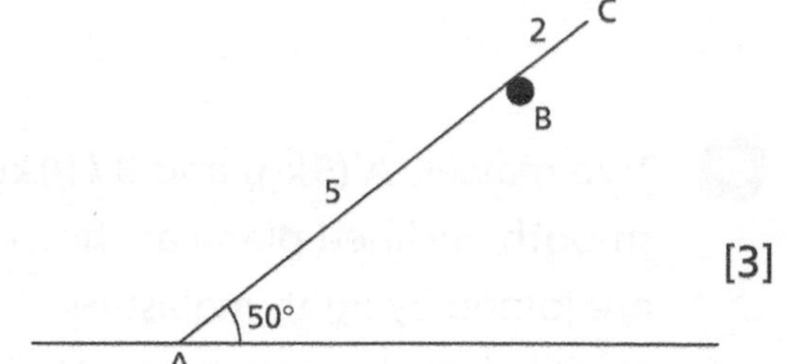

a) Find the reaction force at B. [3]

b) Find the coefficient of friction between the rod and the ground. [5]

3 A uniform ladder, of mass 20 kg and length 8 m, rests on rough, horizontal ground with the other end resting against a smooth wall. The ladder makes an angle of 55° to the horizontal. Find the value of μ, the coefficient of friction between the ladder and the ground, given that the ladder is on the point of slipping. [7]

Total Marks .... / 24

Collins

GCE

Mathematics

Advanced

Paper 1: Pure Mathematics 1

Time: 2 hours

You must have:

Mathematical Formulae and Statistical Tables

Calculator

Calculators must not have the facility for algebraic manipulation, differentiation and integration, or have retrievable mathematical formulae stored in them.

Instructions

- Use **black** ink or ball-point pen.
- If pencil is used for diagrams/sketches/graphs, it must be dark (HB or B).
- Answer **all** the questions and ensure that your answers to parts of questions are clearly labelled.
- Write your answers in the spaces provided.
- Carry out your working on separate sheets of paper where necessary. You should show sufficient working to make your methods clear. Answers without working may not gain full credit.
- Inexact answers should be given to three significant figures unless otherwise stated.

Information

- Mathematical Formulae and Statistical Tables are provided on pages 230–239.
- There are 13 questions in this question paper. The total mark for this paper is 100.
- The marks for each question are shown in brackets
  – *use this as a guide as to how much time to spend on each question.*

Advice

- Read each question carefully before you start to answer it.
- Try to answer every question.
- Check your answers if you have time at the end.
- If you change your mind about an answer, cross it out and put your new answer and any working underneath.

Name: ....

Practice Exam Paper 1

1. The curve C has equation $y = 2x^3 - 6x^2 + 6x + 5$

(a) Find $\frac{dy}{dx}$ **(2)**

(b) Find $\frac{d^2y}{dx^2}$ **(1)**

(c) Find the coordinates of, and determine the nature of, any stationary points on the curve. **(4)**

(Total for Question 1 is 7 marks)

2. Given $f(x) = \ln(x - 3), x > 3, x \in \mathbb{R}$ and $g(x) = 2x + 1, x \in \mathbb{R}$

(a) Find an expression for $f^{-1}(x)$. **(2)**

(b) Find an expression for $fg(x)$. **(2)**

(c) $fg(x)$ may be obtained from the function $f(x)$ by a sequence of two consecutive **translations**. Describe them. **(4)**

(Total for Question 2 is 8 marks)

3. Position vectors $\overrightarrow{OP}$ and $\overrightarrow{OQ}$ are such that $\overrightarrow{OP} = \mathbf{i} + 2\mathbf{j} + 3\mathbf{k}$ and $\overrightarrow{OQ} = -3a\mathbf{i} - 2a\mathbf{j} + a\mathbf{k}$, where a is constant.

Determine the value of a such that $|\overrightarrow{PQ}|$ is minimised. **(6)**

(Total for Question 3 is 6 marks)

4. Use the trapezium rule, with four strips, to obtain an approximation for $\int_1^5 (\ln x)^2 \, dx$ **(4)**

(Total for Question 4 is 4 marks)

5. (a) Sketch the graphs of $y = \left|\frac{1}{3}x - 5\right|$ and $y = |4x - 8|$ on the same axes, showing clearly where the graphs intersect the axes. **(2)**

(b) Hence, or otherwise, solve the equation $\left|\frac{1}{3}x - 5\right| = |4x - 8|$ **(4)**

(Total for Question 5 is 6 marks)

6. Consider the function $f(x) = e^{-2x} + \ln(3x) - x^2 + 5, x > 0, x \in \mathbb{R}$

(a) Show that a root α exists, such that $2 < \alpha < 3$. **(3)**

(b) By using $x_0 = 2.5$ as a first approximation to α, apply the Newton-Raphson iteration once in order to find an estimate for α to 2 decimal places. **(4)**

(Total for Question 6 is 7 marks)

7. Prove from first principles that if $y = \cos 2x$, then $\frac{dy}{dx} = -2\sin 2x$ **(8)**

(Total for Question 7 is 8 marks)

8. A boundary in the shape of a circular sector, of radius r and angle θ, is to be made from a piece of rope, fixed length P.

(a) Show that the area of the sector, A, may be expressed by $A = \frac{Pr}{2} - r^2$ **(4)**

(b) Hence find the value of θ for which the area will be maximised. **(3)**

(Total for Question 8 is 7 marks)

9. (a) Prove the trigonometric identity $\tan 3\theta = \dfrac{3\tan\theta - \tan^3\theta}{1 - 3\tan^2\theta}$ **(5)**

(b) Hence show that the only solutions to $\tan 3\theta = \tan\theta$ occur when $\tan\theta = 0$ **(3)**

(Total for Question 9 is 8 marks)

10. A geometric series has second term $\frac{63}{10}$ and sum to infinity 30.

(a) Show that $100r^2 - 100r + 21 = 0$ **(4)**

(b) Find the two possible values for r. **(2)**

(c) Given r takes the smallest value, find the smallest number of terms for which $S_n > 29.99999$ **(5)**

(Total for Question 10 is 11 marks)

11. (a) Show that $\sin x(\cos^2 x) \approx x - x^3 + \dfrac{x^5}{4}$ for small angles x. **(4)**

(b) In the specific case $x = 0.1$, calculate the percentage error in using this approximation. **(2)**

(Total for Question 11 is 6 marks)

12. Find $\displaystyle\int_0^1 \left(\frac{-3x^3 + 5x^2 + 2x + 4}{2 - x}\right) dx$, giving your answer in the form $p + q \ln r$, where p, q and r are constants to be determined. **(6)**

(Total for Question 12 is 6 marks)

13. Consider the curve defined by the parametric equations given by $x = 1 - \sin 2t$ and $y = \tan t$

(a) Show that the Cartesian form for this equation may be given by $(1 - x)(1 + y^2) = 2y$ **(5)**

(b) Let L be the normal to the curve at the point where $t = 0$.

Find an equation for L. **(6)**

(c) Prove that the line L does not intersect the curve again. **(5)**

(Total for Question 13 is 16 marks)

TOTAL FOR PAPER IS 100 MARKS

Collins

GCE

Mathematics

Advanced

Paper 2: Pure Mathematics 2

Time: 2 hours

You must have:

Mathematical Formulae and Statistical Tables

Calculator

Calculators must not have the facility for algebraic manipulation, differentiation and integration, or have retrievable mathematical formulae stored in them.

Instructions

- Use **black** ink or ball-point pen.
- If pencil is used for diagrams/sketches/graphs, it must be dark (HB or B).
- Answer **all** the questions and ensure that your answers to parts of questions are clearly labelled.
- Write your answers in the spaces provided.
- Carry out your working on separate sheets of paper where necessary. You should show sufficient working to make your methods clear. Answers without working may not gain full credit.
- Inexact answers should be given to three significant figures unless otherwise stated.

Information

- Mathematical Formulae and Statistical Tables are provided on pages 230–239.
- There are 13 questions in this question paper. The total mark for this paper is 100.
- The marks for each question are shown in brackets
  – use this as a guide as to how much time to spend on each question.

Advice

- Read each question carefully before you start to answer it.
- Try to answer every question.
- Check your answers if you have time at the end.
- If you change your mind about an answer, cross it out and put your new answer and any working underneath.

Name: ....

Practice Exam Paper 2

1. Find the equation of the tangent to the curve $y = \frac{x-2}{3x+1}$ at the point $(2, 0)$, giving your answer in the form $ax + by + c = 0$. **(5)**

(Total for Question 1 is 5 marks)

2. (a) Find the first three terms, in ascending powers of x, of the binomial expansion of $\dfrac{1}{\sqrt{4+3x}}$ **(4)**

(b) State the range of values of x for which the expansion is valid. **(2)**

(Total for Question 2 is 6 marks)

3. (a) Show that $(2x - 1)$ is a factor of $p(x) = 2x^3 - 5x^2 - 4x + 3$ **(2)**

(b) Hence write $p(x)$ as the product of three linear factors. **(3)**

(Total for Question 3 is 5 marks)

4. Terry bought a new car in 2009 and had it valued once every two years. He expected the value of the car, V, to be related to the time t (in years) by the relationship $V = Ae^{-kt}$, where A and k are constants.

(a) Interpret the value of the constant A in the context of the question. **(1)**

(b) Show that the expected relationship can be expressed in the form $\ln V = mt + c$, giving m and c in terms of A and/or k. **(3)**

(c) The diagram shows $\ln V$ plotted against t.

By drawing a line of best fit, find an estimate for constants A and k. **(2)**

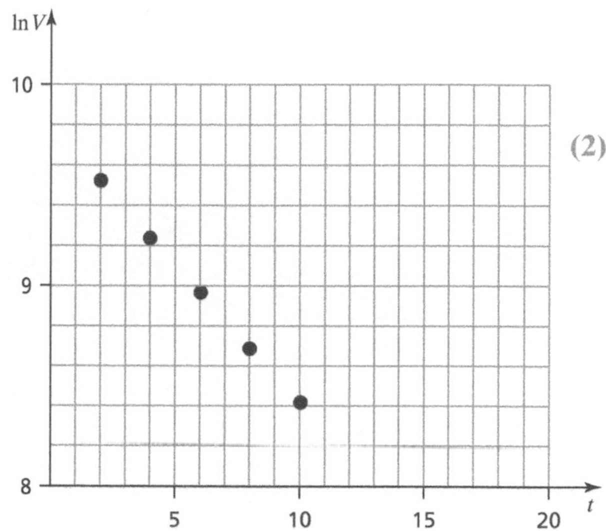

(d) Explain why the model may be unsuitable for estimating the value of Terry's car in 2022. **(1)**

(Total for Question 4 is 7 marks)

5. Three points, $A(6, 5)$, $B(3, -4)$ and $C(-2, 1)$, lie on the circumference of a circle.

(a) Find the equation of the perpendicular bisector of AB. **(4)**

(b) Find the equation of the perpendicular bisector of BC. **(4)**

(c) Hence find the coordinates of the centre of the circle. **(2)**

(Total for Question 5 is 10 marks)

6. Suppose u_n is an arithmetic sequence, with first term a and common difference d.

(a) Given that S_n is defined to be the sum of the first n terms of the sequence u_n,

prove that $S_n = \frac{n}{2}(2a+(n-1)d)$ **(3)**

(b) Determine the value of $\sum_{r=20}^{r=40}(2r+1)$ **(2)**

(Total for Question 6 is 5 marks)

7. (a) On the same axes, sketch the graphs of $y = |x|$ and $y = \sqrt{x+2}$ **(2)**

(b) Find the points of intersection of both graphs. **(3)**

(c) Hence find the exact area bounded by the two curves. **(4)**

(Total for Question 7 is 9 marks)

8. By making the substitution $u = e^x$, solve the inequality $e^{2x} - 3e^x - 10 < 0$ **(5)**

(Total for Question 8 is 5 marks)

9. It is given that $y = 5\sin 2x + \cos 2x$ $(x \geqslant 0)$

(a) Show that $\dfrac{d^2y}{dx^2} = ky$, where k is to be determined. **(4)**

(b) Find the coordinates of the first stationary point on the curve, and show that it is a maximum point. **(4)**

(c) Express $5\sin 2x + \cos 2x$ in the form $R\cos(2x - \alpha)$, where $R > 0$ and $0 < \alpha < \dfrac{\pi}{2}$ **(4)**

(d) Solve the equation $5\sin 2x + \cos 2x = 2$ for $0 \leqslant x < 2\pi$ **(4)**

(Total for Question 9 is 16 marks)

10. By using the substitution $u = \tan x$, or otherwise, show that $\int_0^{\frac{\pi}{3}} \frac{\sec^2 x \, dx}{(2+\tan x)^2} = a + b\sqrt{3}$, where a and b are constants to be determined. **(6)**

(Total for Question 10 is 6 marks)

11. Consider each of the following statements. In each case, say whether the statement is always true, sometimes true, or never true, giving your reasons.

(a) $x\log_{10} a < \log_{10} b \Rightarrow x < \dfrac{\log_{10} b}{\log_{10} a}$ **(2)**

(b) $2^n - 1$ is prime, if n is an odd integer and $n > 1$ **(2)**

(c) $\ln(x + y) = \ln x + \ln y$ **(2)**

(d) $\dfrac{1-\cos^2 2\theta}{\cos^2\theta} = 4\sin^2\theta$ **(2)**

(Total for Question 11 is 8 marks)

12. (a) Split $\dfrac{x^2-13x+34}{(x-3)^2(x+1)}$ into partial fractions. **(5)**

(b) Hence show that $\displaystyle\int_0^1 \frac{x^2-13x+34}{(x-3)^2(x+1)}\,dx = a + b\ln 18$, where a and b are constants to be determined. **(4)**

(Total for Question 12 is 9 marks)

13. Consider the differential equation defined by $\frac{dy}{dx} = -\frac{xy}{\ln y}$

(a) By solving this equation, show that the general solution may be written as $y = e^{\sqrt{k-x^2}}$ **(6)**

(b) Find the domain and the range of y. **(3)**

(Total for Question 13 is 9 marks)

TOTAL FOR PAPER IS 100 MARKS

Collins

GCE

Mathematics

Advanced

Paper 3: Statistics and Mechanics

Time: 2 hours

You must have:

Mathematical Formulae and Statistical Tables

Calculator

Calculators must not have the facility for algebraic manipulation, differentiation and integration, or have retrievable mathematical formulae stored in them.

Instructions

- Use **black** ink or ball-point pen.
- If pencil is used for diagrams/sketches/graphs, it must be dark (HB or B).
- There are **two** sections in this question paper. Answer **all** the questions in Section A and all the questions in Section B.
- Write your answers in the spaces provided.
- Carry out your working on separate sheets of paper where necessary. You should show sufficient working to make your methods clear. Answers without working may not gain full credit.
- Inexact answers should be given to three significant figures unless otherwise stated.

Information

- Mathematical Formulae and Statistical Tables are provided on pages 230–239.
- There are 10 questions in this question paper. The total mark for this paper is 100.
- The marks for each question are shown in brackets
  – *use this as a guide as to how much time to spend on each question.*

Advice

- Read each question carefully before you start to answer it.
- Try to answer every question.
- Check your answers if you have time at the end.
- If you change your mind about an answer, cross it out and put your new answer and any working underneath.

Name: ....

Practice Exam Paper 3

SECTION A: STATISTICS

1. Events A, B and C are such that $\mathrm{P}(A)=\frac{1}{3}$, $\mathrm{P}(B)=\frac{1}{2}$ and $\mathrm{P}(A \mid B)=\frac{1}{9}$

(a) Explain why events A and B are not independent. **(1)**

(b) Find $\mathrm{P}(A \cap B)$ **(2)**

(c) Find $\mathrm{P}(A' \cap B')$ **(2)**

(d) Find $\mathrm{P}(A' \mid B')$ **(2)**

(Total for Question 1 is 7 marks)

2. A student collected values for the hours of daily sunshine and the daily mean visibility (Dm) at Heathrow from the first eight days of July in 1987. The student wished to determine whether there was any correlation between hours of sunshine and mean visibility for the whole of the month.

The values are shown in the table below:

Daily sunshine (hours)	10.6	9.5	13.3	14.6	15.1	6.2	8.4	13.3
Daily mean visibility (Dm)	3400	2400	1800	2600	2200	800	1900	2000

(a) Calculate the product moment correlation coefficient (PMCC) for these values obtained, giving your answer to 3 decimal places. **(1)**

(b) Interpret your calculated result. **(1)**

(c) Stating clear hypotheses, assess at the 5% significance level whether the PMCC for the whole of the month of July is greater than zero. **(4)**

(d) Suggest a reason why the student's conclusion may in fact be unreliable. **(1)**

(Total for Question 2 is 7 marks)

3. The duration of flights from London to Jersey was analysed over the course of one week.

It was found that one-in-six flights took longer than one hour, whereas one-in-ten flights took less than 52 minutes. Let X be the random variable 'duration of flight from London to Jersey'.

(a) Assuming X may be modelled as a normal distribution, find the mean time μ and the standard deviation σ. **(7)**

(b) A flight is classed as being 'on time' if it takes anything up to 58 minutes for its journey.

Find the probability that at least 16 of 20 randomly chosen flights will be on time. **(4)**

(c) By using a suitable approximation, find the probability that, from 100 randomly chosen flights, between 60 and 75 flights (inclusive) can be classed as being 'on time'. **(4)**

(Total for Question 3 is 15 marks)

4. The table shows the number of people, from 100 surveyed, who admit to playing games regularly on their consoles.

Age (x years)	$0 \leqslant x < 10$	$10 \leqslant x < 20$	$20 \leqslant x < 30$	$30 \leqslant x < 40$	$40 \leqslant x < 50$
Frequency (f)	8	23	35	20	14

(a) Estimate the mean age of the people being surveyed. **(2)**

(b) Estimate the standard deviation. **(2)**

(c) Use interpolation to estimate the median of this distribution to 1 decimal place. **(2)**

(d) Suggest two reasons why this data suggests a normal distribution could be used to model the ages of people playing console games. **(2)**

(e) By assuming a normal distribution, find the value of k such that $P(\mu - k < x < \mu + k) = \frac{1}{2}$ **(4)**

(Total for Question 4 is 12 marks)

5. The weekly salaries (\$$x$) of a random sample of 90 workers at a large American company were obtained, and it was found that the sample mean was \$990.35 and the sample standard deviation \$439.80.

(a) Test, at the 2% significance level, whether the mean population salary is less than \$1080. **(5)**

(b) Another test was undertaken, this time at the 5% significance level.

The following hypotheses were tested:

H_0: The mean population salary is equal to M

H_1: The mean population salary is not equal to M

Given that the null hypothesis was accepted, find the possible range of values of M. **(4)**

(Total for Question 5 is 9 marks)

TOTAL FOR SECTION A IS 50 MARKS

SECTION B: MECHANICS

6. A force **F** acts on a mass of 2 kg, such that at time t, $\mathbf{F} = 3t^2\mathbf{i} + 4t\mathbf{j}$

(a) Given that the initial velocity of the mass is $(3\mathbf{i} - \mathbf{j})$, find a general expression for its velocity, **v**, in terms of t. **(4)**

(b) Find the angle α between the direction of the particle's velocity and the direction of the resultant force on the particle after 2 seconds. **(3)**

(Total for Question 6 is 7 marks)

7. Particle A, of mass $5m$, and particle B, of mass $4m$, are attached by light string, hanging over a smooth pulley. Particle A lies on a rough plane, inclined at 30° to the horizontal. The coefficient of friction between particle A and the plane is $\frac{1}{5}$. The particles are initially at rest and the string taut. The particles are then released.

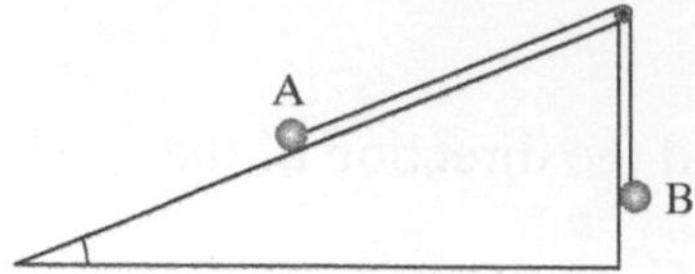

(a) (i) Write down the equation of motion for particle A. **(2)**

(ii) Write down the equation of motion for particle B. **(2)**

(iii) Hence find the acceleration of particle A up the plane. **(4)**

(b) After one second, the string snaps.

Find the further distance that particle A will travel before coming to instantaneous rest. **(6)**

(Total for Question 7 is 14 marks)

8. A uniform rod AB of mass m and length $2l$ is in equilibrium, with A resting on rough ground and a force P acting at B, perpendicular to the rod.

AB is inclined at 60° to the horizontal.

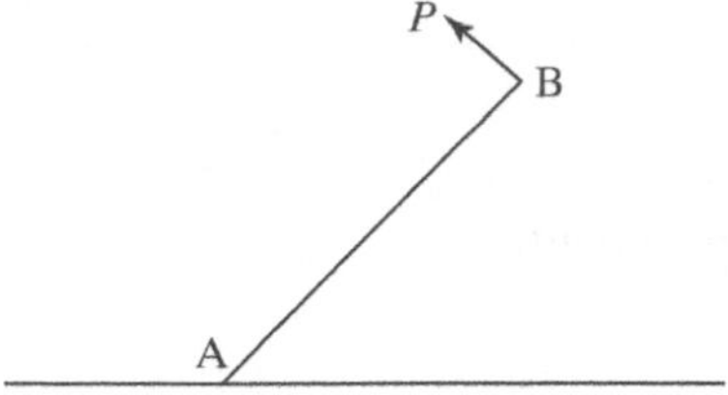

(a) Show that $P = \frac{mg}{4}$ N **(2)**

(b) Given that the rod is on the point of slipping, find the value of μ, the coefficient of friction between the ground and the rod. **(6)**

(c) Find the resultant force the ground exerts on the rod, and the direction in which it acts. **(4)**

(Total for Question 8 is 12 marks)

9. [*In this question,* **i** *and* **j** *are horizontal unit vectors due east and due north respectively.*]

A particle is moving with constant acceleration ($3\mathbf{i} - \mathbf{j}$) ms^{-2}. Its initial velocity is ($-12\mathbf{i} + 5\mathbf{j}$) ms^{-1}.

(a) Find an expression for the particle's general velocity, **v**, in terms of t. **(2)**

(b) Find the time at which the particle is travelling due north. **(2)**

(c) Find the time at which the particle is travelling in a north-west direction. **(2)**

(Total for Question 9 is 6 marks)

10. A bullet in the form of a projectile moves freely under gravity in the plane of a horizontal axis (x) and a vertical axis (y). The particle is projected from point O at time $t = 0$ with speed u and at an angle of elevation α. The bullet is intended to travel an overall horizontal distance of 500 m.

(a) Show that $u > 10\sqrt{5g}$ **(7)**

Now suppose that $u = 12\sqrt{5g}$

(b) (i) Find the angle of projection α. **(2)**

(ii) Hence, or otherwise, find the maximum height reached by the projectile during the course of its motion. **(2)**

(Total for Question 10 is 11 marks)

TOTAL FOR SECTION B IS 50 MARKS

TOTAL FOR PAPER IS 100 MARKS

Mathematical Formulae

Pure Mathematics

Mensuration

Surface area of sphere = $4\pi r^2$

Area of curved surface of cone = $\pi r \times$ slant height

Arithmetic series

$$S_n = \frac{1}{2}n(a + l) = \frac{1}{2}n[2a + (n - 1)d]$$

Binomial series

$$(a+b)^n = a^n + \binom{n}{1}a^{n-1}b + \binom{n}{2}a^{n-2}b^2 + \ldots + \binom{n}{r}a^{n-r}b^r + \ldots + b^n \quad (n \in \mathbb{N})$$

where $\binom{n}{r} = {}^n\mathrm{C}_r = \dfrac{n!}{r!(n-r)!}$

$$(1+x)^n = 1 + nx + \frac{n(n-1)}{1 \times 2}x^2 + \ldots + \frac{n(n-1)\ldots(n-r+1)}{1 \times 2 \times \ldots \times r}x^r + \ldots \quad (|x| < 1, n \in \mathbb{R})$$

Logarithms and exponentials

$$\log_a x = \frac{\log_b x}{\log_b a}$$

$$e^{x \ln a} = a^x$$

Geometric series

$$S_n = \frac{a(1 - r^n)}{1 - r}$$

$$S_\infty = \frac{a}{1 - r} \text{ for } |r| < 1$$

Trigonometric identities

$\sin(A \pm B) = \sin A \cos B \pm \cos A \sin B$

$\cos(A \pm B) = \cos A \cos B \mp \sin A \sin B$

$$\tan(A \pm B) = \frac{\tan A \pm \tan B}{1 \mp \tan A \tan B} \quad \left(A \pm B \neq \left(k + \frac{1}{2}\right)\pi\right)$$

$$\sin A + \sin B = 2\sin\frac{A+B}{2}\cos\frac{A-B}{2}$$

$$\sin A - \sin B = 2\cos\frac{A+B}{2}\sin\frac{A-B}{2}$$

$$\cos A + \cos B = 2\cos\frac{A+B}{2}\cos\frac{A-B}{2}$$

$$\cos A - \cos B = -2\sin\frac{A+B}{2}\sin\frac{A-B}{2}$$

Small angle approximations

$\sin\theta \approx \theta$

$\cos\theta \approx 1 - \frac{\theta^2}{2}$

$\tan\theta \approx \theta$

where θ is measured in radians

Differentiation

First Principles

$$f'(x) = \lim_{h \to 0} \frac{f(x+h) - f(x)}{h}$$

$\mathbf{f(x)}$	$\mathbf{f'(x)}$
$\tan kx$	$k\sec^2 kx$
$\sec kx$	$k\sec kx\tan kx$
$\cot kx$	$-k\operatorname{cosec}^2 kx$
$\operatorname{cosec} kx$	$-k\operatorname{cosec} kx\cot kx$
$\frac{f(x)}{g(x)}$	$\frac{f'(x)g(x) - f(x)g'(x)}{(g(x))^2}$

Mathematical Formulae

Integration (+ constant)

$\mathrm{f}(x)$	$\int \mathrm{f}(x)\,\mathrm{d}x$
$\sec^2 kx$	$\frac{1}{k}\tan kx$
$\tan kx$	$\frac{1}{k}\ln\lvert\sec kx\rvert$
$\cot kx$	$\frac{1}{k}\ln\lvert\sin kx\rvert$
$\operatorname{cosec} kx$	$-\frac{1}{k}\ln\lvert\operatorname{cosec} kx+\cot kx\rvert,\ \frac{1}{k}\ln\left\lvert\tan\left(\frac{1}{2}kx\right)\right\rvert$
$\sec kx$	$\frac{1}{k}\ln\lvert\sec kx+\tan kx\rvert,\ \frac{1}{k}\ln\left\lvert\tan\left(\frac{1}{2}kx+\frac{1}{4}\pi\right)\right\rvert$

$$\int u\frac{\mathrm{d}v}{\mathrm{d}x}\mathrm{d}x = uv - \int v\frac{\mathrm{d}u}{\mathrm{d}x}\mathrm{d}x$$

Numerical Methods

The trapezium rule: $\int_a^b y\ \mathrm{d}x \approx \frac{1}{2}h\{(y_0+y_n)+2(y_1+y_2+\ldots+y_{n-1})\}$, where $h=\frac{b-a}{n}$

The Newton-Raphson iteration for solving $\mathrm{f}(x)=0$: $x_{n+1}=x_n-\frac{\mathrm{f}(x_n)}{\mathrm{f}'(x_n)}$

Statistics

Probability

$\mathrm{P}(A') = 1 - \mathrm{P}(A)$

$\mathrm{P}(A \cup B) = \mathrm{P}(A) + \mathrm{P}(B) - \mathrm{P}(A \cap B)$

$\mathrm{P}(A \cap B) = \mathrm{P}(A)\mathrm{P}(B \mid A)$

$\mathrm{P}(A \mid B) = \frac{\mathrm{P}(B \mid A)\mathrm{P}(A)}{\mathrm{P}(B \mid A)\mathrm{P}(A)+\mathrm{P}(B \mid A')\mathrm{P}(A')}$

For independent events A and B,

$\mathrm{P}(B \mid A) = \mathrm{P}(B)$

$\mathrm{P}(A \mid B) = \mathrm{P}(A)$

$\mathrm{P}(A \cap B) = \mathrm{P}(A)\ \mathrm{P}(B)$

Standard deviation

Standard deviation = $\sqrt{(\text{Variance})}$

Interquartile range = IQR = $Q_3 - Q_1$

For a set of n values $x_1, x_2, \ldots x_i, \ldots x_n$

$$S_{xx} = \sum(x_i - \bar{x})^2 = \sum x_i^2 - \frac{\left(\sum x_i\right)^2}{n}$$

Standard deviation = $\sqrt{\frac{S_{xx}}{n}}$ or $\sqrt{\frac{\sum x^2}{n} - \bar{x}^2}$

Discrete distributions

Distribution of X	$P(X = x)$	Mean	Variance
Binomial $B(n, p)$	$\binom{n}{x} p^x (1 - p)^{n-x}$	np	$np(1 - p)$

Sampling distributions

For a random sample of n observations from $N(\mu, \sigma^2)$

$$\frac{\bar{X} - \mu}{\sigma / \sqrt{n}} \sim N(0,1)$$

Mechanics

Kinematics

For motion in a straight line with constant acceleration:

$v = u + at$

$s = ut + \frac{1}{2}at^2$

$s = vt - \frac{1}{2}at^2$

$v^2 = u^2 + 2as$

$s = \frac{1}{2}(u + v)t$

Statistical Tables

Binomial Cumulative Distribution Function

The tabulated value is $P(X \leqslant x)$, where X has a binomial distribution with index n and parameter p.

$p =$	0.05	0.10	0.15	0.20	0.25	0.30	0.35	0.40	0.45	0.50
$n = 5, x = 0$	0.7738	0.5905	0.4437	0.3277	0.2373	0.1681	0.1160	0.0778	0.0503	0.0312
1	0.9774	0.9185	0.8352	0.7373	0.6328	0.5282	0.4284	0.3370	0.2562	0.1875
2	0.9988	0.9914	0.9734	0.9421	0.8965	0.8369	0.7648	0.6826	0.5931	0.5000
3	1.0000	0.9995	0.9978	0.9933	0.9844	0.9692	0.9460	0.9130	0.8688	0.8125
4	1.0000	1.0000	0.9999	0.9997	0.9990	0.9976	0.9947	0.9898	0.9815	0.9688
$n = 6, x = 0$	0.7351	0.5314	0.3771	0.2621	0.1780	0.1176	0.0754	0.0467	0.0277	0.0156
1	0.9672	0.8857	0.7765	0.6554	0.5339	0.4202	0.3191	0.2333	0.1636	0.1094
2	0.9978	0.9842	0.9527	0.9011	0.8306	0.7443	0.6471	0.5443	0.4415	0.3438
3	0.9999	0.9987	0.9941	0.9830	0.9624	0.9295	0.8826	0.8208	0.7447	0.6563
4	1.0000	0.9999	0.9996	0.9984	0.9954	0.9891	0.9777	0.9590	0.9308	0.8906
5	1.0000	1.0000	1.0000	0.9999	0.9998	0.9993	0.9982	0.9959	0.9917	0.9844
$n = 7, x = 0$	0.6983	0.4783	0.3206	0.2097	0.1335	0.0824	0.0490	0.0280	0.0152	0.0078
1	0.9556	0.8503	0.7166	0.5767	0.4449	0.3294	0.2338	0.1586	0.1024	0.0625
2	0.9962	0.9743	0.9262	0.8520	0.7564	0.6471	0.5323	0.4199	0.3164	0.2266
3	0.9998	0.9973	0.9879	0.9667	0.9294	0.8740	0.8002	0.7102	0.6083	0.5000
4	1.0000	0.9998	0.9988	0.9953	0.9871	0.9712	0.9444	0.9037	0.8471	0.7734
5	1.0000	1.0000	0.9999	0.9996	0.9987	0.9962	0.9910	0.9812	0.9643	0.9375
6	1.0000	1.0000	1.0000	1.0000	0.9999	0.9998	0.9994	0.9984	0.9963	0.9922
$n = 8, x = 0$	0.6634	0.4305	0.2725	0.1678	0.1001	0.0576	0.0319	0.0168	0.0084	0.0039
1	0.9428	0.8131	0.6572	0.5033	0.3671	0.2553	0.1691	0.1064	0.0632	0.0352
2	0.9942	0.9619	0.8948	0.7969	0.6785	0.5518	0.4278	0.3154	0.2201	0.1445
3	0.9996	0.9950	0.9786	0.9437	0.8862	0.8059	0.7064	0.5941	0.4770	0.3633
4	1.0000	0.9996	0.9971	0.9896	0.9727	0.9420	0.8939	0.8263	0.7396	0.6367
5	1.0000	1.0000	0.9998	0.9988	0.9958	0.9887	0.9747	0.9502	0.9115	0.8555
6	1.0000	1.0000	1.0000	0.9999	0.9996	0.9987	0.9964	0.9915	0.9819	0.9648
7	1.0000	1.0000	1.0000	1.0000	1.0000	0.9999	0.9998	0.9993	0.9983	0.9961
$n = 9, x = 0$	0.6302	0.3874	0.2316	0.1342	0.0751	0.0404	0.0207	0.0101	0.0046	0.0020
1	0.9288	0.7748	0.5995	0.4362	0.3003	0.1960	0.1211	0.0705	0.0385	0.0195
2	0.9916	0.9470	0.8591	0.7382	0.6007	0.4628	0.3373	0.2318	0.1495	0.0898
3	0.9994	0.9917	0.9661	0.9144	0.8343	0.7297	0.6089	0.4826	0.3614	0.2539
4	1.0000	0.9991	0.9944	0.9804	0.9511	0.9012	0.8283	0.7334	0.6214	0.5000
5	1.0000	0.9999	0.9994	0.9969	0.9900	0.9747	0.9464	0.9006	0.8342	0.7461
6	1.0000	1.0000	1.0000	0.9997	0.9987	0.9957	0.9888	0.9750	0.9502	0.9102
7	1.0000	1.0000	1.0000	1.0000	0.9999	0.9996	0.9986	0.9962	0.9909	0.9805
8	1.0000	1.0000	1.0000	1.0000	1.0000	1.0000	0.9999	0.9997	0.9992	0.9980
$n = 10, x = 0$	0.5987	0.3487	0.1969	0.1074	0.0563	0.0282	0.0135	0.0060	0.0025	0.0010
1	0.9139	0.7361	0.5443	0.3758	0.2440	0.1493	0.0860	0.0464	0.0233	0.0107
2	0.9885	0.9298	0.8202	0.6778	0.5256	0.3828	0.2616	0.1673	0.0996	0.0547
3	0.9990	0.9872	0.9500	0.8791	0.7759	0.6496	0.5138	0.3823	0.2660	0.1719
4	0.9999	0.9984	0.9901	0.9672	0.9219	0.8497	0.7515	0.6331	0.5044	0.3770

p =	0.05	0.10	0.15	0.20	0.25	0.30	0.35	0.40	0.45	0.50
5	1.0000	0.9999	0.9986	0.9936	0.9803	0.9527	0.9051	0.8338	0.7384	0.6230
6	1.0000	1.0000	0.9999	0.9991	0.9965	0.9894	0.9740	0.9452	0.8980	0.8281
7	1.0000	1.0000	1.0000	0.9999	0.9996	0.9984	0.9952	0.9877	0.9726	0.9453
8	1.0000	1.0000	1.0000	1.0000	1.0000	0.9999	0.9995	0.9983	0.9955	0.9893
9	1.0000	1.0000	1.0000	1.0000	1.0000	1.0000	1.0000	0.9999	0.9997	0.9990
$n = 12$, $x = 0$	0.5404	0.2824	0.1422	0.0687	0.0317	0.0138	0.0057	0.0022	0.0008	0.0002
1	0.8816	0.6590	0.4435	0.2749	0.1584	0.0850	0.0424	0.0196	0.0083	0.0032
2	0.9804	0.8891	0.7358	0.5583	0.3907	0.2528	0.1513	0.0834	0.0421	0.0193
3	0.9978	0.9744	0.9078	0.7946	0.6488	0.4925	0.3467	0.2253	0.1345	0.0730
4	0.9998	0.9957	0.9761	0.9274	0.8424	0.7237	0.5833	0.4382	0.3044	0.1938
5	1.0000	0.9995	0.9954	0.9806	0.9456	0.8822	0.7873	0.6652	0.5269	0.3872
6	1.0000	0.9999	0.9993	0.9961	0.9857	0.9614	0.9154	0.8418	0.7393	0.6128
7	1.0000	1.0000	0.9999	0.9994	0.9972	0.9905	0.9745	0.9427	0.8883	0.8062
8	1.0000	1.0000	1.0000	0.9999	0.9996	0.9983	0.9944	0.9847	0.9644	0.9270
9	1.0000	1.0000	1.0000	1.0000	1.0000	0.9998	0.9992	0.9972	0.9921	0.9807
10	1.0000	1.0000	1.0000	1.0000	1.0000	1.0000	0.9999	0.9997	0.9989	0.9968
11	1.0000	1.0000	1.0000	1.0000	1.0000	1.0000	1.0000	1.0000	0.9999	0.9998
$n = 15$, $x = 0$	0.4633	0.2059	0.0874	0.0352	0.0134	0.0047	0.0016	0.0005	0.0001	0.0000
1	0.8290	0.5490	0.3186	0.1671	0.0802	0.0353	0.0142	0.0052	0.0017	0.0005
2	0.9638	0.8159	0.6042	0.3980	0.2361	0.1268	0.0617	0.0271	0.0107	0.0037
3	0.9945	0.9444	0.8227	0.6482	0.4613	0.2969	0.1727	0.0905	0.0424	0.0176
4	0.9994	0.9873	0.9383	0.8358	0.6865	0.5155	0.3519	0.2173	0.1204	0.0592
5	0.9999	0.9978	0.9832	0.9389	0.8516	0.7216	0.5643	0.4032	0.2608	0.1509
6	1.0000	0.9997	0.9964	0.9819	0.9434	0.8689	0.7548	0.6098	0.4522	0.3036
7	1.0000	1.0000	0.9994	0.9958	0.9827	0.9500	0.8868	0.7869	0.6535	0.5000
8	1.0000	1.0000	0.9999	0.9992	0.9958	0.9848	0.9578	0.9050	0.8182	0.6964
9	1.0000	1.0000	1.0000	0.9999	0.9992	0.9963	0.9876	0.9662	0.9231	0.8491
10	1.0000	1.0000	1.0000	1.0000	0.9999	0.9993	0.9972	0.9907	0.9745	0.9408
11	1.0000	1.0000	1.0000	1.0000	1.0000	0.9999	0.9995	0.9981	0.9937	0.9824
12	1.0000	1.0000	1.0000	1.0000	1.0000	1.0000	0.9999	0.9997	0.9989	0.9963
13	1.0000	1.0000	1.0000	1.0000	1.0000	1.0000	1.0000	1.0000	0.9999	0.9995
14	1.0000	1.0000	1.0000	1.0000	1.0000	1.0000	1.0000	1.0000	1.0000	1.0000
$n = 20$, $x = 0$	0.3585	0.1216	0.0388	0.0115	0.0032	0.0008	0.0002	0.0000	0.0000	0.0000
1	0.7358	0.3917	0.1756	0.0692	0.0243	0.0076	0.0021	0.0005	0.0001	0.0000
2	0.9245	0.6769	0.4049	0.2061	0.0913	0.0355	0.0121	0.0036	0.0009	0.0002
3	0.9841	0.8670	0.6477	0.4114	0.2252	0.1071	0.0444	0.0160	0.0049	0.0013
4	0.9974	0.9568	0.8298	0.6296	0.4148	0.2375	0.1182	0.0510	0.0189	0.0059
5	0.9997	0.9887	0.9327	0.8042	0.6172	0.4164	0.2454	0.1256	0.0553	0.0207
6	1.0000	0.9976	0.9781	0.9133	0.7858	0.6080	0.4166	0.2500	0.1299	0.0577
7	1.0000	0.9996	0.9941	0.9679	0.8982	0.7723	0.6010	0.4159	0.2520	0.1316
8	1.0000	0.9999	0.9987	0.9900	0.9591	0.8867	0.7624	0.5956	0.4143	0.2517
9	1.0000	1.0000	0.9998	0.9974	0.9861	0.9520	0.8782	0.7553	0.5914	0.4119
10	1.0000	1.0000	1.0000	0.9994	0.9961	0.9829	0.9468	0.8725	0.7507	0.5881

Statistical Tables

p =	0.05	0.10	0.15	0.20	0.25	0.30	0.35	0.40	0.45	0.50
11	1.0000	1.0000	1.0000	0.9999	0.9991	0.9949	0.9804	0.9435	0.8692	0.7483
12	1.0000	1.0000	1.0000	1.0000	0.9998	0.9987	0.9940	0.9790	0.9420	0.8684
13	1.0000	1.0000	1.0000	1.0000	1.0000	0.9997	0.9985	0.9935	0.9786	0.9423
14	1.0000	1.0000	1.0000	1.0000	1.0000	1.0000	0.9997	0.9984	0.9936	0.9793
15	1.0000	1.0000	1.0000	1.0000	1.0000	1.0000	1.0000	0.9997	0.9985	0.9941
16	1.0000	1.0000	1.0000	1.0000	1.0000	1.0000	1.0000	1.0000	0.9997	0.9987
17	1.0000	1.0000	1.0000	1.0000	1.0000	1.0000	1.0000	1.0000	1.0000	0.9998
18	1.0000	1.0000	1.0000	1.0000	1.0000	1.0000	1.0000	1.0000	1.0000	1.0000
$n = 25$, $x = 0$	0.2774	0.0718	0.0172	0.0038	0.0008	0.0001	0.0000	0.0000	0.0000	0.0000
1	0.6424	0.2712	0.0931	0.0274	0.0070	0.0016	0.0003	0.0001	0.0000	0.0000
2	0.8729	0.5371	0.2537	0.0982	0.0321	0.0090	0.0021	0.0004	0.0001	0.0000
3	0.9659	0.7636	0.4711	0.2340	0.0962	0.0332	0.0097	0.0024	0.0005	0.0001
4	0.9928	0.9020	0.6821	0.4207	0.2137	0.0905	0.0320	0.0095	0.0023	0.0005
5	0.9988	0.9666	0.8385	0.6167	0.3783	0.1935	0.0826	0.0294	0.0086	0.0020
6	0.9998	0.9905	0.9305	0.7800	0.5611	0.3407	0.1734	0.0736	0.0258	0.0073
7	1.0000	0.9977	0.9745	0.8909	0.7265	0.5118	0.3061	0.1536	0.0639	0.0216
8	1.0000	0.9995	0.9920	0.9532	0.8506	0.6769	0.4668	0.2735	0.1340	0.0539
9	1.0000	0.9999	0.9979	0.9827	0.9287	0.8106	0.6303	0.4246	0.2424	0.1148
10	1.0000	1.0000	0.9995	0.9944	0.9703	0.9022	0.7712	0.5858	0.3843	0.2122
11	1.0000	1.0000	0.9999	0.9985	0.9893	0.9558	0.8746	0.7323	0.5426	0.3450
12	1.0000	1.0000	1.0000	0.9996	0.9966	0.9825	0.9396	0.8462	0.6937	0.5000
13	1.0000	1.0000	1.0000	0.9999	0.9991	0.9940	0.9745	0.9222	0.8173	0.6550
14	1.0000	1.0000	1.0000	1.0000	0.9998	0.9982	0.9907	0.9656	0.9040	0.7878
15	1.0000	1.0000	1.0000	1.0000	1.0000	0.9995	0.9971	0.9868	0.9560	0.8852
16	1.0000	1.0000	1.0000	1.0000	1.0000	0.9999	0.9992	0.9957	0.9826	0.9461
17	1.0000	1.0000	1.0000	1.0000	1.0000	1.0000	0.9998	0.9988	0.9942	0.9784
18	1.0000	1.0000	1.0000	1.0000	1.0000	1.0000	1.0000	0.9997	0.9984	0.9927
19	1.0000	1.0000	1.0000	1.0000	1.0000	1.0000	1.0000	0.9999	0.9996	0.9980
20	1.0000	1.0000	1.0000	1.0000	1.0000	1.0000	1.0000	1.0000	0.9999	0.9995
21	1.0000	1.0000	1.0000	1.0000	1.0000	1.0000	1.0000	1.0000	1.0000	0.9999
22	1.0000	1.0000	1.0000	1.0000	1.0000	1.0000	1.0000	1.0000	1.0000	1.0000
$n = 30$, $x = 0$	0.2146	0.0424	0.0076	0.0012	0.0002	0.0000	0.0000	0.0000	0.0000	0.0000
1	0.5535	0.1837	0.0480	0.0105	0.0020	0.0003	0.0000	0.0000	0.0000	0.0000
2	0.8122	0.4114	0.1514	0.0442	0.0106	0.0021	0.0003	0.0000	0.0000	0.0000
3	0.9392	0.6474	0.3217	0.1227	0.0374	0.0093	0.0019	0.0003	0.0000	0.0000
4	0.9844	0.8245	0.5245	0.2552	0.0979	0.0302	0.0075	0.0015	0.0002	0.0000
5	0.9967	0.9268	0.7106	0.4275	0.2026	0.0766	0.0233	0.0057	0.0011	0.0002
6	0.9994	0.9742	0.8474	0.6070	0.3481	0.1595	0.0586	0.0172	0.0040	0.0007
7	0.9999	0.9922	0.9302	0.7608	0.5143	0.2814	0.1238	0.0435	0.0121	0.0026
8	1.0000	0.9980	0.9722	0.8713	0.6736	0.4315	0.2247	0.0940	0.0312	0.0081
9	1.0000	0.9995	0.9903	0.9389	0.8034	0.5888	0.3575	0.1763	0.0694	0.0214
10	1.0000	0.9999	0.9971	0.9744	0.8943	0.7304	0.5078	0.2915	0.1350	0.0494
11	1.0000	1.0000	0.9992	0.9905	0.9493	0.8407	0.6548	0.4311	0.2327	0.1002

$p =$	0.05	0.10	0.15	0.20	0.25	0.30	0.35	0.40	0.45	0.50
12	1.0000	1.0000	0.9998	0.9969	0.9784	0.9155	0.7802	0.5785	0.3592	0.1808
13	1.0000	1.0000	1.0000	0.9991	0.9918	0.9599	0.8737	0.7145	0.5025	0.2923
14	1.0000	1.0000	1.0000	0.9998	0.9973	0.9831	0.9348	0.8246	0.6448	0.4278
15	1.0000	1.0000	1.0000	0.9999	0.9992	0.9936	0.9699	0.9029	0.7691	0.5722
16	1.0000	1.0000	1.0000	1.0000	0.9998	0.9979	0.9876	0.9519	0.8644	0.7077
17	1.0000	1.0000	1.0000	1.0000	0.9999	0.9994	0.9955	0.9788	0.9286	0.8192
18	1.0000	1.0000	1.0000	1.0000	1.0000	0.9998	0.9986	0.9917	0.9666	0.8998
19	1.0000	1.0000	1.0000	1.0000	1.0000	1.0000	0.9996	0.9971	0.9862	0.9506
20	1.0000	1.0000	1.0000	1.0000	1.0000	1.0000	0.9999	0.9991	0.9950	0.9786
21	1.0000	1.0000	1.0000	1.0000	1.0000	1.0000	1.0000	0.9998	0.9984	0.9919
22	1.0000	1.0000	1.0000	1.0000	1.0000	1.0000	1.0000	1.0000	0.9996	0.9974
23	1.0000	1.0000	1.0000	1.0000	1.0000	1.0000	1.0000	1.0000	0.9999	0.9993
24	1.0000	1.0000	1.0000	1.0000	1.0000	1.0000	1.0000	1.0000	1.0000	0.9998
25	1.0000	1.0000	1.0000	1.0000	1.0000	1.0000	1.0000	1.0000	1.0000	1.0000
$n = 40, x = 0$	0.1285	0.0148	0.0015	0.0001	0.0000	0.0000	0.0000	0.0000	0.0000	0.0000
1	0.3991	0.0805	0.0121	0.0015	0.0001	0.0000	0.0000	0.0000	0.0000	0.0000
2	0.6767	0.2228	0.0486	0.0079	0.0010	0.0001	0.0000	0.0000	0.0000	0.0000
3	0.8619	0.4231	0.1302	0.0285	0.0047	0.0006	0.0001	0.0000	0.0000	0.0000
4	0.9520	0.6290	0.2633	0.0759	0.0160	0.0026	0.0003	0.0000	0.0000	0.0000
5	0.9861	0.7937	0.4325	0.1613	0.0433	0.0086	0.0013	0.0001	0.0000	0.0000
6	0.9966	0.9005	0.6067	0.2859	0.0962	0.0238	0.0044	0.0006	0.0001	0.0000
7	0.9993	0.9581	0.7559	0.4371	0.1820	0.0553	0.0124	0.0021	0.0002	0.0000
8	0.9999	0.9845	0.8646	0.5931	0.2998	0.1110	0.0303	0.0061	0.0009	0.0001
9	1.0000	0.9949	0.9328	0.7318	0.4395	0.1959	0.0644	0.0156	0.0027	0.0003
10	1.0000	0.9985	0.9701	0.8392	0.5839	0.3087	0.1215	0.0352	0.0074	0.0011
11	1.0000	0.9996	0.9880	0.9125	0.7151	0.4406	0.2053	0.0709	0.0179	0.0032
12	1.0000	0.9999	0.9957	0.9568	0.8209	0.5772	0.3143	0.1285	0.0386	0.0083
13	1.0000	1.0000	0.9986	0.9806	0.8968	0.7032	0.4408	0.2112	0.0751	0.0192
14	1.0000	1.0000	0.9996	0.9921	0.9456	0.8074	0.5721	0.3174	0.1326	0.0403
15	1.0000	1.0000	0.9999	0.9971	0.9738	0.8849	0.6946	0.4402	0.2142	0.0769
16	1.0000	1.0000	1.0000	0.9990	0.9884	0.9367	0.7978	0.5681	0.3185	0.1341
17	1.0000	1.0000	1.0000	0.9997	0.9953	0.9680	0.8761	0.6885	0.4391	0.2148
18	1.0000	1.0000	1.0000	0.9999	0.9983	0.9852	0.9301	0.7911	0.5651	0.3179
19	1.0000	1.0000	1.0000	1.0000	0.9994	0.9937	0.9637	0.8702	0.6844	0.4373
20	1.0000	1.0000	1.0000	1.0000	0.9998	0.9976	0.9827	0.9256	0.7870	0.5627
21	1.0000	1.0000	1.0000	1.0000	1.0000	0.9991	0.9925	0.9608	0.8669	0.6821
22	1.0000	1.0000	1.0000	1.0000	1.0000	0.9997	0.9970	0.9811	0.9233	0.7852
23	1.0000	1.0000	1.0000	1.0000	1.0000	0.9999	0.9989	0.9917	0.9595	0.8659
24	1.0000	1.0000	1.0000	1.0000	1.0000	1.0000	0.9996	0.9966	0.9804	0.9231
25	1.0000	1.0000	1.0000	1.0000	1.0000	1.0000	0.9999	0.9988	0.9914	0.9597
26	1.0000	1.0000	1.0000	1.0000	1.0000	1.0000	1.0000	0.9996	0.9966	0.9808
27	1.0000	1.0000	1.0000	1.0000	1.0000	1.0000	1.0000	0.9999	0.9988	0.9917
28	1.0000	1.0000	1.0000	1.0000	1.0000	1.0000	1.0000	1.0000	0.9996	0.9968

Statistical Tables

$p =$	0.05	0.10	0.15	0.20	0.25	0.30	0.35	0.40	0.45	0.50
29	1.0000	1.0000	1.0000	1.0000	1.0000	1.0000	1.0000	1.0000	0.9999	0.9989
30	1.0000	1.0000	1.0000	1.0000	1.0000	1.0000	1.0000	1.0000	1.0000	0.9997
31	1.0000	1.0000	1.0000	1.0000	1.0000	1.0000	1.0000	1.0000	1.0000	0.9999
32	1.0000	1.0000	1.0000	1.0000	1.0000	1.0000	1.0000	1.0000	1.0000	1.0000
$n = 50,\ x = 0$	0.0769	0.0052	0.0003	0.0000	0.0000	0.0000	0.0000	0.0000	0.0000	0.0000
1	0.2794	0.0338	0.0029	0.0002	0.0000	0.0000	0.0000	0.0000	0.0000	0.0000
2	0.5405	0.1117	0.0142	0.0013	0.0001	0.0000	0.0000	0.0000	0.0000	0.0000
3	0.7604	0.2503	0.0460	0.0057	0.0005	0.0000	0.0000	0.0000	0.0000	0.0000
4	0.8964	0.4312	0.1121	0.0185	0.0021	0.0002	0.0000	0.0000	0.0000	0.0000
5	0.9622	0.6161	0.2194	0.0480	0.0070	0.0007	0.0001	0.0000	0.0000	0.0000
6	0.9882	0.7702	0.3613	0.1034	0.0194	0.0025	0.0002	0.0000	0.0000	0.0000
7	0.9968	0.8779	0.5188	0.1904	0.0453	0.0073	0.0008	0.0001	0.0000	0.0000
8	0.9992	0.9421	0.6681	0.3073	0.0916	0.0183	0.0025	0.0002	0.0000	0.0000
9	0.9998	0.9755	0.7911	0.4437	0.1637	0.0402	0.0067	0.0008	0.0001	0.0000
10	1.0000	0.9906	0.8801	0.5836	0.2622	0.0789	0.0160	0.0022	0.0002	0.0000
11	1.0000	0.9968	0.9372	0.7107	0.3816	0.1390	0.0342	0.0057	0.0006	0.0000
12	1.0000	0.9990	0.9699	0.8139	0.5110	0.2229	0.0661	0.0133	0.0018	0.0002
13	1.0000	0.9997	0.9868	0.8894	0.6370	0.3279	0.1163	0.0280	0.0045	0.0005
14	1.0000	0.9999	0.9947	0.9393	0.7481	0.4468	0.1878	0.0540	0.0104	0.0013
15	1.0000	1.0000	0.9981	0.9692	0.8369	0.5692	0.2801	0.0955	0.0220	0.0033
16	1.0000	1.0000	0.9993	0.9856	0.9017	0.6839	0.3889	0.1561	0.0427	0.0077
17	1.0000	1.0000	0.9998	0.9937	0.9449	0.7822	0.5060	0.2369	0.0765	0.0164
18	1.0000	1.0000	0.9999	0.9975	0.9713	0.8594	0.6216	0.3356	0.1273	0.0325
19	1.0000	1.0000	1.0000	0.9991	0.9861	0.9152	0.7264	0.4465	0.1974	0.0595
20	1.0000	1.0000	1.0000	0.9997	0.9937	0.9522	0.8139	0.5610	0.2862	0.1013
21	1.0000	1.0000	1.0000	0.9999	0.9974	0.9749	0.8813	0.6701	0.3900	0.1611
22	1.0000	1.0000	1.0000	1.0000	0.9990	0.9877	0.9290	0.7660	0.5019	0.2399
23	1.0000	1.0000	1.0000	1.0000	0.9996	0.9944	0.9604	0.8438	0.6134	0.3359
24	1.0000	1.0000	1.0000	1.0000	0.9999	0.9976	0.9793	0.9022	0.7160	0.4439
25	1.0000	1.0000	1.0000	1.0000	1.0000	0.9991	0.9900	0.9427	0.8034	0.5561
26	1.0000	1.0000	1.0000	1.0000	1.0000	0.9997	0.9955	0.9686	0.8721	0.6641
27	1.0000	1.0000	1.0000	1.0000	1.0000	0.9999	0.9981	0.9840	0.9220	0.7601
28	1.0000	1.0000	1.0000	1.0000	1.0000	1.0000	0.9993	0.9924	0.9556	0.8389
29	1.0000	1.0000	1.0000	1.0000	1.0000	1.0000	0.9997	0.9966	0.9765	0.8987
30	1.0000	1.0000	1.0000	1.0000	1.0000	1.0000	0.9999	0.9986	0.9884	0.9405
31	1.0000	1.0000	1.0000	1.0000	1.0000	1.0000	1.0000	0.9995	0.9947	0.9675
32	1.0000	1.0000	1.0000	1.0000	1.0000	1.0000	1.0000	0.9998	0.9978	0.9836
33	1.0000	1.0000	1.0000	1.0000	1.0000	1.0000	1.0000	0.9999	0.9991	0.9923
34	1.0000	1.0000	1.0000	1.0000	1.0000	1.0000	1.0000	1.0000	0.9997	0.9967
35	1.0000	1.0000	1.0000	1.0000	1.0000	1.0000	1.0000	1.0000	0.9999	0.9987
36	1.0000	1.0000	1.0000	1.0000	1.0000	1.0000	1.0000	1.0000	1.0000	0.9995
37	1.0000	1.0000	1.0000	1.0000	1.0000	1.0000	1.0000	1.0000	1.0000	0.9998
38	1.0000	1.0000	1.0000	1.0000	1.0000	1.0000	1.0000	1.0000	1.0000	1.0000

Percentage Points of The Normal Distribution

The values z in the table are those which a random variable $Z \sim N(0, 1)$ exceeds with probability p; that is, $P(Z > z) = 1 - \Phi(z) = p$.

p	z	p	z
0.5000	0.0000	0.0500	1.6449
0.4000	0.2533	0.0250	1.9600
0.3000	0.5244	0.0100	2.3263
0.2000	0.8416	0.0050	2.5758
0.1500	1.0364	0.0010	3.0902
0.1000	1.2816	0.0005	3.2905

Answers

TOPIC-BASED QUESTIONS

Pages 166–167: Algebra and Functions

1. a) $fg(x) = \frac{3(x^2-3)}{x^2-3-1} = \frac{3x^2-9}{x^2-4}$ **[1]**

   b) Change the subject of the formula.

   Let $y = \frac{3x}{x-1}$ **[1]**

   $yx - y = 3x$, so $yx - 3x = y$

   $x(y-3) = y$ **[1]**

   $x = \frac{y}{y-3}$

   $f^{-1}(x) = \frac{x}{x-3}$ **[1]**

   The domain is $\{x \in \mathbb{R}: x \neq 3\}$ **[1]**

   The range is $\{y \in \mathbb{R}: y \neq 1\}$ **[1]**

   As $x \to \infty$, $y \to 1$ and as $x \to -\infty$, $y \to 1$. This can also be found by noting the range of $f(x)$ is $\{x \in \mathbb{R}: x \neq 1\}$ and recalling that the domain of $f(x)$ is the range of $f^{-1}(x)$.

2. a) $x + 4 \geqslant 0 \Rightarrow x \geqslant -4$, so the domain is $\{x \in \mathbb{R}: x \geqslant -4\}$ **[1]**

   The range is $\{f \in \mathbb{R}: f \geqslant 0\}$ **[1]**

   b) Change the subject of the formula.

   Let $y = \sqrt{x+4}$ **[1]**

   $y^2 = x + 4$, so $y^2 - 4 = x$

   $f^{-1}(x) = x^2 - 4$ **[1]**

   The domain is the range of $f(x)$ $\therefore$ $\{x \in \mathbb{R}: x \geqslant 0\}$ **[1]**

   The range is the domain of $f(x)$ $\therefore$ $\{f \in \mathbb{R}: f \geqslant -4\}$ **[1]**

   c) $f^{-1}(x) = x^2 - 4$ is easier to sketch, so start by sketching it, then reflect it over the line $y = x$ in order to sketch $f(x) = \sqrt{x+4}$. Remember the domain and the range of each function when sketching the curves.

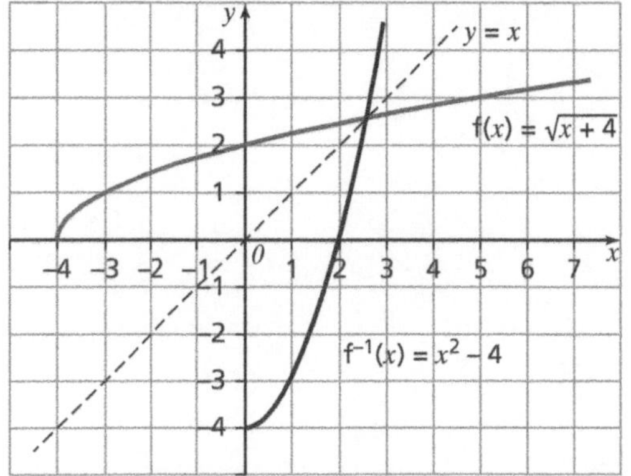

[1 for general shape and correct range of both curves; 1 for correct $y = f(x)$; 1 for correct $y = f^{-1}(x)$]

3. a) $fg(x) = 2(x^2 + 4) - 3 = 2x^2 + 5$ **[1]**

   b) $gf(x) = (2x - 3)^2 + 4 = 4x^2 - 12x + 13$ **[1]**

   c) $f^2(x) = ff(x) = 2(2x - 3) - 3 = 4x - 9$ **[1]**

   d) $4x^2 - 12x + 13 = 29$

   $4x^2 - 12x - 16 = 0$ **[1]**

   $x^2 - 3x - 4 = 0$

   $(x + 1)(x - 4) = 0$

   $x = -1$ and $x = 4$ **[1]**

4. $y = |2x + 3| - 1$ is a translation of $y = |2x + 3|$ by $\begin{pmatrix} 0 \\ -1 \end{pmatrix}$

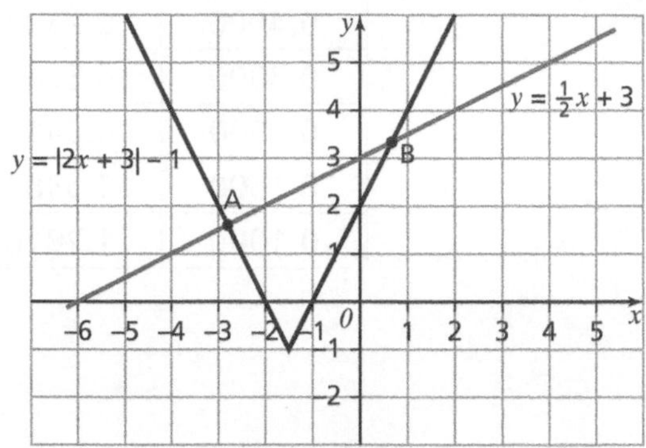

[1 for graph of $y = |2x + 3| - 1$; 1 for $y = \frac{1}{2}x + 3$]

There are two points of intersection.

Point A: $-(2x + 3) - 1 = \frac{1}{2}x + 3$ **[1]**

Point A lies on the part of the modulus graph that has been reflected over the x-axis.

$-\frac{5}{2}x = 7$, so $x = -2\frac{4}{5}$ **[1]**

Point B: $2x + 3 - 1 = \frac{1}{2}x + 3$ **[1]**

$\frac{3}{2}x = 1$, so $x = \frac{2}{3}$ **[1]**

5. $y = 2(f(x - 3)) - 1$

   First do the horizontal translation of $f(x)$ by +3, then the vertical stretch by a factor of 2, then the vertical translation of −1.

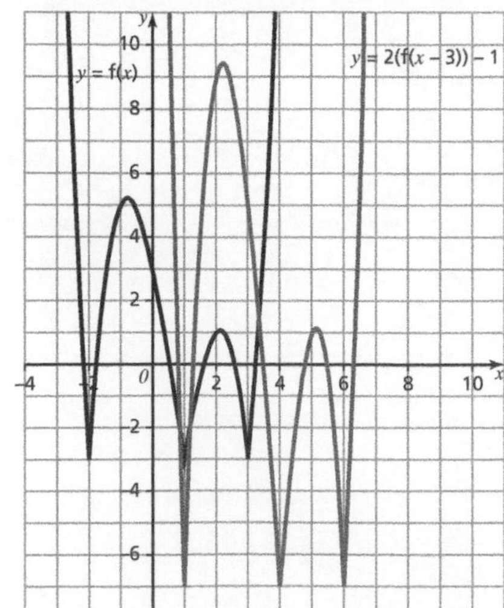

[1 for general shape; 1 for horizontal translation; 1 for stretch; 1 for vertical translation]

6. a) $\frac{4}{x+2}+\frac{2x}{x-1}+5$

$=\frac{4(x-1)}{(x+1)(x-1)}+\frac{2x(x+2)}{(x-1)(x+2)}+\frac{5(x+2)(x-1)}{(x+2)(x-1)}$

$=\frac{4(x-1)+2x(x+2)+5(x+2)(x-1)}{(x+2)(x-1)}$ **[1]**

$=\frac{4x-4+2x^2+4x+5x^2+5x-10}{(x+2)(x-1)}$

$=\frac{7x^2+13x-14}{(x+2)(x-1)}$ **[1]**

Multiply so that each fraction has a common denominator and then write as a single fraction. Then expand and simplify.

b) The question is really asking for the decomposition of $\frac{7x^2+13x-14}{(x+2)(x-1)}\div(x+2)=\frac{7x^2+13x-14}{(x+2)^2(x-1)}$ **[1]** into partial fractions.

$\frac{7x^2+13x-14}{(x+2)^2(x-1)}\equiv\frac{A}{x-1}+\frac{B}{x+2}+\frac{C}{(x+2)^2}$

$\equiv\frac{A(x+2)^2}{(x+2)^2(x-1)}+\frac{B(x-1)(x+2)}{(x+2)^2(x-1)}+\frac{C(x-1)}{(x+2)^2(x-1)}$

$\equiv\frac{A(x+2)^2+B(x-1)(x+2)+C(x-1)}{(x+2)^2(x-1)}$ **[1]**

$7x^2+13x-14=A(x+2)^2+B(x-1)(x+2)+C(x-1)$ **[1]**

To find A, substitute $x=1$.

$7\times1^2+(13\times1)-14=A(1+2)^2$

$6=9A$, so $A=\frac{2}{3}$ **[1]**

To find C, substitute $x=-2$.

$7\times(-2)^2+(13\times-2)-14=C(-2-1)$

$-12=-3C$, so $C=4$ **[1]**

To find B, equate the coefficients.

$A(x+2)^2+B(x-1)(x+2)+C(x-1)$

$=Ax^2+4Ax+4A+Bx^2+Bx-2B-C+Cx$

$=(A+B)x^2+(4A+B+C)x+(4A-2B-C)$ **[1]**

The coefficient of x^2 is 7 and the value of A is $\frac{2}{3}$,

$\frac{2}{3}+B=7$, $B=\frac{19}{3}$ **[1]**

Therefore, $f(x)\div(x+2)$

$=\frac{7x^2+13x-14}{(x+2)^2(x-1)}=\frac{2}{3(x-1)}+\frac{19}{3(x+2)}+\frac{4}{(x+2)^2}$ **[1]**, as required.

7.

$$\begin{array}{r}
2x^2-x-1 \\
x+3\,\overline{)\,2x^3+5x^2-4x+0} \\
-(2x^3+6x^2) \\
\hline
-x^2-4x \\
-(-x^2-3x) \\
\hline
-x+0 \\
-(-x-3) \\
\hline
3
\end{array}$$

[1 for method of algebraic division; 1 for quotient correct; 1 for remainder correct]

$\frac{2x^3+5x^2-4x}{x+3}=2x^2-x-1+\frac{3}{x+3}$ **[1]**

$A=2$, $B=-1$, $C=-1$ and $D=3$ **[1]**

Pages 168–169: Coordinate Geometry

1. Convert $x=2+5\cos(t)$, $y=-4+5\sin(t)$ into a Cartesian equation.

$x=2+5\cos(t)$

$\frac{x-2}{5}=\cos(t)$ **[1]**

$y=-4+5\sin(t)$

$\frac{y+4}{5}=\sin(t)$ **[1]**

$\sin^2(\theta)+\cos^2(\theta)\equiv1$ **[1]**

$\left(\frac{y+4}{5}\right)^2+\left(\frac{x-2}{5}\right)^2=1$

$(x-2)^2+(y+4)^2=25$ **[1]**

The parametric equations describe a circle with centre (2, –4) and radius 5. **[1]**

The range of $x=2+5\cos(t)$, $\{t\in\mathbb{R}: \frac{\pi}{2}\leqslant t\leqslant\pi\}$ and hence the domain of the curve is $\{x\in\mathbb{R}: -3\leqslant t\leqslant2\}$ **[1]**

The range of $y=-4+5\sin(t)$, $\{t\in\mathbb{R}: \frac{\pi}{2}\leqslant t\leqslant\pi\}$ and hence the range of the curve is $\{y\in\mathbb{R}: -4\leqslant t\leqslant1\}$ **[1]**

2. a) $x=3\ln(t)$

$\frac{x}{3}=\ln(t)$

$e^{\frac{x}{3}}=t$ **[1]**

$y=2\ln(t)+4$

$y=2\ln\left(e^{\frac{3}{x}}\right)+4$ **[1]**

$y=2\times\frac{x}{3}+4$

$y=\frac{2}{3}x+4$ **[1]**

The range of $x=3\ln(t)$, $\{t\in\mathbb{R}: t\geqslant1\}$ and hence the domain of $f(x)$ is $\{x\in\mathbb{R}: x\geqslant0\}$. **[1]**

The range of $y=2\ln(t)+4$, $\{t\in\mathbb{R}: t\geqslant1\}$ and hence the range of $f(x)$ is $\{y\in\mathbb{R}: y\geqslant4\}$. **[1]**

Answers

b) 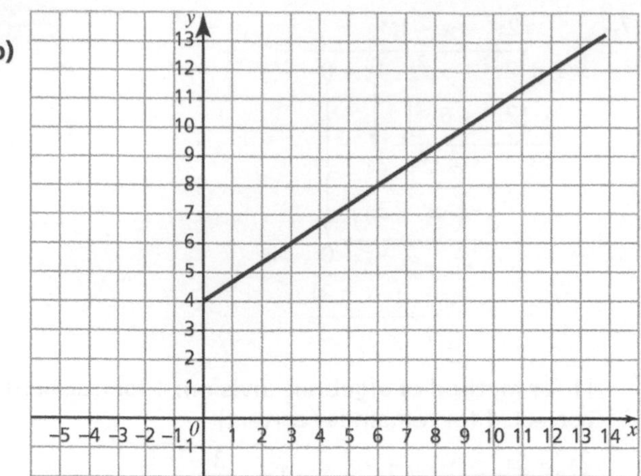

[1 for line correct; 1 for domain and range correct]

3. a)

t	$x = 30\left(\frac{t}{1+t^3}\right)$	$y = 30\left(\frac{t^2}{1+t^3}\right)$
0	0.00	0.00
0.25	7.38	1.85
0.5	13.33	6.67
0.75	15.82	11.87
1	15.00	15.00
1.5	10.29	15.43
2	6.67	13.33
3	3.21	9.64
10	0.30	3.00
100	0.00	0.30

[1 for 10 correct values; 2 for all values correct]

b) 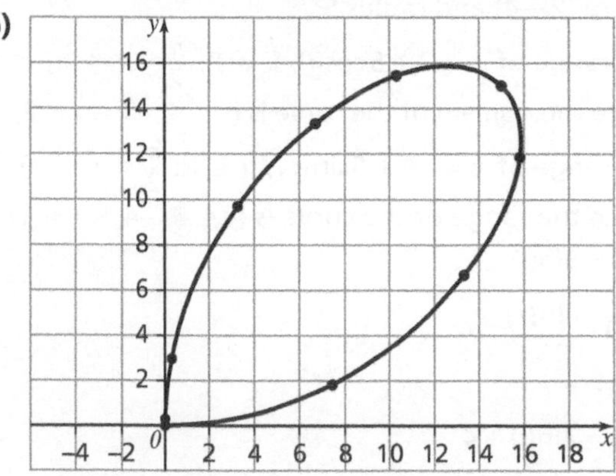

[1 for points plotted correctly; 1 for general shape]

4. a) Using right-triangle trigonometry, $\cos(30°) = \frac{x}{1.5}$ and $\sin(30°) = \frac{y}{1.5}$

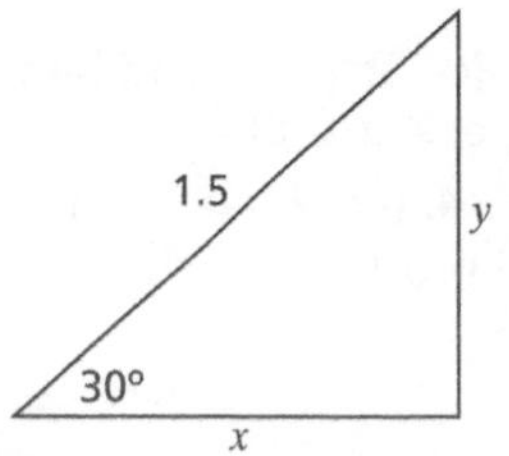

Taking into account the current, the x-component is $x = 1.5\cos(30°)t - t$ **[1 for $1.5\cos(30°)t$; 1 for $-t$]** and the y-component is $y = 1.5\sin(30°)t$ **[1]**.

b) The boat has crossed the river when $y = 30$.

$60 = (1.5\sin(30°))t$ **[1]**

$\frac{60}{1.5\sin(30°)} = t$

$t = 80$ seconds **[2]**

c) $x = 1.5\cos(30°) \times 80 - 80 = 23.92$ m (to 2 d.p.) **[1]**

5. a) The point where the projectile hits the water is $y = 0$.

$0 = 100 + (v\sin(30°))t - 4.9t^2$

$0 = 100 + \left(\frac{v}{2}\right)t - 4.9t^2$ **[1]**

$\frac{v}{2}t = 4.9t^2 - 100$

$v = \frac{9.8t^2 - 200}{t}$ **[1]**

$t \geqslant 5: \quad v \geqslant \frac{(9.8 \times 25) - 200}{5}$

$v \geqslant 9\ \text{ms}^{-1}$ **[1]**

$t \leqslant 6: \quad v \leqslant \frac{(9.8 \times 36) - 200}{6}$

$v \leqslant 25.46\ \text{ms}^{-1}$ **[1]**

b) $x = (v\cos(30°))t$, from part a), $v = 9\ \text{ms}^{-1}$, $t = 5$, so

$x = (9\cos(30°)) \times 5 = \frac{45\sqrt{3}}{2} = 38.97$ m (to 2 d.p.) **[1]**

6. a) $y = a(1 - \cos(t))$

$0.25 = a(1 - \cos\left(\frac{\pi}{2}\right))$ **[1]**

$a = 0.25$ **[1]**

b) $y = 0.25(1 - \cos(t))$, y is maximised when $\cos(t) = -1$ **[1 (may be implied by final answer)]**, which is when $t = \pi, 3\pi, 5\pi, 7\pi$... **[1]**

c) The wheel completes a full revolution every 2π seconds, so find the value of x at $t = 2\pi$ and subtract the value of x at $t = 0$.

$x = 0.25(2\pi - \sin(2\pi)) - 0$ **[1]**

$x = \frac{\pi}{2}$ **[1]**

Pages 170–171: Sequences and Series

1. a) Increasing **[1]**

b) Substitute $n = 1, n = 2, n = 3, n = 4$ and $n = 5$ to get $-6, -4, -2, 0, 2$ **[1]**

c) $u_{(n+1)} = u_n + 2,\ u_1 = -6$ **[1]**

d) $\sum_{1}^{5} u_n = -6 - 4 - 2 + 0 + 2 = -10$ **[1]**

2. a) $42.195 = 10 + 0.2(n - 1)$ **[1]**

$n = (42.195 - 10 + 0.2) \div 0.2 = 161.975$

It takes her 162 days. **[1]**

b) $S_{161} = \frac{161}{2}(10 + 42.195) = 4201.70$ km (to 2 d.p.) **[1]**

3. $4795 = 1295 + 14d$

$d = 250$ **[1]**

$u_{30} = 1295 + (29 \times 250) = 8545$ **[1]**

$S_{30} = \frac{30}{2}(1295 + 8545) = 147600$ **[1]**

4. The odd numbers are $(2n - 1)$ where n is a natural number.

The first odd number is 1 and the 100th odd number is $2 \times 100 - 1 = 199$

$\sum_{1}^{100}(2n - 1) = S_{100} = \frac{100}{2}(1 + 199) = 10000$

[1 for sigma notation; 1 for substitution into S formula; 1 for final answer]

5. a) Substitute $n = 1$, $n = 2$, $n = 3$, $n = 4$ and $n = 5$ to get 7, 14, 28, 56, 112 **[1]**

b) $S_{10} = \frac{7(2^{10} - 1)}{(2 - 1)}$ **[1]** $= 7161$ **[1]**

c) Divergent ($|r| > 1$) **[1]**

Sum is undefined **[1]**

6. a) $a = 5000$, $r = 1.03$ **[1]**

$u_5 = 5000(1.03^5) = £5796.37$ **[1]**

b) $7500 = 5000 \times 1.03^{n-1}$ **[1]**

$\frac{3}{2} < 1.03^{n-1}$

$\log_{10}\left(\frac{3}{2}\right) = (n - 1)\log_{10}(1.03)$ **[1]**

$n = \frac{\log_{10}\left(\frac{3}{2}\right)}{\log_{10}(1.03)} + 1 = 14.7$ **[1]**

After 15 years, the account will have £7500. **[1]**

7. a) $100, 75, \frac{225}{4}, \frac{675}{16}, \frac{2025}{64}$ litres **[1]**

b) $a = 100$, $r = 0.75$ **[1]**

$u_n = 100 \times 0.75^{n-1}$ **[1]**

c) $100 \times 0.75^{n-1} < 1$ **[1]**

$0.75^{n-1} < 0.01$

$(n - 1)(\log_{10}(0.75)) < \log_{10}(0.01)$ **[1]**

$n > \frac{\log_{10}(0.01)}{\log_{10}(0.75)} + 1$

$n = 17.007$ **[1]**

After 18 dilutions, the mixture will be less than 1% dye. **[1]**

8. $\frac{1}{(4x - 3)^3} = (4x - 3)^{-3}$

Rewrite in index form.

$(4x - 3)^{-3} = (-3^{-3})\left(1 + \frac{-3 \times -\frac{4}{3}x}{1!} + \frac{-3(-3 - 1)(-\frac{4}{3}x)^2}{2!} + \frac{-3(-3 - 1)(-3 - 2)\left(-\frac{4}{3}x\right)^3}{3!} + \ldots\right)$ **[1]**

$= -\frac{1}{27}\left(1 + 4x + \frac{32x^2}{3} + \frac{640x^3}{27}\right) + \ldots$ **[1]**

$= -\frac{1}{27} - \frac{4x}{27} - \frac{32x^2}{81} - \frac{640x^3}{729} + \ldots$ **[1]**

Valid for $|x| < \frac{3}{4}$ **[1]**

9. a) $\frac{2}{\sqrt{x + 1}} = 2(x + 1)^{-\left(\frac{1}{2}\right)}$

Rewrite in index form.

$2(x + 1)^{-\left(\frac{1}{2}\right)} = \left(2\left(1^{-\frac{1}{2}}\left(1 + \frac{-\frac{1}{2} \times x}{1!} + \frac{-\frac{1}{2}\left(-\frac{1}{2} - 1\right)(x)^2}{2!} + \frac{-\frac{1}{2}\left(-\frac{1}{2} - 1\right)\left(-\frac{1}{2} - 2\right)(x)^3}{3!} + \cdots\right)\right)\right)$ **[1]**

$= 2\left(1 - \frac{x}{2} + \frac{3x^2}{8} - \frac{5x^3}{16} + \ldots\right)$ **[1]**

$= 2 - x + \frac{3x^2}{4} - \frac{5x^3}{8} + \ldots$ **[1]**

b) The expansion is only valid when $|x| < 1$, so the substitution of $x = 4$ is not valid. **[1]**

10. $\frac{1}{(kx - 1)^5} = (kx - 1)^{-5}$

Rewrite in index form.

Finding the x^2 term in the expansion of $(kx - 1)^{-5}$.

$(-1 + kx)^{-5} = -1^{-5}(1 + \ldots + \frac{-5(-5 - 1)(-kx)^2}{2!} + \ldots$ **[1]**

$= -1(1 + 15k^2x^2 + \ldots)$ **[1]**

$-60 = -15k^2 \Rightarrow k^2 = 4 \Rightarrow k = 2$ **[1]**

Pages 172–175: Trigonometry

1. Angle AOB $= \theta$ is given by $\cos\theta = \frac{14^2 + 14^2 - 25^2}{2 \times 14 \times 14}$ **[1]**

$\Rightarrow \theta = 2.207^c$ **[1]**

Area of segment

$= \frac{1}{2}r^2(\theta - \sin\theta) = \frac{1}{2} \times 14^2 \times (2.207 - \sin 2.207)$ **[1]**

$= 138\,\text{cm}^2$ **[1]**

2. $\angle AOB = \theta = \frac{\pi}{3}$

Area of minor segment

$= \frac{1}{2}r^2(\theta - \sin\theta) = \frac{1}{2} \times 10^2 \times \left(\frac{\pi}{3} - \sin\frac{\pi}{3}\right) = 9.059$ **[2]**

Area of major sector $= \frac{1}{2}r^2\theta = \frac{1}{2} \times 10^2 \times \frac{5\pi}{3} = 261.8$ **[2]**

Ratio $= 9.059 : 261.8 = 1 : 28.9$ **[1]**

3. a) Area of shaded segment

$= \frac{1}{2}r^2((\pi - \theta) - \sin(\pi - \theta)) = \frac{1}{2}r^2((\pi - \theta) - \sin\theta)$ **[2]**

Area of shaded triangle $= \frac{r^2\sin\theta}{2}$ **[1]**

So $\frac{1}{2}r^2((\pi - \theta) - \sin\theta) = \frac{r^2\sin\theta}{2}$ **[1]**

$(\pi - \theta) - \sin\theta = \sin\theta$ **[1]**

$\pi - \theta = 2\sin\theta$

$\theta + 2\sin\theta = \pi$

Answers

b) Let $f(\theta) = \theta + 2\sin\theta - \pi$

$f(1.2) = 1.2 + 2\sin 1.2 - \pi = -0.0775 < 0$ **[2]**

$f(1.3) = 1.3 + 2\sin 1.3 - \pi = 0.0855 > 0$ **[2]**

Sign change, so $1.2^c < \alpha < 1.3^c$ **[1]**

4. $\overrightarrow{AB} = \overrightarrow{OB} - \overrightarrow{OA} = \begin{pmatrix}-1\\1\\3\end{pmatrix} - \begin{pmatrix}2\\-3\\1\end{pmatrix} = \begin{pmatrix}-3\\4\\2\end{pmatrix}$ **[2]**

$\overrightarrow{AB} = \sqrt{(-3)^2 + 4^2 + 2^2} = \sqrt{29}$ **[2]**

5. $|\overrightarrow{OA}|^2 = (2t+4)^2 + (2t)^2 + (1-4t)^2$

$= 4t^2 + 16t + 16 + 4t^2 + 16t^2 - 8t + 1 = 24t^2 + 8t + 17$ **[2]**

$\dfrac{d|\overrightarrow{OA}|^2}{dt} = 48t + 8$ **[2]**

$\dfrac{d|\overrightarrow{OA}|^2}{dt} = 0 \Rightarrow 48t + 8 = 0 \Rightarrow t = -\dfrac{8}{48} = -\dfrac{1}{6}$ **[2]**

So least value of $|\overrightarrow{OA}|$ is

$\sqrt{\left(2\left(-\frac{1}{6}\right)+4\right)^2 + \left(2\left(-\frac{1}{6}\right)\right)^2 + \left(1-4\left(-\frac{1}{6}\right)\right)^2}$

$= \sqrt{\dfrac{147}{9}} = \dfrac{7\sqrt{3}}{3}$ **[1]**

6. $\dfrac{2\tan x}{1-\tan^2 x} - \tan x = 0$ **[1]**

$\dfrac{2\tan x}{1-\tan^2 x} - \dfrac{\tan x(1-\tan^2 x)}{1-\tan^2 x} = 0$

$\dfrac{2\tan x - \tan x(1-\tan^2 x)}{1-\tan^2 x} = 0$ **[1]**

$2\tan x - \tan x\,(1-\tan^2 x) = 0$ **[1]**

$\tan^3 x + \tan x = 0$

$\tan x\,(\tan^2 x + 1) = 0$ **[1]**

$\tan x = 0 \Rightarrow x = 0, x = \pi$ **[2]**

7. $\dfrac{4\sin x - 4x\cos x}{\sin^2 x\tan x} \approx \dfrac{4x - 4x\left(1-\frac{x^2}{2}\right)}{x^3}$ **[3]**

$\approx \dfrac{2x^3}{x^3}$ **[2]**

≈ 2

8. a) $\operatorname{cosec}\left(-\dfrac{\pi}{4}\right) = -\operatorname{cosec}\left(\dfrac{\pi}{4}\right) = -\sqrt{2}$ **[2]**

b) $\cot\left(-\dfrac{5\pi}{4}\right) = \cot\dfrac{3\pi}{4} = -1$ **[2]**

c) $\sec\left(\dfrac{7\pi}{3}\right) = \sec\left(\dfrac{\pi}{3}\right) = 2$ **[2]**

9. $(1-\sec^2 x)(1-\sin^2 x) = 1 + \sec^2 x\sin^2 x - \sec^2 x - \sin^2 x$ **[1]**

$= 1 + \tan^2 x - \sec^2 x - \sin^2 x$ **[1]**

$= \sec^2 x - \sec^2 x - \sin^2 x$ **[1]**

$= -\sin^2 x$

$= -(1-\cos^2 x)$ **[1]**

$= \cos^2 x - 1$

10. $1 + \tan^2\left(x-\dfrac{\pi}{6}\right) - 2\tan\left(x-\dfrac{\pi}{6}\right) - 2 = 0$ **[1]**

$\tan^2\left(x-\dfrac{\pi}{6}\right) - 2\tan\left(x-\dfrac{\pi}{6}\right) - 1 = 0$ **[2]**

$\tan\left(x-\dfrac{\pi}{6}\right) = \dfrac{2\pm\sqrt{4-4(1)(-1)}}{2} = \dfrac{2\pm 2\sqrt{2}}{2} = 1\pm\sqrt{2}$ **[2]**

$\tan\left(x-\dfrac{\pi}{6}\right) = 1+\sqrt{2}$

$x - \dfrac{\pi}{6} = 1.178 \Rightarrow x = 1.70$ **[1]**

$x - \dfrac{\pi}{6} = \pi + 1.178 \Rightarrow x = 4.84$ **[1]**

$\tan\left(x-\dfrac{\pi}{6}\right) = 1-\sqrt{2}$

$x - \dfrac{\pi}{6} = \pi - 0.393 \Rightarrow x = 3.27$ **[1]**

$x - \dfrac{\pi}{6} = -0.393 \Rightarrow x = 0.131$ **[1]**

11. $\cot x(\operatorname{cosec}^2 x - 1) = 3\sqrt{3}$ **[1]**

$\cot x(\cot^2 x) = 3\sqrt{3}$ **[1]**

$\cot^3 x = 3\sqrt{3}$

$\cot x = \sqrt{3}$ **[1]**

$\tan x = \dfrac{1}{\sqrt{3}}$

$x = \dfrac{\pi}{6}$ or $x = \dfrac{7\pi}{6}$ **[2]**

12. a) $\cos 2x = 1 - 2\sin^2 x = 1 - 2\left(-\dfrac{1}{4}\right)^2 = \dfrac{7}{8}$ **[2]**

b) $\sin 2x = 2\sin x\cos x = 2\sin x\sqrt{1-\sin^2 x}$

$= 2\left(-\dfrac{1}{4}\right)\sqrt{1-\left(-\dfrac{1}{4}\right)^2} = -\dfrac{\sqrt{15}}{8}$ **[2]**

$\tan 2x = \dfrac{\sin 2x}{\cos 2x} = -\dfrac{\sqrt{15}}{8}\left(\dfrac{8}{7}\right) = -\dfrac{\sqrt{15}}{7}$ **[2]**

13. $\sin 195° = -\sin 15°$

$= -\sin\left(\dfrac{30}{2}\right)^\circ$ **[1]**

$= -\sqrt{\dfrac{1-\cos 30^\circ}{2}}$ **[1]**

$= -\sqrt{\dfrac{1-\frac{\sqrt{3}}{2}}{2}}$ **[1]**

$\left(= -\dfrac{\sqrt{2-\sqrt{3}}}{2}\right)$

14. $2\sin\dfrac{x}{2}\cos\dfrac{x}{2} + 1 - 2\sin^2\dfrac{x}{2} = 1$ **[2]**

$2\sin\dfrac{x}{2}\cos\dfrac{x}{2} - 2\sin^2\dfrac{x}{2} = 0$

$2\sin\dfrac{x}{2}\left(\cos\dfrac{x}{2} - \sin\dfrac{x}{2}\right) = 0$ **[1]**

$\sin\dfrac{x}{2} = 0$ **[1]**

$\dfrac{x}{2} = 0 \Rightarrow x = 0$ **[1]**

Or:

$\cos\frac{x}{2} - \sin\frac{x}{2} = 0$

$\tan\frac{x}{2} = 1$ **[1]**

$\frac{x}{2} = \frac{\pi}{4}$ or $\frac{x}{2} = \frac{5\pi}{4}$

$\Rightarrow x = \frac{\pi}{2}$ or $x = \frac{5\pi}{2}$ **[2]**

15. $4\cos x + 2\sin x = R\cos(x - \alpha)$

$= R\cos x\cos\alpha + R\sin x\sin\alpha$

$4 = R\cos\alpha$ **[1]**

$2 = R\sin\alpha$ **[1]**

$R^2 = 4^2 + 2^2 = 20 \Rightarrow R = \sqrt{20} = 2\sqrt{5}$ **[2]**

$\frac{R\sin\alpha}{R\cos\alpha} = \tan\alpha = \frac{2}{4} \Rightarrow \alpha = 0.4636^c$ **[2]**

$\cos x - 2\sin x = 2\sqrt{5}\cos\left(x - 0.4636^c\right)$

$4\cos x + 2\sin x = 2$

$2\sqrt{5}\cos\left(x - 0.4636^c\right) = 2$

$\cos\left(x - 0.4636^c\right) = \frac{1}{\sqrt{5}}$ **[1]**

$x - 0.4636 = 1.107 \Rightarrow x = 1.57$ **[2]**

$x - 0.4636 = 2\pi - 1.107 \Rightarrow x = 5.64$ **[2]**

Pages 176–177: Differentiation

1. a) $u = x^2$, $v = \sin 4x$

$\frac{du}{dx} = 2x \qquad \frac{dv}{dx} = 4\cos 4x$ **[1]**

$4x^2\cos 4x + 2x\sin 4x$ **[2]**

b) $u = \cos x$, $v = x$

$\frac{du}{dx} = -\sin x \qquad \frac{dv}{dx} = 1$ **[1]**

$\frac{-x\sin x - \cos x}{x^2}$ **[2]**

c) $4\ln(3x) + 2x^{-1}$ **[1]**

$\frac{4}{x} - 2x^{-2}$ **[2]**

2. $u = e^x$, $v = x$ **[1]**

$\frac{du}{dx} = e^x \qquad \frac{dv}{dx} = 1$ **[1]**

$\frac{xe^x - e^x}{x^2}$ **[1]**

$\frac{xe^x - e^x}{x^2} = 0$ **[1]**

$xe^x - e^x = 0$, $e^x(x - 1) = 0$ **[1]**

$x = 1$ **[1]** and $y = e$ **[1]**

3. a) $\frac{dx}{dt} = -2t(1 + t)^{-2} + 2(1 + t)^{-1}$ **[1]**

$\frac{dy}{dt} = -t^2(1 + t)^{-2} + 2t(1 + t)^{-1}$ **[1]**

$\frac{dy}{dx} = \frac{-t^2(1+t)^{-2} + 2t(1+t)^{-1}}{-2t(1+t)^{-2} + 2(1+t)^{-1}}$ at $x = 1$,

$t = 1 \frac{dy}{dx} = 1.5$ **[1]**

Equation of normal $y - \frac{1}{2} = \frac{-2}{3}(x - 1)$ **[1]**

$6y - 3 = -4x + 4$, $6y = 7 - 4x$ **[1]**

b) $\frac{6t^2}{1+t} = 7 - \frac{8t}{1+t}$ **[1]**

Substitute t into $6y = 7 - 4x$

$6t^2 + t - 7 = 0$ **[1]**

$(6t + 7)(t - 1)\ t = \frac{-7}{6}$ **[1]**

$\left(14, \frac{-49}{6}\right)$ **[1]**

4. a) $3x^2 + 2y\frac{dy}{dx}$ **[1]**

$4x^2\frac{dy}{dx} + 8xy$ **[2]**

$3x^2 + 2y\frac{dy}{dx} + 4x^2\frac{dy}{dx} + 8xy = 0$

$\frac{dy}{dx} = \frac{3x^2 + 8xy}{-2y - 4x^2}$ **[1]**

b) $(1)^3 + y^2 + 4(1)^2y = 2$ **[1]**

$y^2 + 4y - 1 = 0$ **[1]**

c) $y^2 + 4y - 1 = 0$, $(y + 2)^2 - 5 = 0$ **[1]**

$y = \sqrt{5} - 2$ **[1]**

$\frac{dy}{dx} = \frac{8\sqrt{5} - 13}{-2\sqrt{5}}$ **[1]**

Pages 178–179: Integration

1. a) $x^3 - x^2 - 6x + 10 = 3x + 1$ **[1]**

$x^3 - x^2 - 9x + 9 = 0$, $x = 1$ is a solution **[1]**

$(x - 1)(x^2 - 9) = 0$, $(x - 1)(x - 3)(x + 3)$, $x = 1$, $x = 3$ and $x = -3$ **[1]**

(1, 4), (3, 10) and (–3, –11)

b) $\int_1^3 x^3 - x^2 - 6x + 10 = \left[\frac{x^4}{4} - \frac{x^3}{3} - 3x^2 + 10x\right]_1^3$ **[1]** $\frac{22}{6}$ **[1]**

$\int_1^3 3x + 1 = \left[\frac{3x^2}{2} + x\right]_1^3$ **[1]** 14 **[1]**

$14 - \frac{22}{6} = \frac{31}{3}$ **[1]**

2. a) $y^{-2}dy = x\,dx$ **[1]**

$\int y^{-2}dy = \int x\,dx$ **[1]**

$-y^{-1}$ **[1]** $= \frac{x^2}{2} + c$ **[2]**

b) $-(1)^{-1} = \frac{(1)^2}{2} + c$, $c = \frac{-3}{2}$ **[1]**

$\frac{-1}{y} = \frac{x^2}{2} - \frac{3}{2}$ **[1]**

$y = \frac{2}{3 - x^2}$ **[1]**

3. a) $\int \frac{5x^2 - 8x + 1}{2x(x - 1)^2} = \frac{A}{2x} + \frac{B}{x - 1} + \frac{C}{(x - 1)^2}$

Answers

$5x^2 - 8x + 1 = A(x-1)^2 + B(2x)(x-1) + C(2x)$ **[2]**

$x = 1 \quad -2 = 0 + 0 + 2C \quad C = -1$ **[1]**

$x = 0 \quad 1 = A$ **[1]**

$x = 2 \quad 5 = A + 4B + 4C$

$5 = 1 + 4B - 4$

$8 = 4B$

$B = 2$ **[1]**

$y = \frac{1}{2x} + \frac{2}{x-1} - \frac{1}{(x-1)^2}$ **[1]**

b) $\int \frac{1}{2x} + \frac{2}{x-1} - \frac{1}{(x-1)^2}\,dx$

$= \ln(2x)$ **[1]** $+ 2\ln(x-1)$ **[1]** $+ \frac{1}{1-x} + c$ **[1]**

Pages 180–181: Numerical Methods

1. $f(-0.6) = e^{0.6} - \frac{1}{0.6} = 0.16 > 0$ **[2]**

$f(-0.5) = e^{0.5} - \frac{1}{0.5} = -0.35 < 0$ **[2]**

There is a sign change and $f(x)$ is continuous in the interval $-0.6 < x < -0.5$, therefore there exists a root α such that $-0.6 < \alpha < -0.5$ **[1]**

$f'(x) = -e^{-x} - \frac{1}{x^2}$ **[1]**

$x_{n+1} = x_n - \left(\frac{e^{-x} + \frac{1}{x}}{-e^{-x} - \frac{1}{x^2}} \right)$ **[1]**

$x_1 = -0.566$ **[1]**

$\alpha = -0.57$ **[1]**

2. $f(0.9) = -0.31 < 0$ **[2]**

$f(1) = 0.16 > 0$ **[2]**

f is continuous in the interval $0.9 < x < 1.0$ and there is a sign change, so there exists a root α such that $0.9 < \alpha < 1.0$ **[1]**

$x^7 - \sin x = 0$

$x^7 = \sin x$

$x = \sqrt[7]{\sin x}$

So use iteration $x_{n+1} = \sqrt[7]{\sin x_n}$ with $x_0 = 1$:

$x_1 = 0.9756$ **[1]**

$x_2 = 0.9734$

$x_3 = 0.9732$ **[1]**

$\alpha = 0.973$ **[1]**

3. a) $f(-2) = -1.61$, $f(2) = 7.14$ **[2]**

This shows there is an **odd number of roots** in the interval $-2 < x < 2$. **[1]**

b) $f(-0.25) = 0.27 > 0$, $f(0.25) = -0.21 < 0$ **[2]**

So there are roots between $x = -2$ and $x = -0.25$, **[1]**
between $x = -0.25$ and $x = 0.25$, **[1]**
and between $x = 0.25$ and $x = 2$ **[1]**

4. $f(1) = -0.56 < 0$ **[2]**

$f(1.5) = 1.65 > 0$ **[2]**

There is a sign change and $f(x)$ is continuous in the interval $1 < x < 1.5$, therefore there exists a root α such that $1 < \alpha < 1.5$ **[1]**

$f'(x) = \frac{\sec^2 x}{\tan x} = \frac{1}{\sin x \cos x}$ **[1]**

Newton-Raphson iteration is $x_{n+1} = x_n - \sin x_n \cos x_n (\ln(\tan x_n) - 1)$ with $x_0 = 1.25$ **[1]**

$x_1 = 1.2195$ **[1]**

$x_2 = 1.2183$ **[1]**

$\alpha = 1.218$ **[1]**

5. $h = \frac{2}{4} = \frac{1}{2}$ **[1]**

x	0	0.5	1	1.5	2
$\sqrt{e^x + x^2}$	1	1.378	1.928	2.595	3.375

[2]

$\int_0^2 \sqrt{e^x + x^2}\,dx \approx \frac{\left(\frac{1}{2}\right)}{2}[1 + 2(1.378 + 1.928 + 2.595) + 3.375]$ **[2]**

$= 4.04$ units^2 **[1]**

6. a) $f(0) = 1 > 0$ **[2]**

$f(1) = -0.5 < 0$ **[2]**

Sign change, so there is a root α in the range $0 < x < 1$ **[1]**

b) $f'(x) = -\frac{1}{(x+1)^2} - 2x$ **[2]**

$f'(x) < 0$ for all $x > 0$, so f is a decreasing function. **[1]**

Therefore f will cross the x-axis only once.

Pages 182–183: Proof

1. $\frac{1 + \cos x}{\sin x} + \frac{\sin x}{1 + \cos x} = \frac{(1 + \cos x)^2 + \sin^2 x}{\sin x(1 + \cos x)}$ **[1]**

$= \frac{1 + \cos^2 x + 2\cos x + \sin^2 x}{\sin x(1 + \cos x)}$ **[1]**

$= \frac{2 + 2\cos x}{\sin x(1 + \cos x)}$ **[1]**

$= \frac{2(1 + \cos x)}{\sin x(1 + \cos x)}$ **[1]**

$= \frac{2}{\sin x}$

$= 2\operatorname{cosec} x$

2. $\frac{\cos 2x}{\cos x - \sin x} = \frac{\cos^2 x - \sin^2 x}{\cos x - \sin x}$ **[1]**

$= \frac{(\cos x + \sin x)(\cos x - \sin x)}{\cos x - \sin x}$ **[1]**

$= \sin x + \cos x$

3. $\sin 3x = \sin(2x + x) = \sin 2x \cos x + \cos 2x \sin x$
$= 2\sin x \cos^2 x + \sin x(1 - 2\sin^2 x)$
$= 2\sin x \cos^2 x + \sin x - 2\sin^3 x$ **[3]**

$\cos 3x = \cos(2x + x) = \cos 2x \cos x - \sin 2x \sin x$
$= (1 - 2\sin^2 x)\cos x - 2\sin^2 x \cos x$

$= \cos x - 2\sin^2 x \cos x - 2\sin^2 x \cos x$ **[3]**

$$\frac{\sin 3x - \sin x}{\cos x - \cos 3x} = \frac{2\sin x\cos^2 x + \sin x - 2\sin^3 x - \sin x}{\cos x - \left(\cos x - 4\sin^2 x\cos x\right)}$$

$$= \frac{2\sin x\cos^2 x - 2\sin^3 x}{4\sin^2 x\cos x}$$ **[1]**

$$= \frac{2\sin x\left(\cos^2 x - \sin^2 x\right)}{4\sin^2 x\cos x}$$ **[1]**

$$= \frac{\left(\cos^2 x - \sin^2 x\right)}{2\sin x\cos x}$$ **[1]**

$$= \frac{\cos 2x}{\sin 2x}$$ **[1]**

$= \cot 2x$

4. Proof:
   Assume that $n^2 + 2$ is divisible by 4. **[1]**
   Case 1:
   Assume n odd. So $n = 2k + 1$ **[1]**
   $n^2 + 2 = (2k+1)^2 + 2 = 4k^2 + 4k + 1 + 2 = 4(k^2 + k) + 3$ **[1]**
   Now 4 divides $n^2 + 2$ and $4(k^2 + k)$, but not 3. Therefore there is a contradiction. **[1]**
   Case 2:
   Assume n even. So $n = 2k$ **[1]**
   $n^2 + 2 = (2k)^2 + 2 = 4k^2 + 2$ **[1]**
   Now 4 divides $n^2 + 2$ and $4k^2$, but not 2. Therefore there is a contradiction. **[1]**
   Hence $n^2 + 2$ is not divisible by 4.
5. Proof:
   Suppose there exist integers a and b such that $14a + 21b = 1$ **[1]**
   Then dividing through by 7, you obtain $2a + 3b = \frac{1}{7}$ **[1]**
   Since a and b are integers, you have that $2a + 3b$ is an integer. **[1]**
   But $\frac{1}{7}$ is not an integer, so you have a contradiction. **[1]**
   Therefore there are no integers a and b such that $14a + 21b = 1$
6. Proof:
   Suppose OP is not perpendicular to L. **[1]**
   Then there exists a point Q on L such that OQ is perpendicular to L. **[1]**
   Then OPQ is a right-angled triangle, with $\angle OQP = 90°$ **[1]**
   Therefore OP is the hypotenuse of this triangle and so OP > OQ. **[1]**
   But Q lies outside the circle, and so OP < OQ. **[1]**
   Therefore there is a contradiction. **[1]**
   Therefore the tangent line L is perpendicular to OP.
7. Assume there exists an x such that $f(x) = 3$ **[1]**
   Then $\frac{3x+1}{x+2} = 3$ **[1]**
   $\Rightarrow 3x + 1 = 3x + 6$
   $\Rightarrow 1 = 6$ **[1]**
   Therefore, there is a contradiction. **[1]**
   Therefore, for every x, $f(x) \neq 3$

Pages 184–185: Probability

1. a) $P(A \cap B) = P(A) + P(B) - P(A \cup B)$ **[1]**
   $\frac{1}{4} + \frac{1}{3} - \frac{5}{12} = \frac{1}{6}$ **[2]**

   b) 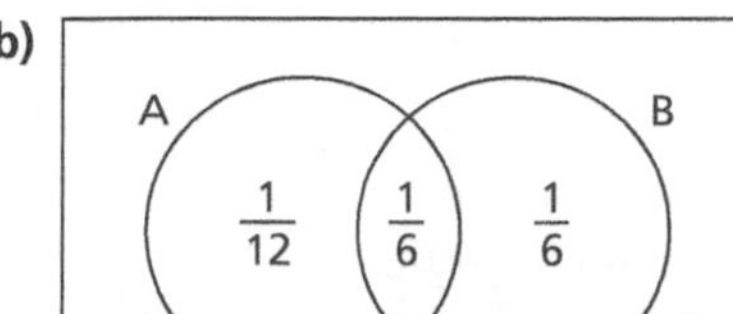

   [1 for $\frac{1}{6}$ in overlap; 1 for $\frac{1}{6}$ and $\frac{1}{12}$ in circles; 1 for $\frac{7}{12}$ outside circles]

   c) $\dfrac{\frac{1}{6}}{\frac{1}{6}+\frac{1}{6}}$ **[1]** $= \dfrac{\frac{1}{6}}{\frac{2}{6}}$ **[1]** $= \frac{1}{2}$ **[1]**

   d) $\frac{1}{12}$ **[1]**

   e) $\frac{7}{12}$ **[1]**

   f) $1 - \frac{1}{6}$ **[1]** $= \frac{5}{6}$ **[1]**

2. a) $\frac{20}{30}$ **[1]** $= \frac{2}{3}$ **[1]**

   b) $\frac{4}{30} \times \frac{20}{29} + \frac{4}{30} \times \frac{3}{29} + \frac{20}{30} \times \frac{19}{29} + \frac{20}{30} \times \frac{4}{29}$ **[2]** $= 0.634$ **[1]**

   c) $\frac{4}{30} \times \frac{3}{29} + \frac{4}{30} \times \frac{26}{29} + \frac{26}{30} \times \frac{4}{29}$ **[1]** $= \frac{22}{87}$
   $= 0.253$ **[1]**

   d) $\frac{6}{18 + 4 + 6}$ **[1]** $= \frac{6}{28} = \frac{3}{14}$ **[1]**

Pages 186–187: Statistical Distributions

1. $P(X < 32) = P\left(Z < \frac{32 - 36}{2.5}\right) = P(Z < -1.6)$ **[1]**
   0.0548 **[1]**
2. a) $\frac{225}{30} = 7.5$ **[1]**
   $\sqrt{\frac{1755}{30} - 7.5^2}$ **[1]** $= 1.5$ **[1]**

   b) The mean will remain unchanged **[1]** as both values average to the mean **[1]**. The standard deviation will increase **[1]**; both values are greater than 2 standard deviations away from the mean so the average distance from the mean will increase. **[1]**

   c) Not a normal **[1]** as 13% of musicians are greater than 2 standard deviations above the mean; for a normal it should be 2.5% **[1]**
3. $X \sim B(200, 0.45)$
   $np = 90$ **[1]**
   $np(1 - p) = 49.5$ **[1]**
   $P(X < 80) = P\left(Z < \frac{79.5 - 90}{\sqrt{49.5}}\right)$ **[1]**
   $P(Z < -1.49)$ **[1]**
   0.0681 **[1]**

4. a) n must be large [1]
p close to 0.5 [1]
b) Mean $= np$ [1]
Variance: $np(1-p)$ [1]

5. $X \sim B(100, 0.4)$ [1]
$np = 40$
$np(1-p) = 24$ [1]
$P(X \geqslant 50) = P\left(Z > \frac{49.5 - 40}{\sqrt{24}}\right)$ [1]
$P(Z > 1.94)$ [1]
0.0262 [1]

Pages 188–189: Statistical Hypothesis Testing

1. $H_0: p = 0$ **[1]** $H_1: p > 0$ **[1]**
One-tailed test [1]
Critical value 0.6215 [1]
$0.572 < 0.6215$ [1]
Therefore insufficient evidence to reject the null hypothesis **[1]**; there is no evidence of positive correlation **[1]**.

2. $H_0: \mu = 60$ **[1]** $H_1: \mu > 60$ **[1]**
$\bar{X} \sim N\left(60, \frac{100}{30}\right)$ **[1]** $P(\bar{X} > 65)$ **[1]**
$P\left(Z > \frac{65-60}{\frac{10}{\sqrt{30}}}\right)$ [1]
0.0031 [1]
$0.0031 < 0.05$ [1]
Therefore sufficient evidence to reject the null hypothesis **[1]**; there is evidence to support the managing director's claim **[1]**.

This is one-tailed.

3. $H_0: \mu = 30$ **[1]** $H_1: \mu > 30$ **[1]**
$\bar{X} \sim N\left(30, \frac{2.25}{10}\right)$ $P(\bar{X} > 31.4)$ [1]
$P\left(Z > \frac{31.4-30}{\frac{1.5}{\sqrt{10}}}\right)$ [1]
0.00159 [1]
$0.00159 < 0.05$ [1]
Therefore sufficient evidence to reject the null hypothesis **[1]**; there is evidence to support the manager's claim. **[1]**

This is one-tailed.

Pages 190–191: Kinematics

1. a) $\mathbf{a} = \frac{d\mathbf{v}}{dt} = -2t\mathbf{i} + 5\mathbf{j}$ [2]
Substituting $t = 2$, $\mathbf{a} = (-4\mathbf{i} + 5\mathbf{j})\,\text{ms}^{-2}$ [1]
b) $\mathbf{r} = \int \mathbf{v}\,dt = \int\left((2-t^2)\mathbf{i} + (5t-1)\mathbf{j}\right)dt$
$= \left(2t - \frac{t^3}{3}\right)\mathbf{i} + \left(\frac{5t^2}{2} - t\right)\mathbf{j} + \mathbf{c}$ [2]
Substituting $t = 0$, $\mathbf{r} = -2\mathbf{i}$ [1]
$\Rightarrow \mathbf{c} = -2\mathbf{i}$ [1]
$\mathbf{r} = \left(2t - \frac{t^3}{3} - 2\right)\mathbf{i} + \left(\frac{5t^2}{2} - t\right)\mathbf{j}$
Substitute $t = 2 \Rightarrow \mathbf{r} = -\frac{2}{3}\mathbf{i} + 8\mathbf{j}$ [1]

2. a) $\mathbf{v} = (\cos 2t)\mathbf{i} - (\cos t)\mathbf{j}$ [2]
Set **i** and **j** components equal: [1]
$\cos 2t = -\cos t$
$\cos 2t + \cos t = 0$
$2\cos^2 t - 1 + \cos t = 0$ [1]
$(2\cos t - 1)(\cos t + 1) = 0$ [1]
$\cos t = \frac{1}{2}$ or $\cos t = -1$ [2]
Smallest value of t is therefore $\cos^{-1}\left(\frac{1}{2}\right) = \frac{\pi}{3}$ [1]
b) $\mathbf{a} = \frac{d\mathbf{v}}{dt} = (-2\sin 2t)\mathbf{i} + (\sin t)\mathbf{j}$ [2]
Substitute $t = \frac{\pi}{3}$ [1]
$\mathbf{a} = \left(-2\sin\frac{2\pi}{3}\right)\mathbf{i} + \left(\sin\frac{\pi}{3}\right)\mathbf{j}$
$\mathbf{a} = -\sqrt{3}\mathbf{i} + \frac{\sqrt{3}}{2}\mathbf{j}$ [1]
$|\mathbf{a}| = \sqrt{\left(\sqrt{3}\right)^2 + \left(\frac{\sqrt{3}}{2}\right)^2} = \frac{\sqrt{15}}{2}\,\text{ms}^{-2}$ [2]

3. a) $\frac{d\mathbf{r}}{dt} = \begin{pmatrix} ak\cos kt \\ -bk\sin kt \end{pmatrix}$ [2]
$\frac{d^2\mathbf{r}}{dt^2} = \begin{pmatrix} -ak^2\sin kt \\ bk^2\cos kt \end{pmatrix}$ [2]
$\frac{d^2\mathbf{r}}{dt^2} = -k^2\begin{pmatrix} a\sin kt \\ b\cos kt \end{pmatrix} = -k^2\mathbf{r}$ as required [1]
b) If P was stationary, then $ak\cos kt = 0$ (1) [1]
and $-bk\sin kt = 0$ (2) [1]
Solving (2) gives $t = \frac{2n\pi}{k}$ [1]
Substituting in (1) gives
$ak\cos k\left(\frac{2n\pi}{k}\right) = ak\cos(2n\pi) \neq 0$ [1]
Therefore equations (1) and (2) are inconsistent, and so P is never stationary. [1]

4. a) Vertical motion upwards: $-65 = (25\sin 30°)T - 4.9T^2$ [2]
$4.9T^2 - (25\sin 30°)T - 65 = 0$
Solve using quadratic formula to give $T = 5.135$ s [2]
Horizontal motion: $x = (25\cos 30°)T$
$= 25\cos 30° \times 5.135\,\text{m}$ [1]
$= 111\,\text{m}$ [1]
b) $v^2 = u^2 + 2as$ vertically upwards gives

$v^2 = (25\sin30°)^2 + 2 \times (-9.8) \times (-65)$ **[2]**

$v = 37.8\,\text{ms}^{-1}$ **[1]**

5.

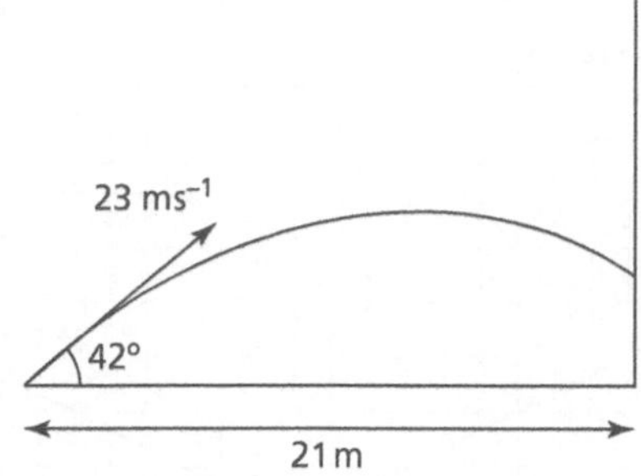

Horizontal motion: $21 = (23\cos42°)t$ **[1]**

Time to collision is: $\dfrac{21}{23\cos 42°} = 1.229\,\text{s}$ **[1]**

Vertical velocity: $23\sin42° - 9.8 \times 1.229 = 3.350$ **[2]**

Horizontal velocity: $23\cos42° = 17.09\,\text{ms}^{-1}$ **[1]**

Calculate speed using Pythagoras:

$\sqrt{3.350^2 + 17.09^2} = 17.4\,\text{ms}^{-1}$ **[2]**

Pages 192–193: Forces

1. Resolving perpendicular to plane: $R = 2g\cos 25°$ (1) **[1]**

Friction limiting: $F = \mu R$ (2) **[1]**

Equation of motion down plane:

$2g\sin25° - F = 2 \times 2.65$ (3) **[2]**

Solving (1), (2), (3) simultaneously $\Rightarrow \mu = 0.168$ **[2]**

2.

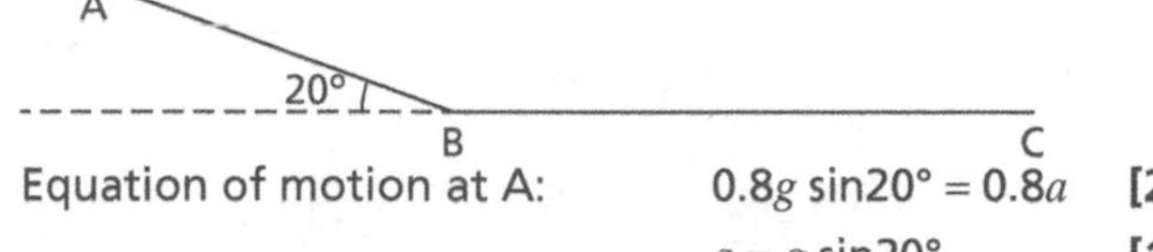

Equation of motion at A: $0.8g\sin20° = 0.8a$ **[2]**

$a = g\sin20°$ **[1]**

Using $v^2 = u^2 + 2as$ from A → B $\quad v^2 = 2g\sin20° \times 2$ **[1]**

Resolving vertically at B: $R = 0.8g$ (1) **[1]**

Friction limiting at B: $F = 0.3R$ (2) **[1]**

Equation of motion at B: $-F = 0.8a$ (3) **[1]**

Solving (1), (2), (3) simultaneously: $a = -0.3g$ **[1]**

Using $v^2 = u^2 + 2as$ from B → C :

$0 = 2g\sin20° \times 2 - 2 \times 0.3gs$ **[2]**

$s = 2.28\,\text{m}$ **[1]**

3. a) Equation of motion of Q: $2.5g - T = 2.5a$ (1) **[2]**

Equation of motion of P:

$T - 3g\sin30° - F = 3a$ (2) **[2]**

Resolve perpendicular to plane: $R = 3g\cos30°$ (3) **[1]**

Friction limiting: $F = 0.2R$ (4) **[1]**

Solving (1), (2), (3), (4) simultaneously gives

$a = 0.860\,\text{ms}^{-2}$ **[1]**

b) Substituting back into equation (1) gives $T = 22.4\,\text{N}$ **[1]**

c)

30°
30°
T
T
F_res

Force on pulley $= 2T\cos30°$ **[1]**

$= 2 \times 22.36 \times \cos30° = 38.7\,\text{N}$ **[1]**

4. a) Resolving parallel to the plane:

$X\cos50° = 4g\sin40°$ **[2]**

$X = \dfrac{4g\sin 40°}{\cos 50°} = 39.2\,\text{N}$ **[1]**

b) Resolving perpendicular to plane:

$R = 4g\cos40° + X\sin50°$ **[2]**

$R = 4g\cos40° + 39.2\sin50° = 60.1\,\text{N}$ **[1]**

5. a) Resolving parallel to plane for A:

$T = 5g\sin 30° = \dfrac{5g}{2} = 24.5\,\text{N}$ **[2]**

b) Resolving parallel to plane for B:

$P = T + 10g\sin30°$ **[2]**

$P = \dfrac{5g}{2} + \dfrac{10g}{2} = \dfrac{15g}{2} = 73.5\,\text{N}$ **[1]**

6. Let the tension in the string attached to A be T_1 and the tension in the string at B be T_2.

Resolving vertically: $T_1\sin35° + T_2\sin65° = 15g$ **[2]**

Resolving horizontally: $T_1\cos35° = T_2\cos65°$ **[2]**

Solving simultaneously to give $T_1 = 63.1\,\text{N}$ **[1]**

$T_2 = 122\,\text{N}$ **[1]**

Page 194: Moments

1. a) Resolving horizontally: $T_1\cos50° = T_2\cos60°$ **[2]**

Resolving vertically: $T_1\sin50° + T_2\sin60° = 2g$ **[2]**

Solving simultaneously to give $T_1 = 10.43\,\text{N}$ **[1]**

$T_2 = 13.41\,\text{N}$ **[1]**

b) Taking moments about A: $2gx = 6T_2\sin60°$ **[2]**

$x = \dfrac{6 \times 13.41\sin 60°}{2g} = 3.56\,\text{m}$ **[1]**

2. a) Taking moments about A: $5R_B = \dfrac{7}{2}\cos 50°(6g)$ **[2]**

$R_B = 26.5\,\text{N}$ **[1]**

b) Resolving horizontally: $F = R_B\sin50°$ **[1]**

Resolving vertically: $R_A + R_B\cos50° = 6g$ **[2]**

Friction limiting: $F = \mu R_A$ **[1]**

Solving simultaneously: $\mu = 0.485$ **[1]**

3.

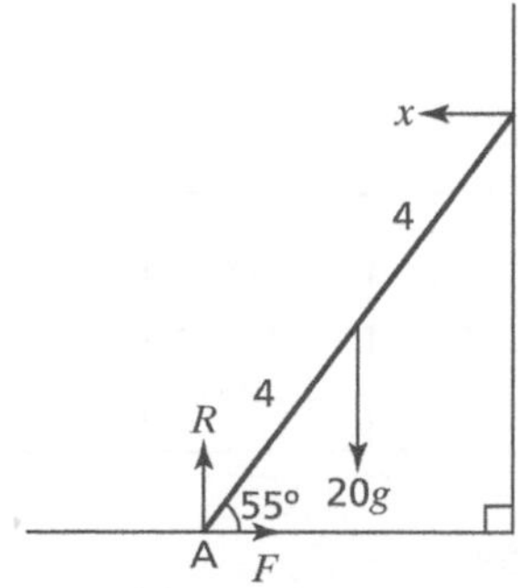

Moments about A: $20g \times 4\cos55° = X(8\sin55°)$ **[2]**

$X = 68.62\,\text{N}$ (1) **[1]**

Resolving horizontally: $F = X$ (2) **[1]**

Resolving vertically: $R = 20g$ (3) **[1]**

Friction limiting: $F = \mu R$ (4) **[1]**

Solving (1), (2), (3), (4) simultaneously: $\mu = 0.350$ **[1]**

Answers

PRACTICE EXAM PAPERS

Types of mark:
M = correct method applied to appropriate numbers
A = accuracy, dependent on **M** mark being gained
B = correct final answer, partially correct answer or correct intermediate stage

PRACTICE EXAM PAPER 1: PURE MATHEMATICS 1

1. (a) 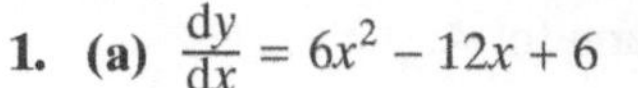
$\frac{dy}{dx} = 6x^2 - 12x + 6$ **M1 A1**

(b) $\frac{d^2y}{dx^2} = 12x - 12$ **A1**

(c) $\frac{dy}{dx} = 6x^2 - 12x + 6 = 0$ **A1**
$\Rightarrow x^2 - 2x + 1 = 0$
$\Rightarrow (x-1)^2 = 0$ **M1**
$\Rightarrow x = 1, y = 7$ **A1**
At $x = 1$, $\frac{d^2y}{dx^2} = 0 \left(\text{and } \frac{d^3y}{dx^3} \neq 0\right)$, so this is a point of inflection. **A1**

2. (a) $y = \ln(x-3)$
$e^y = x - 3$ **M1**
$x = 3 + e^y$
$\Rightarrow f^{-1}(x) = 3 + e^x$ **A1**

(b) $fg(x) = \ln(2x + 1 - 3) = \ln(2x - 2)$ **M1 A1**

(c) $\ln(2x-2) = \ln(2(x-1)) = \ln 2 + \ln(x-1)$ **M1**
$= \ln 2 + \ln((x-3)+2)$ **A1**
So translations are –2 in the x-axis direction and ln2 in the y-axis direction. **A1 A1**

3. $\overrightarrow{PQ} = \overrightarrow{OQ} - \overrightarrow{OP}$ **M1**
$-3a\mathbf{i} - 2a\mathbf{j} + a\mathbf{k} - \mathbf{i} + 2\mathbf{j} + 3\mathbf{k}$
$= (-3a-1)\mathbf{i} + (-2a-2)\mathbf{j} + (a-3)\mathbf{k}$ **A1**
$|\overrightarrow{PQ}|^2 = (-3a-1)^2 + (-2a-2)^2 + (a-3)^2$ **M1**
$= 14a^2 + 8a + 14$ **A1**
Differentiate and set to zero: **M1**
$28a + 8 = 0$
$\Rightarrow a = -\frac{2}{7}$ **A1**

4.

x	1	2	3	4	5
y	0	0.48045	1.2069	1.92181	2.59029

B2

$\int_1^5 (\ln x)^2\,dx \approx \frac{1}{2} \times 1 \times \begin{bmatrix} 0 + 2.59029 + \\ 2(0.48045 + 1.2069 + 1.92181) \end{bmatrix}$ **M1**
$= 4.90$ units^2 **A1**

5. (a)

Cross axes at (15, 0), (0, 5) **B1**
and (2, 0), (0, 8) **B1**

(b) At A: $-\left(\frac{1}{3}x - 5\right) = -(4x - 8)$ **M1**
$\Rightarrow x = \frac{9}{11}, y = \frac{52}{11}$ **A1**
At B: $-\left(\frac{1}{3}x - 5\right) = 4x - 8$ **M1**
$\Rightarrow x = 3, y = 4$ **A1**

6. (a) $f(2) = 2.81 > 0$ **B1**
$f(3) = -1.80 < 0$ **B1**
Sign change, and f continuous between $x = 2$ and $x = 3$, so $2 < \alpha < 3$. **B1**

(b) $f'(x) = -2e^{-2x} + \frac{1}{x} - 2x$ **M1 A1**
$x_1 = 2.5 - \frac{f(2.5)}{f'(2.5)} = 2.5 - \left(\frac{0.7716\ldots}{-4.6134\ldots}\right)$ **M1**
$= 2.67$ **A1**

7. Consider $\frac{\delta y}{\delta x} \approx \frac{\cos(2(x+h)) - \cos 2x}{h}$ **M1**
$= \frac{\cos 2x \cos 2h - \sin 2x \sin 2h - \cos 2x}{h}$ **M1 A1**
$= \frac{\cos 2x \cos 2h - (2\sin x \cos x)(2 \sin h \cos h) - \cos 2x}{h}$ **A1**
$= \frac{\cos 2x}{h}\cos 2h - (4 \sin x \cos x)\left(\frac{\sin h}{h}\cos h\right) - \frac{\cos 2x}{h}$ **A1**
As $h \to 0$, $\cos h \to 1$, $\cos 2h \to 1$ and $\frac{\sin h}{h} \to 1$ **M1 A1**
So $\frac{dy}{dx} = -4 \sin x \cos x$ **A1**
$= -2\sin 2x$

8. (a) $P = 2r + r\theta \Rightarrow \theta = \frac{P - 2r}{r}$ **B1**
$A = \frac{1}{2}r^2\theta$ **B1**
$A = \frac{1}{2}r^2\left(\frac{P-2r}{r}\right)$ **M1**
$= \frac{Pr}{2} - r^2$ **A1**

(b) $\frac{dA}{dr} = 0$ **M1**

$\frac{P}{2} - 2r = 0 \Rightarrow P = 4r$ **A1**

$4r = 2r + r\theta$

$\Rightarrow \theta = 2^c$ **A1**

9. **(a)** $\tan 3\theta = \tan(2\theta + \theta)$ **M1**

$= \frac{\tan 2\theta + \tan\theta}{1 - \tan 2\theta \tan\theta}$ **A1**

$= \frac{\frac{2\tan\theta}{1-\tan^2\theta} + \tan\theta}{1 - \frac{2\tan\theta}{1-\tan^2\theta}\tan\theta}$ **A1**

$= \frac{\frac{\tan\theta(1-\tan^2\theta) + 2\tan\theta}{1-\tan^2\theta}}{\frac{(1-\tan^2\theta) - 2\tan^2\theta}{1-\tan^2\theta}}$ **A1**

$= \frac{\tan\theta(1-\tan^2\theta) + 2\tan\theta}{(1-\tan^2\theta) - 2\tan^2\theta}$ **A1**

$= \frac{3\tan\theta - \tan^3\theta}{1 - 3\tan^2\theta}$

(b) $\frac{3\tan\theta - \tan^3\theta}{1 - 3\tan^2\theta} = \tan\theta$ **M1**

$2\tan^3\theta + 2\tan\theta = 0$

$2\tan\theta(\tan^2\theta + 1) = 0$ **M1**

$\tan^2\theta \neq -1$ **A1**

So $\tan\theta = 0$ only solution.

10. **(a)** $ar = \frac{63}{10}$ **B1**

$\frac{a}{1-r} = 30 \Rightarrow a = 30 - 30r$ **B1**

$(30 - 30r)r = \frac{63}{10}$ **M1**

$(30 - 30r)10r = 63$

$300r^2 - 300r + 63 = 0$ **A1**

$100r^2 - 100r + 21 = 0$

(b) $(10r - 7)(10r - 3) = 0$ **M1**

$r = \frac{7}{10}$ or $r = \frac{3}{10}$ **A1**

(c) Taking $r = \frac{3}{10}$, then $a = \frac{63}{10r} = \frac{63}{3} = 21$

$\frac{a(1-r^n)}{1-r} = \frac{21}{0.7}(1 - 0.3^n) = 29.99999$ **M1 A1**

$1 - 0.3^n = \frac{29.99999}{30}$

$-0.3^n = \frac{29.99999}{30} - 1$

$0.3^n = 1 - \frac{29.99999}{30}$

$n\ln(0.3) = \ln\left(1 - \frac{29.99999}{30}\right)$ **M1**

$n = \frac{\ln\left(1 - \frac{29.99999}{30}\right)}{\ln(0.3)}$

$n = 12.39$ **A1**

So require 13 terms. **A1**

11. **(a)** $\sin x \approx x$, $\cos x \approx 1 - \frac{x^2}{2}$ **B1 B1**

$\sin x(\cos^2 x) \approx x\left(1 - \frac{x^2}{2}\right)^2 = x\left(1 - x^2 + \frac{x^4}{4}\right)$ **M1**

$= x - x^3 + \frac{x^5}{4}$ **A1**

(b) Percentage error $= \frac{\text{actual} - \text{approx}}{\text{actual}} \times 100$ **M1**

Error $= \frac{\sin 0.1(\cos^2 0.1) - 0.0990025}{\sin 0.1(\cos^2 0.1)} \times 100$

$= -0.166\%$ **A1**

12. Attempt algebraic long division (or equivalent) **M1**

$\frac{-3x^3 + 5x^2 + 2x + 4}{2 - x} = 3x^2 + x + \frac{4}{2-x}$ **A1**

$\int_0^1 \left(3x^2 + x + \frac{4}{2-x}\right)dx = \left[x^3 + \frac{x^2}{2} - 4\ln|2 - x|\right]_0^1$ **M1 A1**

$= \frac{3}{2} - (-4\ln 2)$ **M1**

$= \frac{3}{2} + 4\ln 2$ **A1**

13. **(a)** LHS $= (1 - x)(1 + y^2) = (1 - (1 - \sin 2t))(1 + \tan^2 t)$

$= \sin 2t \sec^2 t$ **M1 A1**

$= 2\sin t\cos t\left(\frac{1}{\cos^2 t}\right)$ **A1**

$= \frac{2\sin t}{\cos t}$ **A1**

$= 2\tan t$

$= 2y$ **A1**

(b) At $t = 0$:

$x = 1, y = 0$ **B1**

$\frac{dy}{dx} = \frac{\frac{dy}{dt}}{\frac{dx}{dt}} = \frac{\sec^2 t}{-2\cos 2t} = \frac{-1}{2\cos 2t\cos^2 t}$ **M1 A1**

At $t = 0$, $\frac{dy}{dx} = -\frac{1}{2}$ **A1**

So gradient of normal $= 2$ **M1**

Equation is $y = 2(x - 1)$ **A1**

(c) Normal intersects curve when

$\tan t = 2 - 2\sin 2t - 2$ **M1**

$\tan t = -2\sin 2t$

$\tan t = -4\sin t\cos t$ **A1**

$\frac{\sin t}{\cos t} = -4\sin t\cos t$ **A1**

$\sin t = -4\sin t\cos^2 t$

$\sin t + 4\sin t\cos^2 t = 0$

$\sin t(1 + 4\cos^2 t) = 0$ **M1**

$\sin t = 0$ (or $\cos^2 t = -\frac{1}{4}$) **A1**

So $\sin t = 0 \Rightarrow t = 2n\pi$ gives only one coordinate.

Answers

PRACTICE EXAM PAPER 2: PURE MATHEMATICS 2

1. $\frac{dy}{dx} = \frac{3x + 1 - 3(x - 2)}{(3x + 1)^2} = \frac{7}{(3x + 1)^2}$ **M1 A1**

At $x = 2, \frac{dy}{dx} = \frac{1}{7}$ **A1**

Equation is $y = \frac{1}{7}(x - 2)$ **M1**

$\Rightarrow x - 7y - 2 = 0$ **A1**

2. **(a)** $\frac{1}{\sqrt{4 + 3x}} = (4 + 3x)^{-\frac{1}{2}}$

$= \frac{1}{2}\left(1 + \frac{3x}{4}\right)^{-\frac{1}{2}}$ **M1**

$= \frac{1}{2}\left(1 + \left(-\frac{1}{2}\right)\frac{3x}{4} + \frac{\left(-\frac{1}{2}\right)\left(-\frac{3}{2}\right)}{2!}\left(\frac{3x}{4}\right)^2 + \cdots\right)$ **M1 A1**

$= \frac{1}{2}\left(1 - \frac{3}{8}x + \frac{27}{128}x^2 + \cdots\right)$

$= \frac{1}{2} - \frac{3}{16}x + \frac{27}{256}x^2 + \ldots$ **A1**

(b) Valid for $\left|\frac{3x}{4}\right| < 1$ **M1**

$\Rightarrow |x| < \frac{4}{3}$ or $-\frac{4}{3} < x < \frac{4}{3}$ **A1**

3. **(a)** $p\left(\frac{1}{2}\right) = 2\left(\frac{1}{2}\right)^3 - 5\left(\frac{1}{2}\right)^2 - 4\left(\frac{1}{2}\right) + 3$ **M1**

$= \frac{1}{4} - \frac{5}{4} - 2 + 3 = 0$ **A1**

(b) $p(x) = (2x - 1)(x^2 + kx - 3)$

Equate $x^2 : -1 + 2k = -5 \Rightarrow k = -2$ (or use long division) **M1 A1**

$p(x) = (2x - 1)(x^2 - 2x - 3)$
$= (2x - 1)(x - 3)(x + 1)$ **A1**

4. **(a)** A is the price Terry paid for the car when new. **B1**

(b) $V = Ae^{-kt}$

$\ln V = \ln(Ae^{-kt})$ **M1**

$\ln V = \ln A + \ln e^{-kt}$

$\ln V = \ln A - kt \ln e$

$\ln V = \ln A - kt$ **A1**

$\Rightarrow m = -k$ and $c = \ln A$ **A1**

(c) From the graph, $m = -0.14$

$\Rightarrow k = 0.14$ **B1**

$c = 9.80$

$\Rightarrow A = e^{9.80} \approx 18000$ **B1**

(d) 2022 relies on extrapolating the graph, and the model may not be accurate for values of t outside this range. **B1**

5. **(a)** Midpoint of AB is $\left(\frac{9}{2}, \frac{1}{2}\right)$ **B1**

Gradient of AB is $\frac{-4 - 5}{3 - 6} = 3$ **M1**

Gradient of normal is $-\frac{1}{3}$ **M1**

Equation is $y - \frac{1}{2} = -\frac{1}{3}\left(x - \frac{9}{2}\right)$ or $3y + x - 6 = 0$ **A1**

(b) Midpoint of BC is $\left(\frac{1}{2}, -\frac{3}{2}\right)$ **B1**

Gradient of AB is $\frac{1 - (-4)}{-2 - 3} = -1$ **M1**

Gradient of normal is 1 **M1**

Equation is $y + \frac{3}{2} = x - \frac{1}{2}$, or $y - x + 2 = 0$ **A1**

(c) Solve simultaneously: **M1**

$y = 1$ and $x = 3$ **A1**

Centre is $(3, 1)$

6. **(a)** Writing: $S_n = a + (a + d) + (a + 2d) + \ldots + (a + (n - 1)d)$
and in reverse: $S_n = (a + (n - 1)d) + (a + (n - 2)d) + (a + (n - 3)d) + \ldots + a$ **B1**

Adding: $2S_n = 2a + (n - 1)d$ **M1**

Hence: $S_n = \frac{n}{2}(2a + (n - 1)d)$ **A1**

(b) Considering $\sum_{r=20}^{r=40}(2r + 1)$, you have $a = 41$, $d = 2$ and $n = 21$.

$S_n = \frac{n}{2}(2a + (n - 1)d)$

$= \frac{21}{2}(82 + 20 \times 2)$ **M1**

$= 1281$ **A1**

7. **(a)**

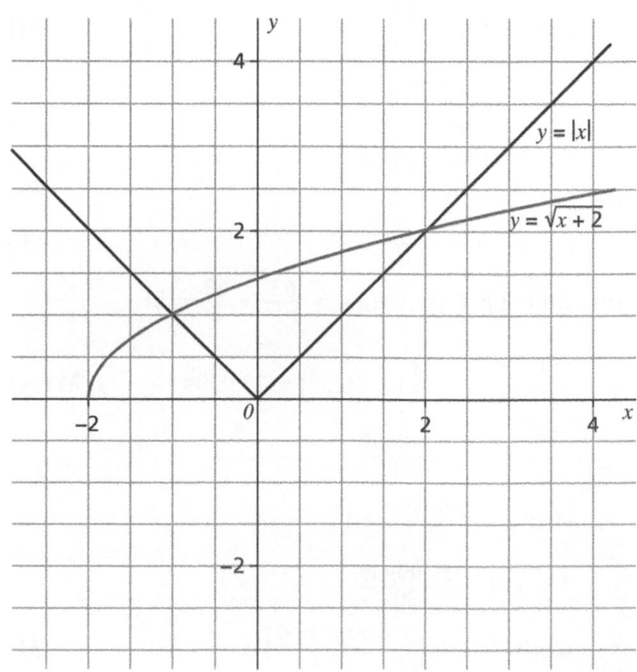

B1 B1

(b) $|x| = \sqrt{x+2}$

Squaring gives $x^2 = x + 2$ **M1**

(or read off intersection points from **accurate** graph)

$x^2 - x - 2 = 0$

$(x-2)(x+1) = 0$

$\Rightarrow x = -1, y = 1$ **A1**

and $x = 2, y = 2$ **A1**

(c) Area under 'modulus' function (considering as two triangles) $= \frac{1}{2} + 2 = \frac{5}{2}$ **B1**

Area under $y = \sqrt{x+2}$ is given by

$$\int_{-1}^{2} \sqrt{x+2}\, dx = \left[\frac{2}{3}(x+2)^{\frac{3}{2}}\right]_{-1}^{2}$$ **M1**

$$= \frac{16}{3} - \frac{2}{3}$$

$$= \frac{14}{3}$$ **A1**

So required area is $\frac{14}{3} - \frac{5}{2} = \frac{13}{6}$ units^2 **A1**

8. Substitute $u = e^x$

$u^2 - 3u - 10 < 0$ **M1**

$(u+2)(u-5) < 0$ **M1**

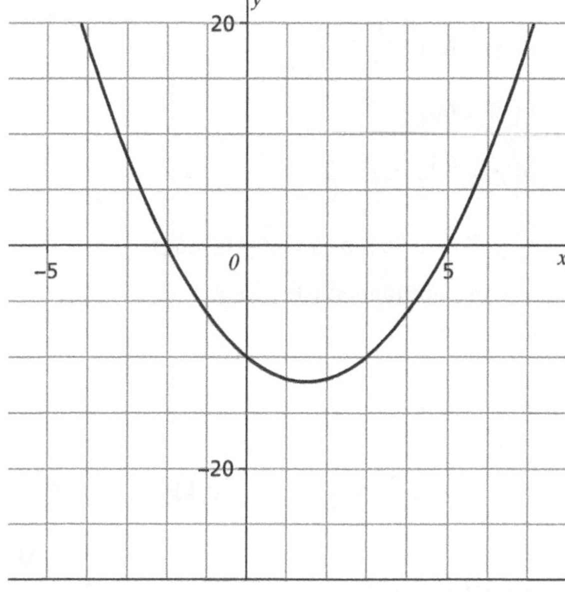

$\Rightarrow -2 < u < 5$ **A1**

But $u = e^x > 0$

$\Rightarrow 0 < e^x < 5$ **A1**

$\Rightarrow x < \ln 5$ **A1**

9. (a) $\frac{dy}{dx} = 10\cos 2x - 2\sin 2x$ **M1 A1**

$\frac{d^2y}{dx^2} = -20\sin 2x - 4\cos 2x$ **A1**

$\frac{d^2y}{dx^2} = -4y$ **A1**

(b) $\frac{dy}{dx} = 0 \Rightarrow 10\cos 2x = 2\sin 2x$ **M1**

$\tan 2x = 5$

$x = \frac{1}{2}\arctan 5 = 0.687$ **A1**

$y = 5.10$ **A1**

$\frac{d^2y}{dx^2} = -4y < 0$ hence maximum point **B1**

(c) $5\sin 2x + \cos 2x \equiv R\cos(2x - \alpha)$

$= R\cos 2x \cos\alpha + R\sin 2x \sin\alpha$ **M1**

$R\cos\alpha = 1$

and $R\sin\alpha = 5$ **A1**

$\tan\alpha = 5 \Rightarrow \alpha = 1.373$ **A1**

$R = \sqrt{1^2 + 5^2} = \sqrt{26}$ **A1**

$\Rightarrow 5\sin 2x + \cos 2x \equiv \sqrt{26}\cos(2x - 1.373)$

(d) $\sqrt{26}\cos(2x - 1.373) = 2$

$\cos(2x - 1.373) = \frac{2}{\sqrt{26}}$

$2x - 1.373 = 1.1677 \quad \Rightarrow x = 1.27$ **M1A1**

$2x - 1.373 = 2\pi - 1.1677 \quad \Rightarrow x = 3.244$ **A1**

$2x - 1.373 = 2\pi + 1.1677 \quad \Rightarrow x = 4.412$ **A1**

10. $u = \tan x \Rightarrow du = \sec^2 x\, dx$ **B1**

$$\int_0^{\frac{\pi}{3}} \frac{\sec^2 x\, dx}{(2+\tan x)^2} = \int_0^{\sqrt{3}} \frac{du}{(2+u)^2}$$ **M1 A1**

$$= \left[\frac{-1}{2+u}\right]_0^{\sqrt{3}}$$ **A1**

$$= \frac{-1}{2+\sqrt{3}} - \left(-\frac{1}{2}\right)$$ **A1**

$$= \frac{-2+2+\sqrt{3}}{2\left(2+\sqrt{3}\right)}$$

$$= \frac{\sqrt{3}}{2\left(2+\sqrt{3}\right)}\left(\frac{2-\sqrt{3}}{2-\sqrt{3}}\right)$$

$$= -\frac{3}{2} + \sqrt{3}$$ **A1**

11. (a) Sometimes true, i.e. when $\log_{10} a$ is positive. **B2**

(b) Sometimes true, **B1**

e.g. $n = 3 \Rightarrow 2^3 - 1 = 7$ which is prime, but $2^9 - 1 = 511 = 73 \times 7$, so not prime. **B1**

(c) Sometimes true, e.g. when $x = y = 2$ **B2**

(d) Always true. **B1**

$$\text{LHS} = \frac{1-\cos^2 2\theta}{\cos^2\theta} = \frac{\sin^2 2\theta}{\cos^2\theta} = \frac{(2\sin\theta\cos\theta)^2}{\cos^2\theta}$$

$$= 4\sin^2\theta = \text{RHS}$$ **B1**

12. (a) $\dfrac{x^2 - 13x + 34}{(x-3)^2(x+1)} = \dfrac{A}{(x-3)^2} + \dfrac{B}{x-3} + \dfrac{C}{x+1}$ **M1**

$x^2 - 13x + 34 \equiv A(x+1) + B(x-3)(x+1) + C(x-3)^2$

$x = 3 \Rightarrow 4 = 4A \Rightarrow A = 1$ **M1 A1**

$x = -1 \Rightarrow 48 = 16C \Rightarrow C = 3$ **A1**

$x = 0 \Rightarrow 34 = A - 3B + 9C$

$34 = 1 - 3B + 27 \Rightarrow B = -2$ **A1**

$\dfrac{x^2 - 13x + 34}{(x-3)^2(x+1)} = \dfrac{1}{(x-3)^2} - \dfrac{2}{x-3} + \dfrac{3}{x+1}$

(b)

$\displaystyle\int_0^1 \frac{x^2 - 13x + 34}{(x-3)^2(x+1)}\,dx = \int_0^1 \left(\frac{1}{(x-3)^2} - \frac{2}{x-3} + \frac{3}{x+1}\right)dx$

$= \left[\dfrac{-1}{x-3} - 2\ln|x-3| + 3\ln|x+1|\right]_0^1$ **M1 A1**

$= \left(\left(\tfrac{1}{2}\right) - 2\ln 2 + 3\ln 2\right) - \left(\tfrac{1}{3} - 2\ln 3\right)$ **M1**

$= \dfrac{1}{6} + \ln 2 + \ln 9$

$= \dfrac{1}{6} + \ln 18$ **A1**

13. (a) Attempt to separate the variables: **M1**

$\displaystyle\int \frac{\ln y}{y}\,dy = \int -x\,dx$ **A1**

Integrate LHS by sight or use substitution $u = \ln y$

$\dfrac{(\ln y)^2}{2} = c - \dfrac{x^2}{2}$ **M1 A2**

$(\ln y)^2 = k - x^2$

$\ln y = \sqrt{k - x^2}$

$y = e^{\sqrt{k-x^2}}$ **A1**

(b) For domain, $k - x^2 > 0$ **M1**

$\Rightarrow x^2 - k < 0$

$\Rightarrow -\sqrt{k} < x < \sqrt{k}$

So domain is $\Rightarrow -\sqrt{k} < x < \sqrt{k}, x \in \mathbb{R}$ **A1**

Range is $0 < y \leq e^{\sqrt{k}}, y \in \mathbb{R}$ **A1**

PRACTICE EXAM PAPER 3: STATISTICS AND MECHANICS

Section A: Statistics

1. (a) A and B are not independent since $P(A|B) \neq P(A)$ **B1**

(b) $P(A \cap B) = P(A \mid B)P(B) = \frac{1}{9} \times \frac{1}{2} = \frac{1}{18}$ **M1 A1**

(c) $P(A' \cap B') = P\left[(A \cup B)'\right] = 1 - \begin{pmatrix} P(A) + P(B) - \\ P(A \cap B) \end{pmatrix}$ **M1**

$= 1 - \left(\frac{1}{3} + \frac{1}{2} - \frac{1}{18}\right) = \frac{2}{9}$ **A1**

(d) $P(A' \mid B') = \dfrac{P(A' \cap B')}{P(B')} = \dfrac{\left(\frac{2}{9}\right)}{\left(\frac{9}{18}\right)} = \dfrac{4}{9}$ **M1 A1**

2. (a) Using calculator

PMCC = 0.420 **B1**

(b) PMCC is between 0 and 0.5, suggesting there is weak positive correlation between the hours of sunshine and the mean visibility. **B1**

(c) $H_0: \rho = 0$ **B1**

$H_1: \rho > 0$ **B1**

Use of critical value 0.6215 **M1**

$0.420 < 0.6215$ so accept H_0. **A1**

(There is insufficient evidence to suggest a correlation between hours of daily sunshine and mean visibility.)

(d) Data was taken from only the first 8 days in July, rather than a random sample from the whole month. **B1**

3. (a) $X \sim N(\mu, \sigma^2)$

$P(X < 52) = 0.1$ **M1**

$P\left(Z < \dfrac{52 - \mu}{\sigma}\right) = 0.1$

$\dfrac{52 - \mu}{\sigma} = -1.281$ or $\mu = 52 + 1.281\sigma$ **A1**

$P(X > 60) = \frac{1}{6}$ **M1**

$P(X < 60) = \frac{5}{6}$

$P\left(Z < \dfrac{60 - \mu}{\sigma}\right) = \frac{5}{6}$

$\dfrac{60 - \mu}{\sigma} = 0.966$, or $\mu = 60 - 0.966\,\sigma$ **A1**

Solving simultaneously gives: **M1**

$\mu = 56.56$ **A1**

$\sigma = 3.56$ **A1**

(b) $X \sim N(\mu, \sigma^2)$

$P(X < 58)$

$P\left(Z < \dfrac{58 - 56.56}{3.56}\right) = P(Z < 0.40449) = 0.6572$

M1 A1

$Y \sim B(20, 0.6572)$ **M1**

$P(Y = 16, 17, 18, 19, 20) = 0.132$ **A1**

(c) $T \sim N(np, npq)$

$T \sim N(100 \times 0.6572, 100 \times 0.6572 \times 0.3428) \sim N(65.72, 22.53)$ **M1**

Apply continuity correction to both 60 and 75: **M1**

$P(T < 75.5) - P(T < 59.5)$ **A1**

$= 0.98032 - 0.09503$

$= 0.88529$ **A1**

4. (a) Using mid-values (may be implied) **M1**

$\bar{x} = \dfrac{\sum fx}{\sum f} = 25.9$ years old **A1**

(b) $\sigma = \sqrt{\frac{\sum fx^2}{\sum f} - \left(\frac{\sum fx}{\sum f}\right)^2} = 11.41$ years old **M1 A1**

(c) $20 + \left(\frac{50 - 31}{35}\right) \times 10 = 25.4$ **M1 A1**

(d) Data is approximately symmetrical about the mean. **B1**

Median and mean are approximately the same. **B1**

(e) $P(X < \mu + k) - P(X < \mu - k) = 0.5$ **M1**

$2P(X < \mu + k) - 1 < 0.5$

$P(X < \mu + k) < 0.75$

$P\left(Z < \left(\frac{\mu + k - 25.9}{11.41}\right)\right) < 0.75$ **M1**

$\frac{\mu + k - 25.9}{11.41} = 0.674$ **A1**

$k = 0.674 \times 11.41 = 7.69$ **A1**

5. (a) One-tailed test:

$H_0: \mu = 1080$ and $H_1: \mu < 1080$ **B1**

$\overline{X} \sim N\left(\mu, \frac{\sigma^2}{n}\right)$ **M1**

$P(\overline{X} < 990.35) = P\left(Z < \frac{990.35 - 1080}{\left(\frac{439.80}{\sqrt{90}}\right)}\right)$ **A1**

$= P(Z < -1.9338) = 0.0266$ **A1**

> 0.02

So accept H_0, i.e. there is insufficient evidence to suggest that $\mu < 1080$. **A1**

(b) Two-tailed test:

$Z_{\frac{\alpha}{2}} = 1.96$ **B1**

$\overline{X} \sim N\left(\mu, \frac{\sigma^2}{n}\right)$

$-1.96 < \frac{\overline{X} - \mu}{\frac{\sigma}{\sqrt{n}}} < 1.96$

$\Rightarrow \overline{X} - 1.96\frac{\sigma}{\sqrt{n}} < \mu < \overline{X} + 1.96\frac{\sigma}{\sqrt{n}}$ **M1**

Critical values: $\mu = 990.35 \pm 1.96 \times \frac{439.80}{\sqrt{90}}$ **A1**

H_0 accepted $\Rightarrow$ $\$899.49 < M < \$1081.21$ **A1**

Section B: Mechanics

6. (a) $\mathbf{a} = \frac{\mathbf{F}}{m} = \frac{3t^2\mathbf{i} + 4t\mathbf{j}}{2} = \frac{3t^2}{2}\mathbf{i} + 2t\mathbf{j}$ **B1**

$\mathbf{v} = \int \mathbf{a}\ \mathrm{d}t = \int\left(\frac{3t^2}{2}\mathbf{i} + 2t\mathbf{j}\right)\mathrm{d}t = \left(\frac{t^3}{2}\right)\mathbf{i} + t^2\mathbf{j} + \mathbf{c}$ **M1 A1**

At $t = 0$, $\mathbf{v} = 3\mathbf{i} - \mathbf{j} \Rightarrow \mathbf{c} = 3\mathbf{i} - \mathbf{j}$ **A1**

$\Rightarrow \mathbf{v} = \left(\frac{t^3}{2} + 3\right)\mathbf{i} + (t^2 - 1)\mathbf{j}$

(b) At $t = 2$

$\mathbf{v} = 7\mathbf{i} + 3\mathbf{j}$ and $\mathbf{F} = 12\mathbf{i} + 8\mathbf{j}$ **B1**

$\alpha = \arctan\left(\frac{8}{12}\right) - \arctan\left(\frac{3}{7}\right)$ **M1**

$= 10.5°$ **A1**

7. (a) **(i)** $T - F - 5mg \sin 30 = 5ma$ **M1 A1**

(ii) $4mg - T = 4ma$ **M1 A1**

(iii) $F = \frac{R}{5}$ **B1**

$R = 5mg \cos 30$ **B1**

$\Rightarrow F = \frac{mg\sqrt{3}}{2}$

Solving simultaneously: **M1**

$a = \left(\frac{3 - \sqrt{3}}{18}\right)g = 0.690\ \text{ms}^{-2}$ **A1**

(b) After 1 second, for particle A, $v = u + at \Rightarrow$ $v = 0 + 0.690 \times 1 = 0.690\ \text{ms}^{-1}$ **B1**

New equation of motion for A:

$-\frac{mg\sqrt{3}}{2} - \frac{5mg}{2} = 5ma$ **M1 A1**

$\Rightarrow a = -3.20\ \text{ms}^{-2}$ **A1**

Applying $v^2 = u^2 + 2as$: $0 = 0.690^2 - 2 \times 3.20 \times s$ **M1**

$\Rightarrow s = 0.0744$ m **A1**

8. (a) Taking moments about A: $mgl \cos 60 = 2lP$ **M1 A1**

$\Rightarrow P = \frac{mg}{4}$

(b) Horizontal: $F = P\cos 30 = \frac{mg\sqrt{3}}{8}$ **M1 A1**

Vertical: $P \sin 30 + R = mg$ **M1 A1**

$\Rightarrow R = mg - \frac{mg}{8} = \frac{7mg}{8}$

Friction limiting: $\mu = \frac{F}{R} = \frac{\sqrt{3}}{7} = 0.247$ **M1 A1**

(c) Resultant force $= mg\sqrt{\frac{3}{64} + \frac{49}{64}} = mg\sqrt{\frac{52}{64}} = \frac{mg\sqrt{13}}{4}$ **M1 A1**

Direction $= \arctan\left(\frac{R}{F}\right) = \arctan\left(\frac{7}{\sqrt{3}}\right)$

$= 76.1°$ from horizontal **M1 A1**

9. (a) Using $v = u + at$ **M1**

$v = -12\mathbf{i} + 5\mathbf{j} + (3\mathbf{i} - \mathbf{j})t = (-12 + 3t)\mathbf{i} + (5 - t)\mathbf{j}$ **A1**

(b) Set coefficient of $\mathbf{i}$ to equal 0 **M1**

$-12 + 3t = 0$

$t = 4$ s (then velocity $= 4\mathbf{j}$ so north) **A1**

(c) Set coefficient of $\mathbf{j}$ to equal negative coefficient of $\mathbf{i}$ **M1**

$5 - t = -(-12 + 3t)$

$5 - t = 12 - 3t$

$2t = 7$

$t = \frac{7}{2}$ seconds **A1**

Answers

10. (a) Horizontal: $x = ut\cos\alpha$ **M1**

Vertical: $y = ut\sin\alpha - \frac{1}{2}gt^2$ **M1**

For maximum range, $y = 0$

$ut\sin\alpha - \frac{1}{2}gt^2 = 0$ **A1**

$t\left(u\sin\alpha - \frac{gt}{2}\right) = 0$

$\Rightarrow t = \frac{2u\sin\alpha}{g}$ **A1**

$\Rightarrow x = u\cos\alpha\left(\frac{2u\sin\alpha}{g}\right) = \frac{u^2(2\sin\alpha\cos\alpha)}{g}$

$= \frac{u^2\sin 2\alpha}{g}$ **A1**

$\sin 2\alpha = \frac{500g}{u^2}$

$\alpha < 90$, so $0 < \sin 2\alpha < 1$ **M1**

$\Rightarrow u^2 > 500g$

$\Rightarrow u > 10\sqrt{5g}$ **A1**

(b) (i) $u = 12\sqrt{5g} \Rightarrow \sin 2\alpha = \frac{500g}{720g}$ **M1**

$\Rightarrow \alpha = 22.0°$ **A1**

(ii) Using $v^2 = u^2 + 2as$ vertically:

$0 = \left(12\sqrt{5g}\sin 22.0°\right)^2 - 2gH$ **M1**

$\Rightarrow H = 50.5$ m **A1**

Acknowledgements

The authors and publisher are grateful to the copyright holders for permission to use quoted materials and images.

Page 221: Contains public sector information licensed under the Open Government Licence v1.0
All other images © Shutterstock.com

Every effort has been made to trace copyright holders and obtain their permission for the use of copyright material. The author and publisher will gladly receive information enabling them to rectify any error or omission in subsequent editions. All facts are correct at time of going to press.

Published by Collins
An imprint of HarperCollins*Publishers* Ltd
1 London Bridge Street
London SE1 9GF

HarperCollins*Publishers*
Macken House, 39/40 Mayor Street Upper,
Dublin 1, D01 C9W8, Ireland

ISBN 9780008268527

First published 2018
This edition published 2020

10 9 8 7 6 5

British Library Cataloguing in Publication Data.

A CIP record of this book is available from the British Library.

Authors: Rebecca Evans, Phil Duxbury and Leisa Bovey
Project management: Richard Toms
Commissioning: Katherine Wilkinson, Clare Souza and Kerry Ferguson
Cover Design: Sarah Duxbury and Kevin Robbins
Inside Concept Design: Ian Wrigley
Text Design and Layout: Jouve India Private Limited
Production: Natalia Rebow
Printed in United Kingdom

This book contains FSC™ certified paper and other controlled sources to ensure responsible forest management.

For more information visit: www.harpercollins.co.uk/green